Numerical Heat Transfer and Fluid Flow 2023

Numerical Heat Transfer and Fluid Flow 2023

Editor

Artur Bartosik

MDPI

Basel • Beijing • Wuhan • Barcelona • Belgrade • Novi Sad • Cluj • Manchester

Editor
Artur Bartosik
Kielce University of
Technology
Kielce
Poland

Editorial Office
MDPI AG
Grosspeteranlage 5
4052 Basel, Switzerland

This is a reprint of articles from the Special Issue published online in the open access journal *Energies* (ISSN 1996-1073) (available at: https://www.mdpi.com/journal/energies/special_issues/8X00CA2HL2).

For citation purposes, cite each article independently as indicated on the article page online and as indicated below:

Lastname, A.A.; Lastname, B.B. Article Title. *Journal Name* **Year**, *Volume Number*, Page Range.

ISBN 978-3-7258-2155-6 (Hbk)
ISBN 978-3-7258-2156-3 (PDF)
doi.org/10.3390/books978-3-7258-2156-3

Contents

About the Editor

Artur Bartosik

Artur Bartosik has been appointed as a researcher and academic teacher at the Kielce University of Technology since 1983 until now. In 1993 he established the Centre for Continuing Education at the University and then in 2001 the Regional Centre for Innovation and Technology Transfer, which he managed as President of the Board from 2001 to 2007. Currently, he is the head of the Department of Production Engineering and the head of the Industrial Laboratory for Low Emission and Renewable Energy Sources. Together with his team, he works to establish a structure for university-wide multidisciplinary collaboration in terms of energy self-sufficiency. His scientific interest is focused on experiments, modelling, and simulation of single- and two-phase flows, including convective heat transfer. Artur, as a researcher, spent 2 years at the University of Saskatchewan in Canada and several weeks at Ilmenau Technical University in Germany. He is the Editor-in-Chief of WSEAS Transactions on Fluid Mechanics and an Associate Editor of several scientific journals. He was a chairman of several national and international conferences and is a member of professional scientific bodies. He serves lectures for students on fluid mechanics; heat transfer; fluid-flow machinery; hydrotransport; renewable energy sources; transnational technology transfer; applied fluid mechanics. He participated as an expert or coordinator in national and international projects.

energies

MDPI

Editorial

Advances in Numerical Heat Transfer and Fluid Flow

Artur S. Bartosik

Department of Production Engineering, Faculty of Management and Computer Modelling, Kielce University of Technology, Al. Tysiaclecia P.P. 7, 25-314 Kielce, Poland; artur.bartosik@tu.kielce.pl

1. Introduction

Scientists continuously are looking for new methods that allow them to better understand the flow and heat transfer phenomena. They try to optimise the parameters of fluid flow and heat exchange in order to increase the efficiency of specific technological processes and enhance environmental sustainability. The aim of this article is to provide an overview of the latest results in experiments and modelling of heat exchange between laminar or turbulent flow and surroundings, which were submitted to the Special Issue "Numerical Heat Transfer and Fluid Flow 2023".

The starting points in the prediction of heat exchange between the moving fluid and the surroundings are conservation equations, like the continuity, Navier–Stokes, and energy equations. This set of equations can be solved numerically using direct or indirect methods (DNS). The direct method requires high computing power. Thanks to the theory of turbulence, the time-averaged Navier–Stokes equations, called Reynolds-averaged Navier–Stokes equations (RANS), can be solved by applying additional equations, which allow the components of turbulent stress tensors to be designated. Components of the turbulent stress tensor can be calculated using a direct or indirect method [1–9]. Again, the direct method requires high computer power. In the literature, an indirect approach is commonly used in engineering applications because it requires less computing power. In such a case, turbulence models are applied to calculate the components of the turbulent stress tensor. One can find that hybrid models are continuously developed; they use different approaches to the near-wall region, where a small scale of turbulence exists, and to the outer region, where a large scale of turbulence dominates. For instance, the turbulence model can be applied to the region close to the wall, while large eddy simulation (LES) can be applied to the outer region. If non-Newtonian flow is considered, additional equations are needed in order to include a proper relation between shear stress and shear rate.

Computational methods that allow us to decrease the computation time are still developing. As a result, we already have considerable resources of software packages which allow scientists and engineers to simulate the heat transfer between fluid and the surroundings in specific applications. Some of the software packages were applied in this Special Issue.

The articles gathered in the Special Issue are useful for researchers, engineers, and students who deal with fluid flow and heat transfer.

Citation: Bartosik, A.S. Advances in Numerical Heat Transfer and Fluid Flow. *Energies* **2024**, *17*, 2108. https://doi.org/10.3390/en17092108

Received: 21 April 2024
Accepted: 25 April 2024
Published: 28 April 2024

2. Review of New Advances

Measurements of fluid thermal dynamics are still required in order to increase the knowledge and understanding of flow phenomena and to build an experimental base for the validation of mathematical models. Holešová et al. [10] focused on the challenges associated with visualising air flow over a heating source in an open laboratory environment. The authors performed an experimental visualisation and numerical simulation of air flow and heat transfer between the heating source and the environment using natural convection. They applied particle image velocimetry to visualise air flow. The data from the measurements were used as input for numerical simulations performed using Ansys Fluent software 18. These results demonstrated the effectiveness of combining measurements and numerical simulation to develop an accurate 3D model of heat transfer in laminar air flow [10]. The authors demonstrated that their 3D model had well-identified vorticities and had the potential to be used to simulate thermal comfort in the laboratory environment. It can be further extended by using and adjusting parameters such as the output of the heater and the heating source temperature [10].

Gas turbines are commonly used in power generation. Better turbine cooling can lead to the design of a turbine with better efficiency. Ahn [11] reviewed 50 articles published over the past 20 years on the use of large eddy simulation (LES) for the internal cooling passage of gas turbines. The author focused mainly on the mid-chord ribbed channels, collected validated LES models, and presented the simulation conditions of the models, then formulated findings that are difficult to obtain experimentally and discussed the effects of rotation, buoyancy, and heat transfer on the rib. The author discussed LES studies related to the shape of the ribbed channels and the LES contribution in the prediction of internal cooling of gas turbines and those with ribbed channels [11].

Kim et al. [12] applied large eddy simulation to predict forced convection around wavy cylinders with different axes for Re = 3000. The authors considered four types of undulated cylinders: streamwise undulation, transverse undulation, in-phase undulation, and antiphase undulation. The authors found that the friction and Nusselt number for cylinders with transverse undulation and in-phase undulation were significantly influenced by the wavelength and wave amplitude, whereas cylinders with streamwise undulation and antiphase undulation showed a very weak influence. The authors showed that proper modification of the geometry can improve both heat transfer and aerodynamic performance [12].

For a fully developed turbulent channel flow in the Reynolds number range of friction of 550–5200, Usov et al. [13] recalibrated the differential Reynolds stress model of Jakirlić and Maduta [14] using direct numerical simulation data in order to use the model in the hybrid RANS/LES framework. In their hybrid method, the authors used the RANS turbulence model only for a thin, near-wall layer in which small-scale turbulence dominated, while for the outer part of the boundary layer, in which large-scale turbulence dominated, they used the LES model. The authors performed comprehensive simulations of mean stream-wise and transverse velocity profiles, distribution of friction coefficient, and Reynolds stress profiles. The authors concluded that the recalibrated hybrid model ensured more accurate profiles of mean velocity, inner Reynolds stress peak, and characteristic turbulence frequency compared to the base model of Jakirlić and Maduta [14], and the model error was reduced by 58% [13].

The efficiency of heat exchange inside a shell and a tube heat exchanger is one of the main challenges of CFD. Estupinan-Campos et al. [15] numerically analysed the influence of geometric variations in baffle angles and the tube profile on heat transfer. They considered the turbulent flow of hot water in the inner tube and cold water in the outer tube. The flow of hot and cold liquids was countercurrent. The 3D simulations for the flow of water inside a shell and a tube heat exchanger were performed using time-averaged mass, momentum, and energy equations, while the problem of closing the set of equations was solved by introducing the standard k-ε model. The results of numerical predictions using Ansys software were compared with the measurements by Bicer et al. [16], achieving an average error equal to 3.15%. The authors recommended parameters such as the height of the baffle walls and the inclination of the deflector, which are given the highest efficiency of heat exchange inside a shell and a tube heat exchanger [15].

Pillow plate heat exchangers provide an alternative to conventional shell and tube heat exchangers. The proper waviness of the channels promotes fluid mixing in the boundary layers, resulting in the intensity of the heat transfer. Heat transfer efficiency can be increased by applying secondary surface structuring. Afsahnoudeh et al. [17] performed numerical studies on the influence of various surface structures in a heat exchanger on the efficiency of heat transfer. The mathematical model consisted of the conservation form of the continuity, momentum, and energy equations, while the closure problem was solved by applying the standard k-ε turbulence model. The authors considered a symmetric flow. Using software OpenFOAM No. 8, numerical simulations were performed for turbulent single-phase flow. As a result of the simulations, the authors determined the pressure drop, the heat transfer coefficient, and the overall thermohydraulic efficiency. The numerical predictions were validated only for the pressure drop, and the relative error was below 5%. The authors confirmed that the application of secondary structure has an impact on the thermohydraulic characteristics in both the outer and inner channels and results in an increase in the efficiency of the pillow plate heat exchanger [17].

On the basis of experiments and simulations, Bartosik [18] analysed the efficiency of heat exchange between the vertical pipe and turbulent slurry flow. The author presented experimental data of four slurries, with a narrow range of solid particle diameters, i.e., two with glass spheres with diameters of 0.125 mm and 0.240 mm, respectively, and two with sand particles with diameters of 0.470 mm and 0.780 mm, respectively, in the range of volume concentration of solid particles from 10% to 40%. The author showed that the experimental data demonstrated a substantial influence of solid particle diameter on the suppression of turbulence, which was the highest for the slurry with particle diameters equal to 0.47 mm. The author developed a mathematical model for heat exchange between pipe and slurry, taking into account the RANS approach with the k-ε turbulence model, and a specially designed turbulence damping function (wall function) depending on the diameter and concentration of solid particles. Numerical simulations of heat exchange between the slurry and the vertical pipe demonstrated a semisinusoidal dependence of the Nusselt number on the average particle diameter. The maximum Nusselt number appeared for a particle diameter of 0.125 mm, while the minimum was 0.470 mm [18].

The main challenge of the modern economy is the reduction in greenhouse gas emissions and the transition toward a zero-carbon economy. The greenhouse gas emissions of air conditioning are one of the main contributors to environmental pollution. A reduction in gas emission can be achieved by applying the caloric effect. One can recognise the caloric effect depending on the applied external field. If the external field is generated due to mechanical stress, such as compression [19], tension [20], or torsion [21], it is called the elastocaloric effect [22]. This effect occurs in specific materials known as shape memory alloys [23,24]. However, if the applied field is the result of hydrostatic pressure, it is referred to as the barocaloric effect [25,26]. Cirillo et al. [27] conducted a study on the application of barocaloric solid-state technology to the cold food supply chain. The authors examined the energy efficiency of a barocaloric cooling system designed to function as a refrigeration machine for cold products. The authors used a heat transfer liquid consisting of 50% ethylene glycol and 50% water mixture, while the solid-state refrigerant was acetoxy silicone rubber. As a reference, they used a vapour compression refrigerator commonly used in industrial settings. This analysis was carried out for incompressible laminar flow using the Navier–Stokes equations and the energy equation in 2D form, and the set of equations was solved using the finite element method. The authors concluded that, for a liquid velocity greater than 0.06 m/s, the barocaloric system outperforms the vapour compression system in terms of cooling power [27].

The gain of energy from deep geothermal resources requires the determination of thermal conductivity and heat capacity; prediction of heat exchange between geothermal wells and transported liquid; sustained power output; and, in some situations, chemical reactions. Such a process is extremely complex due to the dynamics of spatiotemporal evolution, which involves changes in the temperature field, seepage field, and mechanical field. Yang et al. [28] performed simulations of the heat production capacity of an underground reservoir, taking into account time-dependent mass conservation, heat conduction, and fluid heat transfer equations. The authors analysed the sensitivity of parameters related to the thermal performance and mining life of the deep thermal reservoir to thermal output. The results of the simulations showed that the injection temperature and the spacing of the well have little effect on the thermal efficiency and useful life of the mine, while the permeability, number, and length of the fracture have a significant influence. The authors emphasised the theoretical and practical importance of such studies [28].

Photovoltaic modules (PVs) often use cooling mediums, mainly liquid or air. The cooling medium flows over the backside surface of the cells and decreases the temperature of the cells to increase the efficiency of energy production. The cooling temperature is not high, so the use of such heat is limited. In the literature, some researchers have conducted studies on spectral beam splitting technology for sunlight. This technology uses different wavelengths of light from the Sun separately. Part of wavelength of the light is allocated to photoelectric conversion, and the other to photothermal conversion [29,30]. Lu et al. [31] performed simulations of the flow characteristics using a spectral beam-splitting module. The authors applied Ansys software using time-averaged continuity, momentum, energy, and k-ε equations. The authors noted that, for certain parameters of the nanofluid flow, it was possible to obtain a uniform temperature distribution on the surface of the photovoltaic cell while obtaining considerable heat and improving the integrated use of the Sun's energy. The authors stated that their study could provide a reference to optimise nanofluid flow within a spectral beam-splitting module [31].

3. Conclusions

The articles contributing to the Special Issue "Numerical Heat Transfer and Fluid Flow 2023" demonstrate the progress in using CFD for the prediction of heat exchange between laminar, turbulent, single-phase, or two-phase flow and the surroundings. The results of the measurements and several CFD models are presented. The authors formulated a set of conservation equations for a variety of applications and solved them numerically, taking into account the convergence criteria and ensuring a mesh-independent solution. All simulations were performed using commercial software, and most of the mathematical models have been validated.

The reader of the Special Issue "Numerical Heat Transfer and Fluid Flow 2023" can find a description of various phenomena of heat and fluid flow, experimental data, and various approaches to solving engineering problems. The collected articles will allow one to contribute to a better understanding of some phenomena and the interpretation of computed and measured quantities.

Funding: This research received no external funding.

Acknowledgments: The author thanks the contributors of the Special Issue Numerical Heat Transfer and Fluid Flow 2023 for the valuable articles, and thanks for the invitation to act as a guest editor.

Conflicts of Interest: The author declares no conflicts of interest.

References

1. Reynolds, O. On the dynamical theory on incompressible viscous fluids and the determination of the criterion. *Philos. Trans. R. Soc. Lond.* **1895**, *186*, 123–164. [CrossRef]
2. Prandtl, L. Bemerkung uber den warmeubergang in rohr. *Phys. Z.* **1928**, *29*, 487–489.
3. Karman, T. Mechanische ähnlichkeit und turbulenz. *Nachr. Gesselsch. Wiss. Göttingen Math. Phys.* **1930**, *322*, 58–76.
4. Karman, T. Some aspects of the theory of turbulent motion. In Proceedings of the Fourth International Congress for Applied Mechanics, Cambridge, UK, 3–9 July 1934.
5. Kolmogorov, A.N. Equations of turbulent motion of an incompressible fluid. *Izviestia AN SSSR Ser. Fiz.* **1942**, *VI*, 56–58.
6. Van Driest, E.R. On turbulent flow near a wall. *J. Aeronaut. Sci.* **1956**, *23*, 1007–1011. [CrossRef]
7. Patankar, S.V.; Spalding, D.B. *Heat and Mass Transfer in Boundary Layers*; Morgan-Grampian: London, UK, 1967.
8. Spalding, D.B. *Turbulence Models for Heat Transfer*; Report HTS/78/2; Department of Mechanical Engineering, Imperial College London: London, UK, 1978.
9. Spalding, D.B. *Turbulence Models—A Lecture Course*; Report HTS/82/4; Department of Mechanical Engineering, Imperial College London: London, UK, 1983.
10. Holešová, N.; Lenhard, R.; Kaduchová, K.; Holubčík, M. Application of Particle Image Velocimetry and Computational Fluid Dynamics Methods for analysis of natural convection over a horizontal heating source. *Energies* **2023**, *16*, 4066. [CrossRef]
11. Ahn, J. Large Eddy Simulation of flow and heat transfer in a ribbed channel for the internal cooling passage of a gas turbine blade: A Review. *Energies* **2023**, *16*, 3656. [CrossRef]

12. Kim, M.-K.; Chang, C.-H.; Nam, S.-H.; Yoon, H.-S. Large Eddy Simulation of forced convection around wavy cylinders with different axes. *Energies* **2024**, *17*, 894. [CrossRef]
13. Usov, L.; Troshin, A.; Anisimov, K.; Sabelnikov, V. Calibration of a near-wall differential Reynolds Stress Model using the updated Direct Numerical Simulation data and its assessment. *Energies* **2023**, *16*, 6826. [CrossRef]
14. Jakirlić, S.; Maduta, R. Extending the bounds of 'steady' RANS closures: Toward an instability-sensitive Reynolds stress model. *Int. J. Heat Fluid Flow* **2015**, *51*, 175–194. [CrossRef]
15. Estupinan-Campos, J.; Quitiaquez, W.; Nieto-Londono, C.; Quitiaquez, P. Numerical simulation of the heat transfer inside a shell and tube heat exchanger considering different variations in the geometric parameters of the design. *Energies* **2024**, *17*, 691. [CrossRef]
16. Biçer, N.; Engin, T.; Yaşar, H.; Büyükkaya, E.; Aydın, A.; Topuz, A. Design optimization of a shell-and-tube heat exchanger with novel three-zonal baffle by using CFD and Taguchi method. *Int. J. Therm. Sci.* **2020**, *155*, 106417. [CrossRef]
17. Afsahnoudeh, R.; Wortmeier, A.; Holzmüller, M.; Gong, Y.; Homberg, W.; Kenig, E.Y. Thermo-hydraulic performance of pillowplate heat exchangers with secondary structuring: A numerical analysis. *Energies* **2023**, *16*, 7284. [CrossRef]
18. Bartosik, A.S. Effect of the solid particle diameter on frictional loss and heat exchange in a turbulent slurry flow: Experiments and predictions in a vertical pipe. *Energies* **2023**, *16*, 6451. [CrossRef]
19. Chen, J.; Zhang, K.; Kan, Q.; Yin, H.; Sun, Q. Ul-tra-high fatigue life of NiTi cylinders for compression-based elastocaloric cooling. *Appl. Phys. Lett.* **2019**, *115*, 093902. [CrossRef]
20. Tušek, J.; Engelbrecht, K.; Mikkelsen, L.P.; Pryds, N. Elastocaloric effect of Ni-Ti wire for application in a cooling device. *J. Appl. Phys.* **2015**, *117*, 124901. [CrossRef]
21. Wang, R.; Fang, S.; Xiao, Y.; Gao, E.; Jiang, N.; Li, Y.; Mou, L.; Shen, Y.; Zhao, W.; Li, S.; et al. Torsional refrigeration by twisted, coiled, and supercoiled fibers. *Science* **2019**, *366*, 216–221. [CrossRef]
22. Greibich, F.; Schwödiauer, R.; Mao, G.; Wirthl, D.; Drack, M.; Baumgartner, R.; Kogler, A.; Stadlbauer, J.; Bauer, S.; Arnold, N.; et al. Elastocaloric heat pump with specific cooling power of 20.9 Wg–1 exploiting snap-through instability and strain-induced crystallization. *Nat. Energy* **2021**, *6*, 260–267. [CrossRef]
23. Ahcin, Ž.; Kabirifar, P.; Porenta, L.; Brojan, M.; Tušek, J. Numerical modeling of shell-and-tube-like elastocaloric regenerator. *Energies* **2022**, *15*, 9253. [CrossRef]
24. Zhu, Y.; Tsuruta, R.; Gupta, R.; Nam, T. Feasibility investigation of attitude control with shape memory alloy actuator on a tethered wing. *Energies* **2023**, *16*, 5691. [CrossRef]
25. Cirillo, L.; Greco, A.; Masselli, C. Cooling through barocaloric effect: A review of the state of the art up to 2022. *Therm. Sci. Eng. Prog.* **2022**, *33*, 101380. [CrossRef]
26. Aprea, C.; Greco, A.; Maiorino, A.; Masselli, C. The use of barocaloric effect for energy saving in a domestic refrigerator with ethylene-glycol based nanofluids: A numerical analysis and a comparison with a vapor com-pression cooler. *Energy* **2020**, *190*, 116404. [CrossRef]
27. Cirillo, L.; Greco, A.; Masselli, C. The Application of Barocaloric Solid-State Cooling in the Cold Food Chain for Carbon Footprint Reduction. *Energies* **2023**, *16*, 6436. [CrossRef]
28. Yang, Y.; Fu, G.; Zhao, J.; Gu, L. Heat production capacity simulation and parameter sensitivity analysis in the process of thermal reservoir development. *Energies* **2023**, *16*, 7258. [CrossRef]
29. Hu, P.; Zhang, Q.; Liu, Y.; Sheng, C.; Cheng, X.; Chen, Z. Optical analysis of a hybrid solar concentrating Photovoltaic/Thermal (CPV/T) system with beam splitting technique. *Sci. China Technol. Sci.* **2013**, *56*, 1387–1394. [CrossRef]
30. Liu, M.; Du, M.; Long, G.; Wang, H.; Qin, W.; Zhang, D.; Ye, S.; Liu, S.; Shi, J.; Liang, Z.; et al. Iron/Quinone-based all-in-one solar rechargeable flow cell for highly efficient solar energy conversion and storage. *Nano Energy* **2020**, *76*, 104907. [CrossRef]
31. Lu, L.; Tian, R.; Han, X. Optimization of nanofluid flow and temperature uniformity in the spectral beam splitting module of PV/T system. *Energies* **2023**, *16*, 4666. [CrossRef]

energies

MDPI

Article

Application of Particle Image Velocimetry and Computational Fluid Dynamics Methods for Analysis of Natural Convection over a Horizontal Heating Source

Natália Holešová, Richard Lenhard, Katarína Kaduchová and Michal Holubčík *

Department of Power Engineering, Faculty of Mechanical Engineering, University of Žilina, Univerzitná 8215/1, 010 26 Zilina, Slovakia; natalia.holesova@fstroj.uniza.sk (N.H.); richard.lenhard@fstroj.uniza.sk (R.L.); katarina.kaduchova@fstroj.uniza.sk (K.K.)

* Correspondence: michal.holubcik@fstroj.uniza.sk

Abstract: The objective of this article is to address the challenges associated with visualizing air flow over a heating source in an open laboratory environment. The study uses a combination of experimental visualization and numerical simulation techniques to generate a 3D model of the air flow and heat transfer between the heating source and the environment via natural convection. The Particle Image Velocimetry method is used to experimentally visualize the air flow, which is known for its benefits of high speed and accuracy, and for its ability to avoid disturbing the flow of the fluid being investigated. The data obtained from this experimental method are used as input for numerical simulations using the Ansys Fluent program. The numerical simulations identify air vortices and other elements that disrupt the airflow in the laboratory environment. The resulting 3D model accurately represents the actual situation in the laboratory and could be further optimized by adjusting parameters such as the output of the heater and the heating source temperature. These parameters play a crucial role in ensuring thermal comfort in the laboratory environment, which is of utmost importance for user comfort. In conclusion, the study provides valuable insights into the visualization of air flow over a heating source and demonstrates the effectiveness of combining experimental and numerical simulation techniques to generate accurate 3D models of air flow and heat transfer.

Keywords: natural convection; numerical simulation; PIV method; CFD method; heating source

Citation: Holešová, N.; Lenhard, R.; Kaduchová, K.; Holubčík, M. Application of Particle Image Velocimetry and Computational Fluid Dynamics Methods for Analysis of Natural Convection over a Horizontal Heating Source. *Energies* **2023**, *16*, 4066. https://doi.org/10.3390/en16104066

Academic Editor: Artur Bartosik

Received: 28 March 2023
Revised: 26 April 2023
Accepted: 27 April 2023
Published: 12 May 2023

1. Introduction

The quality of the environment has become a significant concern in recent times due to its impact on human health. The air flow within interiors, where people spend a large portion of their day, is influenced by various sources, such as heating, air conditioning, and ventilation [1]. The airflow affects thermal comfort and the concentration of pollutants in the air [2].

Optical methods, which has seen an unprecedented increase in computational and acquisition technologies in recent years, such as Particle Image Velocimetry (PIV), can be used to analyse the air flow within interiors. PIV is a non-contact method that captures the instantaneous velocity and properties of fluid flow by recording the time position of trace particles in the air flow.

This article focuses on visualizing the airflow over the surface of a heating source that uses natural convection to transfer heat to the laboratory environment. The heat from the heating foil affects the thermal comfort in the laboratory. The aim of the study is to measure the air speed above the surface of the heating foil experimentally at its maximum performance using PIV optical method and to create a numerical simulation based on these experimental data. The simulation is then used to optimize the air parameters to ensure user thermal comfort within the laboratory.

Natural convective flow occurs in fluids due to density gradients that are in turn developed due to temperature differences in the fluid. Accordingly, free or natural convection has been studied using similarity theory [3], integral theory [4], and by using experiments [5].

The analysis of natural convection from oriented heat exchange walls is the basis of heat transfer books and forms the starting point for research in this field [6–8]. The way of solution of natural convection occurs over a horizontal plate is different from how it occurs around a vertical plate. The mechanism for this is known as indirect natural convection [9], which is explained in detail for example in the work of Guha and Pradhan [10].

Thermal and velocity boundary layers are typical for convective flow. The presence of boundary layers in this type of flow was confirmed by Stewartson [11]. Researchers, such as Chen et al. [12] and Ghiaasiaan [6], have studied the dependencies of the Nusselt number as a function of Prandtl and Grashof numbers. They found that the velocity boundary layer is thicker than the thermal boundary layer only when the Prandtl number is large ($Pr > 1$). If the Prandtl number is close to 1 ($Pr < 1$, $Pr = 1$)), the boundary layers are of comparable thickness.

Numerous papers on computational fluid dynamics (CFD), which uses basic transport, state equations of fluid mechanics and criterion equations [13], have focused on natural convection [14] over a heated horizontal plate. These papers delve into the fundamental physical and thermodynamic understanding of natural convection. Guha and Sengupta [15] used post-processing of the CFD simulations to show the quantitative details of the 3D velocity and temperature fields, depending on the model geometry. In the work of Guha, Jain, and Pradhan [16], a CFD study of natural convection around heated inclined plates was conducted. The study provided an understanding of the contours of velocity, temperature, and pressure as a function of inclination angle over the entire range from vertical to horizontal and the calculation of the Nusselt number.

In addition, natural convection is a crucial phenomenon that appears in many technological devices, particularly in heat sources and climatization. Research in these areas needs to continue to ensure user comfort and improve the quality of the environment.

Natural Convection and Its Boundary Layers

Convective heat transfer is the transfer of heat through a fluid in motion, also called convection. The process of convective heat transfer involves the transfer of both enthalpy and kinetic energy resulting from the large-scale movement of the fluid- macroscopic movement [17].

Typically, convective heat transfer happens in the presence of a solid surface and a moving fluid, wherein a layer of fluid near the surface experiences a temperature change either by heating or cooling owing to the difference in temperature between it and the surrounding fluid. The disparity in density produced due to the temperature change creates a buoyancy force, which propels the movement of fluid flow close to the surface. This process is called free or natural convection, where the fluid behind the heated or cooled layer remains stationary (zero velocity) with a constant temperature [8,18].

When studying natural convection, the boundary layer theory is utilized, which employs comparable physical laws and techniques as those used in forced convection analysis. However, it is essential to factor in buoyancy forces while creating the momentum equation, and the boundary conditions are distinct [19,20]. During natural convection, the solid surface and the boundary layer's edge remain motionless, while the bulk fluid remains at rest [8].

The nature of the flow in natural convection is determined based on the Grashof number, which is a dimensionless quantity. This number, represented by Gr, is calculated by comparing the buoyancy force to the viscous force in a fluid and is given by the following equation:

$$\mathrm{Gr} = \frac{L^3 g \beta \Delta T}{\nu^2} \tag{1}$$

The Grashof number is used to predict the onset of natural convection in a fluid. When the Grashof number is less than a critical value, the flow in the fluid is dominated by the viscosity of the fluid, and conduction is the primary mode of heat transfer. When th Grashof number is greater than the critical value, buoyancy forces become significant, an natural convection occurs.

The Grashof number is an important parameter in many engineering application including heat exchangers, cooling systems, and geothermal energy.

The velocity field of the fluid in natural convection is reliant on the temperature, whic necessitates iterative computations for analysis [21].

Four different categories of natural convection can transpire around a horizontal plat and the classification of each group depends on the relative temperature of the plate' surface with respect to the flowing fluid around it. It is essential to adapt the relations o natural convection for each type, as well as determining the direction of fluid flow toward the surface [22].

Figure 1 illustrates the four types of natural convection around a horizontal plate. I cases where the surface is hot and isolated from the upper side, the fluid tends to rise, bu the plate's presence hinders the upward flow, forcing the fluid to move around the plat before flowing freely around the edges (Figure 1a). Similarly, in Figure 1b, the temperatur of the plate is cooler than the surrounding fluid, causing the flow to be directed downward Conversely, in Figure 1c,d, the plate's temperature is warmer than the fluid, causing th fluid to move upwards and downwards, respectively.

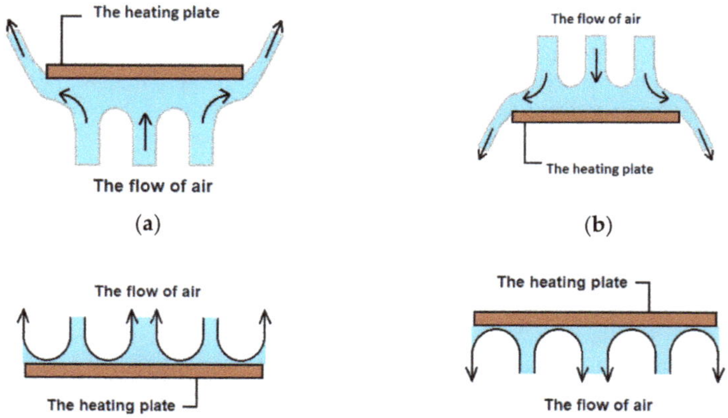

(**a**) (**b**) (**c**) (**d**)

Figure 1. Categories of natural convection occurring around a horizontal plate. (**a**) Heated horizonta facing down ($T_{plate} > T_{air}$) (**b**) Cooled horizontal facing up ($T_{plate} < T_{air}$) (**c**) Heated horizontal facing up ($T_{plate} > T_{air}$) (**d**) Cooled horizontal facing down ($T_{plate} < T_{air}$).

The focus of this work is on a specific situation where a heating foil is situated o a desk, and the air flows over this heating source. It is crucial to determine the type o natural convection occurring in this scenario to optimize the heating efficiency and ensur proper ventilation. By analysing the temperature of the heating source relative to th surrounding fluid and the direction of fluid flow towards the surface, one can determin the appropriate relations for natural convection and make informed decisions regarding heating optimization.

2. Visualization Methods

The text elucidates two approaches to observe the airflow above a heating plate: experimental method called Particle Image Velocimetry (PIV) and a computational method involving numerical simulation using Ansys Fluent. The objective was to collect information on the velocity vectors of air present above the heating foil.

2.1. Measuring Airflow through Particle Image Velocimetry

The non-contact optical technique, PIV, is utilized to compute fluid velocity. This method necessitates injecting minute tracer particles into the fluid stream and then exposing the fluid to intense illumination.

The tracer particles move along with the fluid and reflect light, which is captured by imaging devices. The PIV method determines the rate of motion of the tracer particles instead of the fluid stream, as the tracer particles function as data carriers. The accuracy of the PIV technique depends on the selection and incorporation of the tracer particles into the fluid, as the particles are recorded in images and their local velocities are analysed using computer software [23,24].

The size of the tracer particles is critical to the accuracy of the PIV method, as they must be large enough to monitor the fluid flow and to reliably capture the small turbulences of the current structure. However, the particles must not be too small, as their scattered light may not be captured by the scanning technique. To sum up, the PIV technique includes the introduction of tracer particles into a fluid stream, lighting up the fluid with a powerful light source, and employing imaging devices and computer programs to gauge the velocity of the tracer particles. The accuracy of the PIV technique depends on the careful selection and incorporation of tracer particles into the fluid [22,25–27].

2.2. An Experiment

A laboratory experiment involved placing a heating foil on a worktable and creating a measuring plane around it using a high-powered laser with a wavelength of 532 nm, energy of 1200 mJ, and maximum power of 2000 W. The laser used was a two-chamber Nd: YAG pulsating laser CLASS 4. To capture the scattered light from tracer particles, a Flowsense EO high-speed CCD camera equipped with a matrix detector was used. The measuring plane was transformed into a light section by employing cylindrical optics. The FT700 CE particle generator from Dantec was used to create tracer particles in the form of oil droplets with sizes between 1 μm and 5 μm, using a maximum pressure of 4 bar and a volume of 4.5 l. However, the oil droplets alone were insufficient to capture the air flow around the heating foil in detail. To address this, a fog generator F2010 Plus was employed to create smaller particles (in size 1 μm) that provided more accurate information about the air flow.

The PIV technique involved taking images of the tracer particles' starting and ending positions, which were analyzed using computer software to produce vector maps representing the velocity fields. The study utilized the PIV method to capture the flow of the fluid around the heating foil, which is demonstrated in Figure 2 of the research.

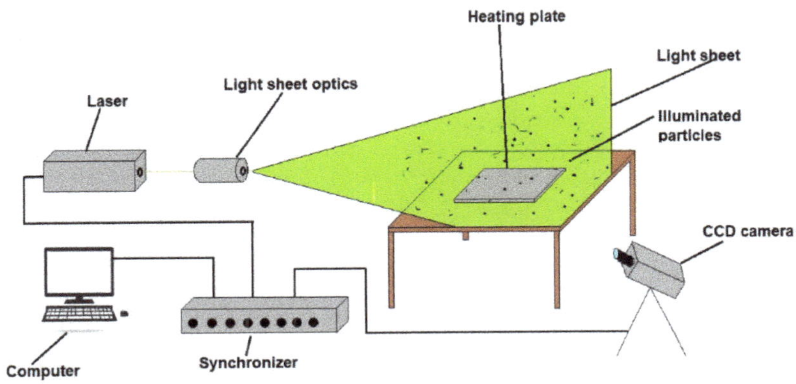

Figure 2. The scheme of PIV method.

In summary, the laboratory experiment involved using the PIV method to measure the fluid flow around a heating foil. Tracer particles were produced with a particle generator and their motion was recorded using a high-speed CCD camera. The resulting images were analyzed with software to generate velocity maps of the fluid flow. For better visualization, a fog generator was employed to produce smaller particles. The PIV method process is illustrated in Figure 2.

Figure 3 depicts the experimental setup, which includes a heating self-adhesive silicone foil (TF542) with specific dimensions (0.3 m × 0.2 m × 0.0013 m) and temperature capabilities—maximum temperature +200 °C. The foil generates a certain amount of power and affects the surrounding air flow through natural convection.

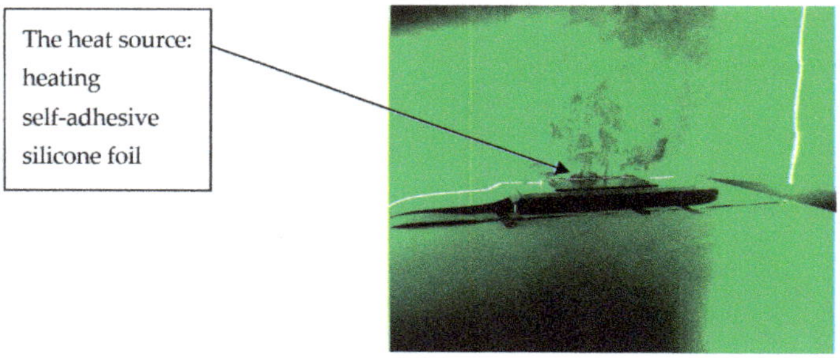

Figure 3. Image of the Experimental Process and Selected Heating Element.

2.3. Data Processing

Data processing is a critical aspect of PIV analysis as it involves extracting meaningful information from the raw data obtained from the PIV experiment. The data processing steps involve various image processing, signal processing, and data analysis techniques to obtain accurate and reliable velocity measurements.

Before conducting any PIV experiments, it is necessary to perform a camera calibration to establish the scale between the object and image planes. In this study, the scale factor was determined, and 135 images were captured by CCD camera using the *Image Run* function within a very short time interval of 1000 μs, after successful calibration. To avoid measurement errors caused by the onset and closing of the laser, images that could have a large measurement error were filtered out [28,29].

Subsequently, the *Image Masking* function was then used to eliminate interfering elements and unnecessary areas from the images. In this experiment, a disturbing area from the bottom of the heating foil was removed because the foil had a glossy surface that reflected the laser beam.

The subsequent step in the procedure was to analyse and correlate the evaluation images using an adaptive correlation multi-stage algorithm based on cross-correlations. The investigation area was set at 64 × 64, with a *Grid Step Size* of 16 × 16 pixels. The *Range Validation* function was implemented to limit the size range of the vectors to be considered as accurate, within the speed range of 0 $m \cdot s^{-1}$ to 0.5 $m \cdot s^{-1}$. Vectors that did not fall within this range were removed from the evaluation. The *Moving Average* function was then applied to adjust the sizes of the vectors according to the neighbouring ones, creating a smooth velocity field without protruding vectors.

At the end of the evaluation, the *Peak Validation* function was used, setting a limit ratio of 1.2 px. The resulting vector field of air velocities after using the *Peak Validation* function is displayed in Figure 4.

Overall, data processing is a critical step in PIV analysis as it determines the accuracy and reliability of the velocity measurements. The quality of the PIV data can be improved by optimizing the processing parameters and using appropriate data filtering and smoothing techniques.

Figure 4. Vector field of axial air velocities after using the function Peak Validation in experimental method PIV.

2.4. The Numerical Simulation of the Air Flow over the Heating Foil

The ANSYS Fluent software was used to perform a numerical simulation of the air flow around the heating foil in an open laboratory environment. The simulation involved a heating foil placed on a worktable, with length, width, and height dimensions of 300 mm, 200 mm, and 10 mm, respectively. The working fluid used in the simulation was air, and a simplified 3D model was employed.

The geometric parameters were inputted to create the 3D model in Design Modeler, which is depicted in Figure 5 and shows the model's geometry and the surfaces under investigation. The dimensions of the model are presented in Table 1.

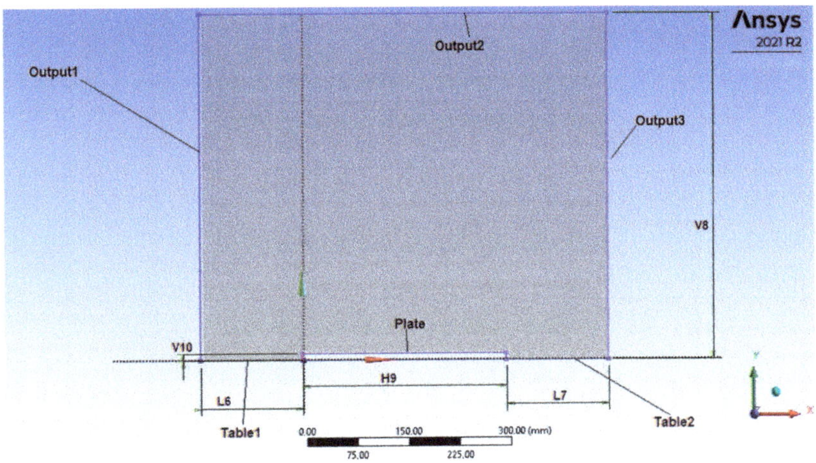

Figure 5. Geometry of CFD model, in the axial direction.

Table 1. Dimensions of surfaces of model.

Dimensions	(m)
H9	0.300 m
L7	0.150 m
L6	0.150 m
V10	0.010 m
V8	0.500 m

In order to model heat transfer, particularly natural convection, a mesh was created and adapted for this purpose. The Inflation function was used to increase the number of cells around the heating foil's surface, with cell sizes ranging from 0.001 m to 0.005 m and a total of 3,343,815 cells [30].

As the mathematical model in the ANSYS Fluent program was used a viscous model with laminar flow. This mathematical model is a commonly used model in ANSYS Fluent, a computational fluid dynamics software package. It is used to simulate the flow of a fluid that is both viscous and laminar. The viscous laminar model is generally suitable for simulating natural convection, which is the flow that occurs when a fluid is heated and rises due to buoyancy effects. In natural convection, the fluid flow is typically slow and laminar, which makes the viscous laminar model a good choice for simulating this type of flow. It is important to consider the specific characteristics of the flow being simulated and to choose the modelling approach that is best suited for the particular situation.

In this model, the fluid is assumed to be incompressible, and the flow is governed by the Navier-Stokes equations, which describe the motion of fluids. The model assumes that the flow is laminar, meaning that the fluid particles move in parallel layers without mixing or turbulence. This is in contrast to turbulent flow, which is characterized by chaotic swirling motion.

The viscous laminar model also takes into account the viscosity of the fluid, which is a measure of its resistance to deformation. Viscosity is important in determining the amount of frictional forces between the layers of fluid, which in turn affects the overall flow behaviour.

Therefore, a laminar viscous model was chosen to simulate this type of solved problem with using a Solver based on pressure with steady time and gravitational acceleration on the Y direction.

To use the viscous laminar model in ANSYS Fluent, it is typically needed to specify the properties of the fluid being simulated, such as its density, viscosity, and other relevant parameters. Very important is also to define the geometry of the simulation domain and set up appropriate boundary conditions to simulate the desired flow behaviour.

The equations that this mathematical model uses are:

1. Continuity equation, which expresses the conservation of mass, stating that the rate of change of mass in any given volume of the fluid must be balanced by the rate of flow of mass into or out of that volume.

$$\nabla \cdot \mathrm{v} = 0 \tag{2}$$

2. Momentum equation, which expresses the conservation of momentum, stating that the rate of change of momentum of any given volume of the fluid must be balanced by the sum of the forces acting on that volume. The first term on the right-hand side represents the pressure gradient, which drives the fluid flow. The second term represents the viscous forces, which oppose the flow and are proportional to the rate of change of velocity with respect to distance.

$$\rho \cdot \left(\frac{\partial \mathrm{v}}{\partial t} + \mathrm{v} \cdot \nabla \mathrm{v} \right) = -\nabla \mathrm{p} + \mu \nabla^2 \mathrm{v} \tag{3}$$

Tables 2–6 provide details on the boundary, operational, material, and general conditions, as well as the set parameters and additional simulation settings.

Table 2. Material conditions.

Material	Density ($kg \times m^{-3}$)	Specific Heat ($J \times kg^{-1} \times K^{-1}$)	Thermal Conductivity ($W \times m^{-1} \times K^{-1}$)	Viscosity ($kg \times m^{-1} \times s^{-1}$)	Thermal Expansion Coefficient (K^{-1})
air	1.2047	1006.43	0.0242	sutherland	0.00343
aluminum	2719	871	202.4	–	–

Table 3. Boundary conditions.

Name	Type of Condition	Set Parameters	
		Velocity ($m \cdot s^{-1}$)	**Temperature (°C)**
plate 1	No slip wall	0	124
plate 2	No slip wall	0	124
plate 3	No slip wall	0	124
Table 1	No slip wall	0	23.20
Table 2	No slip wall	0	23.20
output 1	Pressure-outlet	–	26.85
output 2	Pressure-outlet	–	26.85
output 3	Pressure-outlet	–	26.85

Table 4. Solution Methods.

Pressure-Velocity Coupling	
Scheme	coupled
Spatial Discretization	
Gradient	Green-Gauss Node Based
Pressure	Second Order
Momentum	Second Order Upwind
Energy	Second Order Upwind

Table 5. Solution controls.

Pseudo Transient Explicit Relaxation Factors	
Pressure	0.5
Momentum	0.5
Density	1
Body Forces	1
Energy	0.75

Table 6. Run calculation.

Pseudo Transient Settings	
Time step method	Automatic
Time scale factor	1
Length scale method	Conservative
Verbosity	0

The settings and inputs listed in the tables can be employed to conduct a numerical simulation of heat and mass transfer in an open space, using either a 2D or 3D model. In this particular case, the simulation was run for 500 iterations, but the boundary, material, and operational conditions can be modified to suit the specific characteristics of the system being modelled.

3. Results

To observe the airflow surrounding a heating foil, an experimental PIV (Particle Image Velocimetry) method was implemented. The approach involved using a high-speed camera to capture images of the light scattered from tracer particles within the fluid stream. The resulting images were subsequently analyzed using computer software to generate velocity maps along the entirety of the heating foil, pinpointing areas of peak velocity and temperature values of the airflow. The average temperature of the airflow was 26.85 °C, while the heating foil had 124 °C and the average air velocity was approximately 0.2 $m \cdot s^{-1}$.

These values were then used as inputs for a numerical simulation, which resulted in obtaining contours of temperature and airflow velocity for a 3D model of natural convection. Figure 6 shows the temperature contour, and Figure 7 shows the airflow velocity contour. The velocity contour demonstrates the creation of disturbing elements and air vortices due to the open space in the laboratory, which is expected in a real situation where there are other equipment influencing the heat transfer of the experiment. The details of these disturbing elements are shown in Figure 8.

The contours obtained from the numerical simulation show that the maximum temperature of 124 °C is on the surface of the heating foil, and the temperature gradually decreases with increasing distance from the surface until it stabilizes at the ambient air temperature of 26.85 °C. The velocity distribution map indicates that the airflow is stagnant at the top of the heating foil and in its immediate surroundings, but gradually gains speed as the distance from the surface of the heating foil increases. The highest velocity value obtained in this model is 0.250 m/s.

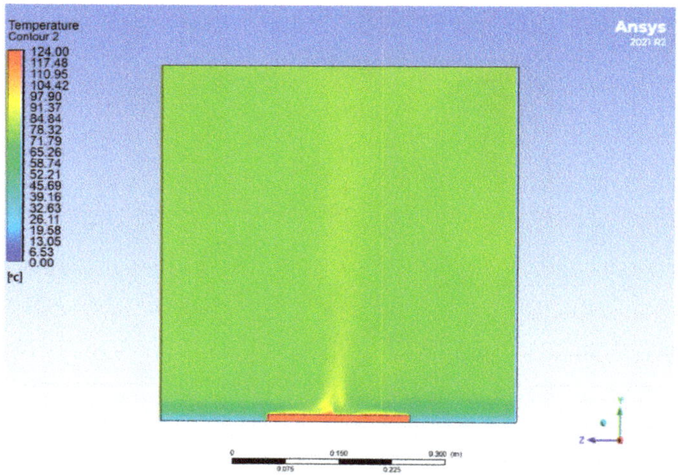

Figure 6. The contour of temperature in CFD in axial direction.

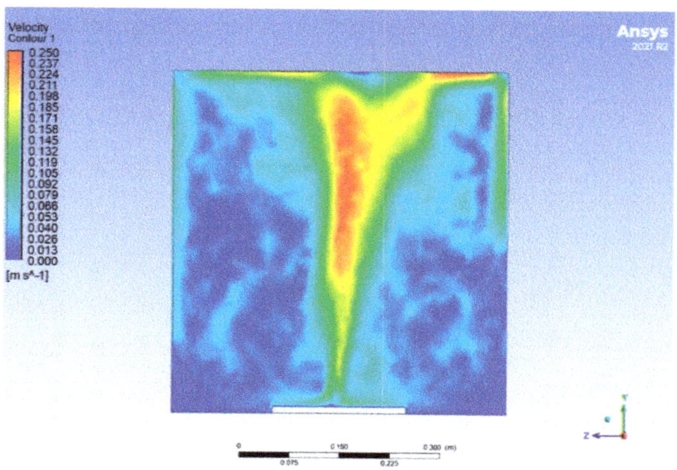

Figure 7. The contour of axial air velocity around the heating foil in CFD.

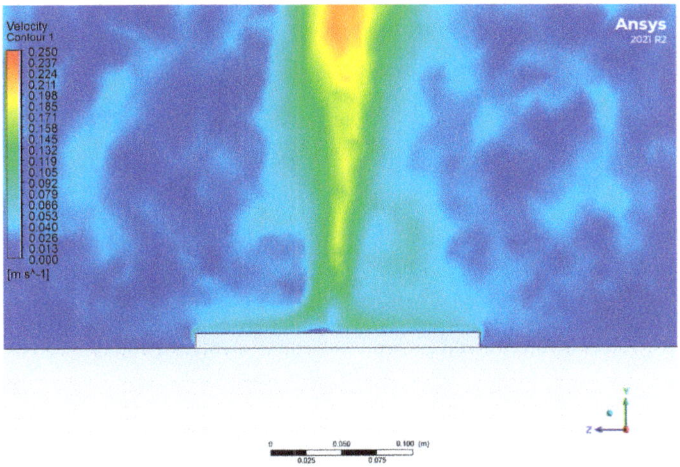

Figure 8. Detail of disturbing elements and air vortices from CFD.

Straight lines were drawn on the images generated by the PIV method using an internal function, which allowed for the plotting of air velocities at predetermined heights above the surface of the heating foil. The straight lines were plotted at several distances above the surface of the heating foil in the range from 0.02 m to 0.4 m in different increments—0.02 m, 0.03 m, 0.04 m, 0.05 m, 0.06 m, 0.08 m, 0.1 m, 0.15 m, 0.2 m, 0.3 m, and 0.4 m.

According to natural convection theory, a boundary layer should exist which implies that there will be no velocity in the immediate vicinity above the heating foil. Figures 9 and 10 exhibit the correlation between air velocity and particular heights above the heating foil's entire surface.

Figure 11 illustrates a comparison between the velocity curves derived from experimental measurements using PIV and the CFD simulation, taken at a height of 150 mm above the heating foil surface. It is apparent that the experimental speeds reach lower values than the simulation, likely due to various external influences that are difficult to account for in a numerical model. These external factors can cause random phenomena that occur in a real environment but are difficult to predict in a simulated one. The maximum air velocities at this height are 0.1720 m·s^{-1} in the CFD simulation, located in the middle of the plate, and 0.1614 m·s^{-1} in the PIV measurement, with the peak shifting more towards the wall due to interfering elements.

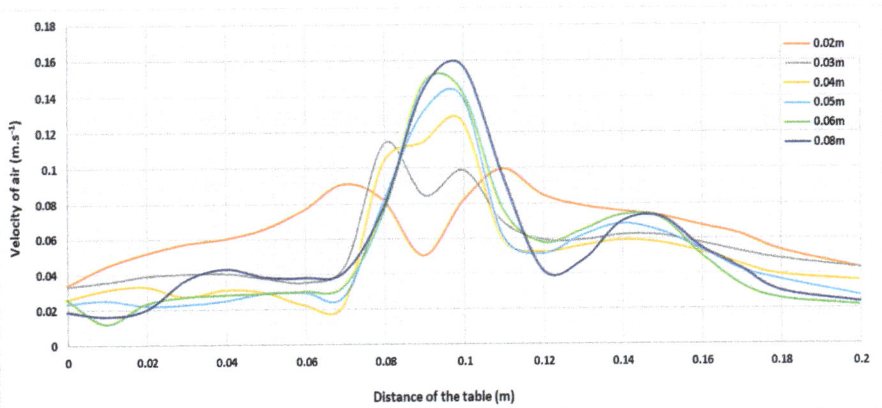

Figure 9. The axial air velocity dependence along the entire heating foil by PIV method.

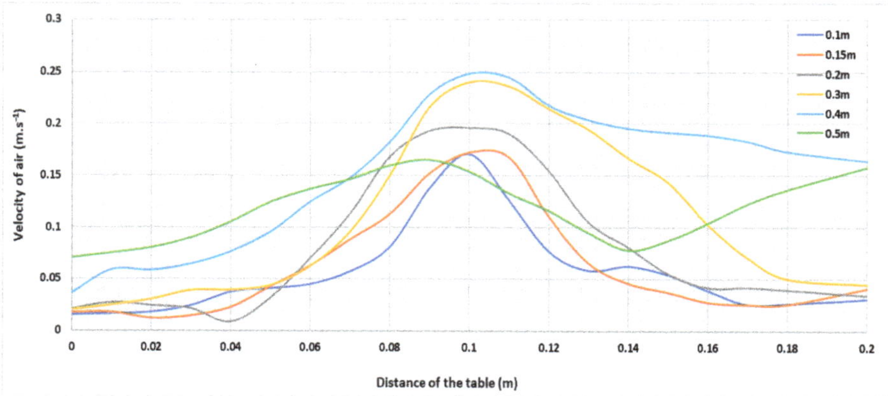

Figure 10. The axial air velocity dependence along the entire heating foil by PIV metho (another case).

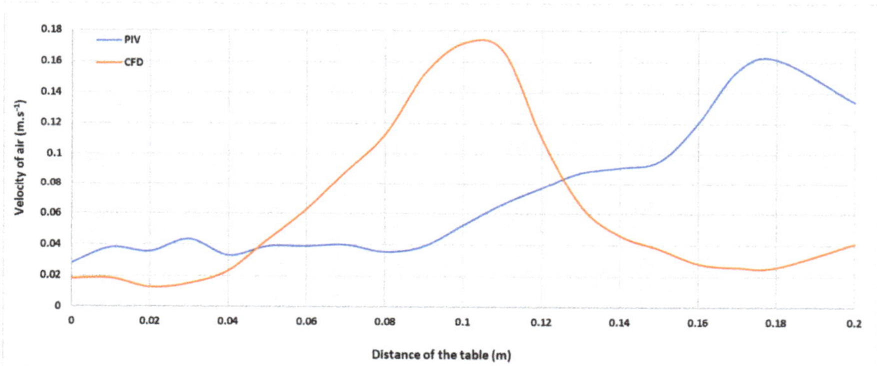

Figure 11. The axial air velocity at a height of 150 mm above the heating foil by PIV compared wit CFD method in the same height.

The velocity of the air around a heating horizontal source due to natural convectio depends on several factors, including the size and shape of the source, the temperatur difference between the source and the surrounding fluid, and the properties of the fluic such as its density, viscosity, and thermal conductivity.

In general, as the temperature of the source increases, the air around the sourc becomes less dense and rises, creating an upward flow of air. As this warm air rise it creates a region of low pressure, which draws in cooler air from the surroundings t replace it. This cooler air is then heated by the source and rises in turn, creating a cycle o fluid motion.

The velocity of the air in this convection cell depends on the temperature differenc between the source and the surrounding fluid, as well as the size and shape of the sourc The velocity of the air can be calculated using equations derived from fluid mechanic such as the Grashof number, which relates the buoyancy force to the viscosity and therma diffusivity of the fluid. It is worth noting that the velocity of the air due to natural convectio around a heating source is generally much lower than that due to forced convection, whic is fluid flow that is driven by an external force, such as a fan or pump.

4. Conclusions and Discussion

This article discusses the visualization of airflow in the vicinity of a horizontal heatin source subjected to natural convection in an open laboratory environment. Two meth ods were used to visualize the airflow: experimental Particle Image Velocimetry (PIV) and numerical simulation in Ansys Fluent. The velocity of air was the main parameter monitored in both methods. The experimental results were used as input data for the numerical simulation.

The simulation's findings indicated the existence of air swirls and disruptive components. These were recognized by charting the correlation between air speed across the whole heating foil at specific elevations above it. Additionally, the simulation recorded the speed boundary layer near the heating foil, with the highest air velocities situated in the middle of the heating foil.

To sum up, the successful visualization of airflow was accomplished by utilizing both PIV and numerical simulation methods. The implementation of PIV allowed for the detection of air vortices and other environmental issues, while numerical simulation enabled the optimization of the airflow. Additionally, by modifying the input data, the 3D model of natural convection over the horizontal heating foil to open space of the laboratory can be used for similar simulations by changing the input data.

The visualization results are important for achieving optimal parameters, such as the output of the heater and the heating source temperature, which affect the thermal comfort in the interior and ensure comfort for the user. The results can also help in better understanding the vorticity of flow in the environment and prevent disturbing elements that create this vorticity. The research was influenced by many external factors, which would also need to be tested at different values. The vision would be the creation of a calibration graph in which the dependences of the performance of such internal heating sources on external parameters would be plotted, so that the thermal comfort would be achieved through the proper calibration. The presence of air vortices together with dust from surrounding areas can cause discomfort in the workplace and human health problems. Therefore, this visualization can contribute to improving the overall environment quality and well-being of the occupants.

Author Contributions: N.H. conceptualization, investigation, writing—original draft preparation, data curation; R.L.; writing—review and editing, funding acquisition, methodology; K.K. data curation, resources; M.H. data curation, methodology, funding acquisition. All authors have read and agreed to the published version of the manuscript.

Funding: The work was created with the support of the project KEGA 047ŽU-4/2022 Innovative approaches to fluid dynamics modelling in energy systems, KEGA 048ŽU-4/2019: Visualization of flow in environmental engineering and KEGA 046ŽU-4/2021: Innovative methods of increasing the thermal efficiency of small heat sources by heat recovery through phase transformations.

Data Availability Statement: Data available in a publicly accessible repository.

Conflicts of Interest: The authors declare no conflict of interest.

Nomenclature

Nomenclature and related units for abbreviations.

CFD	computational fluid dynamics
g	gravitational constant ($m \cdot s^{-2}$)
Gr	Grashof number
L	characteristic dimension (m)
Nu	Nusselt number (−)
PIV	Particle Image Velocimetry
Pr	Prandtl number (−)
T	temperature of fluid (K)
T_∞	temperature of wall (K)
ΔT	difference between wall and fluid temperature (K)
u, v	velocity components in the direction of the x, y, z axis ($m \cdot s^{-1}$)
x, y	axes in 2D coordinate system (m)
β	coefficient of volumetric thermal expansion at reference temperature (K^{-1})
ν	kinematic viscosity ($m^2 \cdot s^{-1}$)

References

1. Orman, Ł.J.; Majewski, G.; Radek, N.; Pietraszek, J. Analysis of Thermal Comfort in Intelligent and Traditional Buildings. *Energies* **2022**, *15*, 6522. [CrossRef]
2. Čajová Kantová, N.; Sładek, S.; Jandačka, J.; Čaja, A.; Nosek, R. Simulation of Biomass Combustion with Modified Flue Gas Tract. *Appl. Sci.* **2021**, *11*, 1278. [CrossRef]
3. Guha, A.; Samanta, S. A similarity theory for natural convection from a horizontal plate for prescribed heat flux or wall temperature. *Int. J. Heat Mass Transf.* **2012**, *55*, 3857–3868.
4. Guha, A.; Samanta, S. Closed-form analytical solutions for laminar natural convection on horizontal plates. *J. Heat Transf.* **2013**, *135*, 102501. [CrossRef]
5. Rotem, Z.; Claassen, L. Natural convection above unconfined horizontal surfaces. *J. Fluid Mech.* **1969**, *39*, 173–192. [CrossRef]
6. Ghiaasiaan, S.M. *Convective Heat and Mass Transfer*, 2nd ed.; CRC Press: Boca Raton, FL, USA, 2018.
7. Kothandaraman, C.P. *Fundamentals of Heat and Mass Transfer*; New Age International: New Delhi, India, 2006.
8. Incropera, F. *Fundamentals of Heat and Mass Transfer*; Wiley: New York, NY, USA, 2007.
9. Schlichting, H.; Gersten, K. *Boundary-Layer Theory*; Springer: Berlin/Heidelberg, Germany, 2000; ISBN 3-540-66270-7.
10. Guha, A.; Pradhan, K. A unified integral theory of laminar natural convection over surfaces at arbitrary inclination from horizontal to vertical. *Int. J. Therm. Sci.* **2011**, *111*, 475–490. [CrossRef]
11. Stewardson, K. On the free convection from a horizontal plate. *Z. Angew. Math. Phys. ZAMP* **1958**, *9*, 276–282. [CrossRef]
12. Chen, T.S.; Tien, H.C.; Armaly, B.F. Natural convection on horizontal, inclined and vertical plates with variable surface temperature or heat flux. *Int. J. Heat Mass Transf.* **1986**, *29*, 1465–1478. [CrossRef]
13. Brestovič, T.; Čarnogurská, M.; Příhoda, M.; Lázár, M.; Pyszko, R.; Jasminská, N. A Similarity Model of the Cooling Process of Fluids during Transportation. *Processes* **2021**, *9*, 802. [CrossRef]
14. Sheremet, M.A. Numerical Simulation of Convective Heat Transfer. *Energies* **2023**, *16*, 1761. [CrossRef]
15. Guha, A.; Sengupta, S. Effects of finiteness on the thermo-fluid-dynamics of natural convection above horizontal plates. *Phys. Fluids* **2016**, *28*, 063603. [CrossRef]
16. Guha, A.; Jain, A.; Pradhan, K. Computation and physical explanation of the thermo-fluid-dynamics of natural convection around heated inclined plates with inclination varying from horizontal to vertical. *Int. J. Heat Mass Transf.* **2019**, *135*, 1130–1151. [CrossRef]
17. Verein Deutscher Ingenieure. *VDI Heat Atlas*; Springer: Dusseldorf, Germany, 2010; ISBN 978-3-540-77876-9.
18. Pengying, Y.; Can, H.; Zhilang, Z.; Ting, L.; Moubin, L. Simulating natural convection with high Rayleigh numbers using the Smoothed Particle Hydrodynamics method. *Int. J. Heat Mass Transf.* **2021**, *166*, 120758.
19. Siddiqa, S.; Begum, N.; Hossain, M.A.; Abrar, M.N.; Gorla, R.S.R.; Al-Mdallal, Q. Effect of thermal radiation on conjugate natural convection flow of a micropolar fluid along a vertical surface. *Comput. Math. Appl.* **2021**, *83*, 74–83. [CrossRef]
20. Nemati, H.; Moradaghay, M.; Moghimi, M.A.; Meyer, J.P. Natural convection heat transfer over horizontal annular elliptical finned tubes. *Int. Commun. Heat Mass Transf.* **2020**, *118*, 104823. [CrossRef]
21. Lenhard, R. Natural Convection Heat Transfer around a Horizontal Circular Cylinder Near an Isothermal Vertical Wall. In *EPJ Web of Conferences*; EDP Sciences: Les Ulis, France, 2018; Volume 180, p. 02077.
22. Dholey, S. Natural Convection Flow of Power-Law Fluids over a Heated Horizontal Plate Surface. *Z. Nat. A* **2019**, *74*, 971–987. [CrossRef]
23. Ansari, A.; Nili-Ahmadabadi, M.; Tavakoli, M.R.; Minaeian, A. Experimental study of the effects of a vertical channel on the natural convection of a horizontal cylinder. *Therm. Sci.* **2021**, *25*, 1935–1948. [CrossRef]
24. Raffel, M. *Particle Image Velocimetry—A Practical Guide*; Springer: Berlin/Heidelberg, Germany, 2007.
25. Pavelek, M. *Vizualizační a Optické Měřicí Metody*; Fond rozvoje vysokých škol: Brno, Czech Republic, 2007.
26. Patsch, M. Simulation of Combustion Air Flow in the Gasification Biomass Boiler. In *MATEC Web of Conferences*; EDP Sciences: Les Ulis, France, 2018; Volume 168, p. 02015.
27. Yifan, F.; Yuguo, L. TIV and PIV based natural convection study over a square flat plate under stable stratification. *Int. J. Heat Mass Transf.* **2019**, *140*, 660–670.
28. Lin, T.; Zargar, O.A.; Lin, K.Y.; Juiña, O.; Sabusap, D.L.; Hu, S.C.; Leggett, G. An experimental study of the flow characteristics and velocity fields in an operating room with laminar airflow ventilation. *J. Build. Eng.* **2020**, *29*, 101184. [CrossRef]
29. Kopecký, V. *Laserová Anemometrie v Mechanice Tekutin*; Tribun EU: Liberec, Czech Republic, 2008.
30. Kozúbková, M. *Modelování Proudení Těkutin FLUENT, CFX*; VŠB—Technická Univerzita Ostrava: Ostrava, Czech Republic, 2008.

energies

MDPI

eview

.arge Eddy Simulation of Flow and Heat Transfer in a Ribbed Channel for the Internal Cooling Passage of a Gas Turbine Blade: A Review

on Ahn

School of Mechanical Engineering, Kookmin University, 77 Jeongneung-ro, Seongbuk-gu, Seoul 02707, Republic of Korea; jahn@kookmin.ac.kr

Abstract: Herein, 50 articles published over the past 20 years on using large eddy simulation (LES) for the internal cooling passage of a gas turbine, especially the mid-chord ribbed channel, are reviewed for the first time. First, the numerical challenges of performing LES on a ribbed channel and experimental verification are summarized. Next, LES data and the major engineering findings that are difficult to obtain experimentally or using Reynolds-averaged Navier–Stokes simulation (RANS) are covered, and heat transfer on and inside the rib, and the effects of rotation and buoyancy are discussed. Next, recent LES studies related to the shape of the ribbed channel are reviewed, and finally, the contribution of using LES for research on the internal cooling of gas turbines in the future, including those with ribbed channels, is anticipated.

Keywords: large eddy simulation (LES); local heat transfer; conjugate heat transfer; rotation; rib geometry

itation: Ahn, J. Large Eddy mulation of Flow and Heat Transfer a Ribbed Channel for the Internal ooling Passage of a Gas Turbine ade: A Review. *Energies* **2023**, *16*, 56. https://doi.org/10.3390/ 16093656

cademic Editor: Artur Bartosik

eceived: 6 March 2023
evised: 21 April 2023
ccepted: 21 April 2023
ublished: 24 April 2023

1. Introduction

Gas turbines are widely used for land-based power generation as well as aircraft and ship propulsion due to their large power-to-weight ratio. Increasing the turbine inlet total temperature (TITT) and compression ratio can improve the cycling efficiency and power output of gas turbines [1]. Advancements in turbine cooling have enabled engine design to exceed the temperature limitation of the commonly used materials [2]. As shown in Figure 1a, a turbine is cooled by extracting some of the compressor air while maximizing the cooling effect with the minimum amount of air [3].

As shown in Figure 1a, the coolant is supplied to the blade to cool the inside and then discharged through a hole in the surface to form an insulating film. As shown in Figure 1b, the cooling channel consists of multiple passes, and components that promote heat transfer are applied to each part. In most regions, ribs are installed as heat transfer promoters. An impinging jet is applied to the leading edge, and a pin fin or dimple is sometimes installed on the trailing edge where it is difficult to install a rib [4,5].

Recently, studies on heat transfer in a triangular channel simulating a trailing edge or pin fin have been conducted [6], and research on heat transfer in an impinging jet or a semicircular channel at a leading edge has also been published [7]. However, most of the middle portion can be seen as a ribbed channel [8], which is the most commonly studied among the gas turbine internal cooling methods [9].

The cooling passage of a gas turbine blade can be modeled as a ribbed channel (Figure 2a) [10]. The ribs promote heat transfer while separating/reattaching the flow in the passage (Figure 2b) [11,12]. Since the heat transfer augmentation effect of the ribs decreases as the heat flows downstream, several of them are installed periodically. To enhance the heat transfer effect, the downstream rib must be located after the reattachment point, so the installation period (p) is usually around 10 times the height of the rib (e) [13].

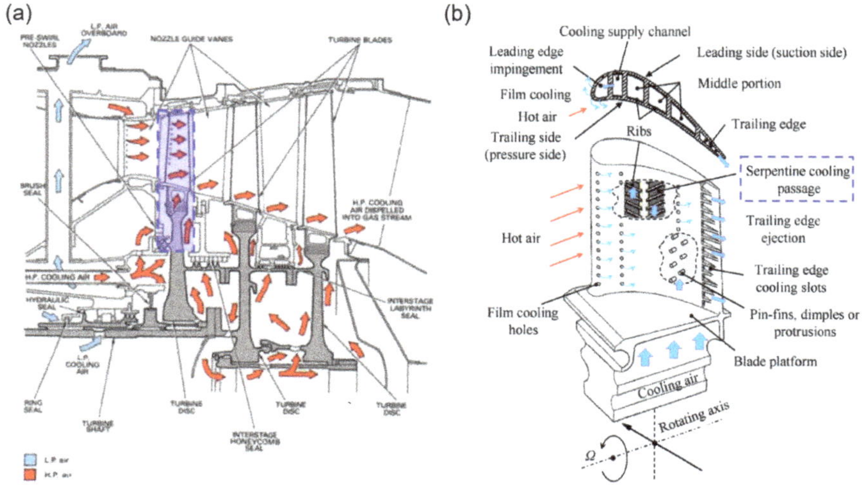

Figure 1. Cooling of gas turbine blades: (**a**) The air supply system for gas turbine cooling [3]; (**b**) ga turbine blades with cooling applied [4].

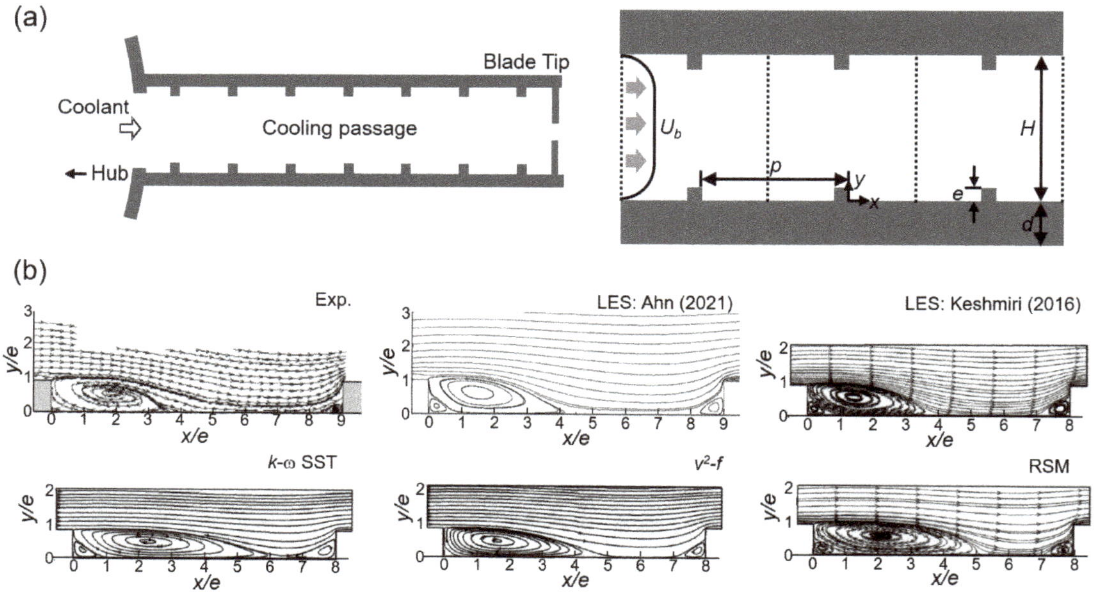

Figure 2. Modeling and flow patterns of gas turbine internal cooling passages: (**a**) the ribbed channe modeling cooling passage [11]; (**b**) streamlines in ribbed channels obtained experimentally [11] an using computational fluid dynamics (CFD) [10,12]. LES: large eddy simulation; *p*: installation perioc *e*: rib height; U_b: bulk velocity; *H*: channel height; *d*: wall thickness; *x*: streamwise coordinat *y*: wall-normal coordinate; SST: shear stress transport model; v', the velocity fluctuation normal t the streamlines; *f*: an elliptic relaxation function; RSM: Reynolds stress model.

As shown in Figure 2b, there is a recirculation area behind the rib where heat transfe is poor, and rib turbulators with various geometries were developed to improve this [14 Measurement techniques such as thermochromic liquid crystal (TLC) [15] or infrarec cameras [16] are applied to obtain the local heat transfer distribution. When applying these techniques to the experimental setup to consider the rotation effect, it is sometimes difficult to rotate the optical equipment such as a camera along with the test section [17].

The rib also serves as a fin that expands the heat exchange area, while promoting heat transfer in the experimentally obtained channel heat transfer data on ribs is very rare because it is difficult to impose boundary conditions on them while gaining optical access from the front and back sides of the ribs [18]. By performing computational fluid dynamics (CFDs) simulation, the heat transfer distribution of the rib can be obtained and the effect of rotation, which is also difficult to deal with experimentally, can also be observed [19].

The Reynolds-averaged Navier–Stokes simulation (RANS) technique depending on a semi-empirical model is most often used to analyze the fluid flow and heat transfer in the gas turbine blades' cooling channel [19]. However, a variation of around 250% was found between different RANS models when predicting ribbed channel flow heat transfer [20]. In recent years, large eddy simulation (LES) has been used to study the internal cooling of turbine blades as it reduces the uncertainty of heat transfer prediction using RANS to within 1/4 [21].

Research on ribbed channels applied to gas turbine cooling passages began in the 1970s, with the first review papers being published in this recently [22], mostly including experimental results [7,17]. In the most recent reviews of CFD data on heat transfer in ribbed channels [23], the target was a solar heater rather than a gas turbine blade, and thus analysis of the rotation effect, etc., was not included and only RANS results were reported. CFD studies on the ribbed channel of a gas turbine blade were reviewed in 1995 [24], which is before using LES started in earnest.

The first articles on using LES to analyze gas turbine cooling passages were published in the early 2000s [25], with over 50 appearing over the past 20 years. LES can be used to predict the heat transfer mechanism more closely to the experimental findings than RANS, and provides information that is difficult to obtain experimentally, such as on instantaneous temperature fields and turbulent heat transfer. The internal cooling passage problem in a gas turbine includes the heat transfer mechanism that is difficult to predict using RANS. Since there are relatively few difficulties in the numerical analysis using LES, we anticipate that it will become more widely used in the industry in the future [20,21].

The LES results for ribbed channels from 50 articles are summarized in Table 1. In this review, the processes that are not easy to predict using RANS but are so using LES are highlighted. Subsequently, the LES results for heat transfer and conjugate heat transfer in the rib, for which experimental data are scarce, are summarized. Next, LES data on the rotation effect and rib geometry of the internal cooling passage are reviewed [3,5]. Finally, the future direction for research using LES to analyze the cooling passage of a gas turbine is covered.

Table 1. The simulation conditions for performing LES analysis of ribbed channels by research institution.

Institution	Country	Year [Ref]	Software	Reynolds Number	p/e	e/H	Aspect Ratio	Rib Geometry
Tokyo Univ. Agriculture and Mechanics	Japan	2000 [25]	In-house	4000–9000	10	0.1	1, 2, 4	90°
		2001 [26]	In-house	1000, 4000	10	0.1	1	60°, 90°
		2001 [27]	In-house	4000	10	0.1	1	60°, 90°
		2001 [28]	In-house	4000	10	0.1	1	60°, 90°
		2003 [29]	In-house	4000	10	0.1	0.25, 1, 4	60°, 90°
		2004 [30]	In-house	4000	10	0.1	1	60°, 90°
		2004 [31]	In-house	4000	10	0.1	1	60°, 90°
		2008 [32]	In-house	1000, 4000	10	0.1	1	60°

Table 1. *Cont.*

Institution	Country	Year [Ref]	Software	Reynolds Number	p/e	e/H	Aspect Ratio	Rib Geometry
Denken	Japan	2002 [33]	In-house	100,000	10	0.1	∞	90°
		2005 [34]	In-house	50,000	10	0.1	2	60°
Univ. Iowa	USA	2003 [35]	In-house	10,020	1, 5, 10	0.1	∞	90°
Iowa State Univ.	USA	2004 [36]	In-house	5600	10	0.2	∞	90°
Virginia Tech	USA	2004 [37]	In-house	20,000	10	0.1	1	90°
		2005 [38]	In-house	20,000	10	0.1	1	90°
		2006 [39]	In-house	20,000	10	0.1	1	90°
		2006 [40]	In-house	20,000	10	0.1	1	90°
		2006 [41]	In-house	20,000	10	0.1	1	90°
		2008 [42]	In-house	20,000	10	0.1	1	90°
		2018 [43]	In-house	100,000	10	0.1	1	90°
		2021 [44]	In-house	10,000	10	0.3		90°
		2021 [45]	In-house	20,000	10	0.1	1	90°, BS, FS
		2022 [46]	In-house	20,000	10	0.1	1	90°, BS, FS
Seoul Nat'l Univ.	Republic of Korea	2005 [47]	In-house	30,000	10	0.1	∞	90°, semicircle
		2007 [48]	In-house	30,000	10	0.1	∞	90°
		2010 [49]	In-house	30,000	10	0.1	∞	90°, detached
Louisiana State Univ.	USA	2005 [50]	In-house	12,500	10	0.1	1	90°
		2005 [51]	In-house	25,000	10	0.1	0.25, 1, 4	90°
		2007 [52]	In-house	25,000, 100,000	10	0.1	0.25, 1, 4	90°
Cambridge	UK	2005 [53]	In-house	14,200	20	0.1	1	90°
		2015 [54]	In-house	20,000	10	0.1	1	90°
		2021 [55]	In-house	14,200	20	0.1	1	90°
Von Karman Institute	Belgium	2006 [56]	Fluent6.1	40,000	10	0.3	1	90°
		2015 [57]	In-house	40,000	10	0.3	1	90°
		2016 [58]	In-house	40,000	10	0.3	1	90°
IIT	India	2012 [59]	In-house	2053	10	0.1	1	90°
Univ. Manchester	UK	2015 [12]	In-house	30,000	9	0.1	∞	90°
Sapienza Univ. Roma	Italy	2015 [60]	In-house	15,000	10	0.1	1	90°
		2017 [61]	In-house	15,000	10	0.1	1	90°
ONERA	France	2016 [62]	In-house	40,000	10	0.3	1	90°
Peking Univ.	China	2016 [63]	In-house	30,000	9	0.1	1	90°
Univ, Manitoba	Canada	2015 [64]	PIV	13,000	8	0.1	1	90°, V(30,45,60°)
		2017 [65]	In-house	5600	8	0.1	1	90°, V(45,60°)
		2021 [66]	In-house	5600	8	0.1	1	90°

Table 1. *Cont.*

Institution	Country	Year [Ref]	Software	Reynolds Number	p/e	e/H	Aspect Ratio	Rib Geometry
Niigata Univ.	Japan	2016 [67]	In-house	5000	2, 4, 8, 16	0.1	∞	90°
		2020 [68]	In-house	5000	2, 4, 8, 16	0.1	∞	90°
Kookmin Univ.	Republic of Korea	2017 [69]	In-house	30,000	10	0.1	∞	90°
		2021 [10]	In-house	30,000	10	0.1	∞	90°
		2021 [70]	In-house	30,000	10	0.1	∞	90°
		2022 [71]	In-house	30,000	10	0.1	∞	90°
Univ. Stuttgart	Germany	2018 [72]	o-FOAM	30,000	10	0.1	1	
Karlsruhe Institute of Tech.	Germany	2018 [73]	Fluent v.15	100,000	10	0.1	1	90, V(60°)
CEFRACS	France	2020 [74]	In-house	15,000	10	0.1	1	90°
		2021 [75]	In-house	15,000	10	0.1	1	90°

PIV: particle image velocimetry; BS: backward step; FS: forward step.

2. Numerical Challenges and Comparative Studies

In the 50 papers analyzed in this review paper (see Table 1), the governing equations comprise the following incompressible Navier–Stokes equations and energy equations [10].

$$\frac{\partial \overline{u}_i}{\partial x_i} = 0 \quad (1)$$

$$\frac{\partial \overline{u}_i}{\partial t} + \frac{\partial \overline{u}_i \overline{u}_j}{\partial x_j} = -\frac{d\overline{p}}{dx_i} + \frac{1}{\mathrm{Re}} \frac{\partial^2 \overline{u}_i}{\partial x_j \partial x_j} + \frac{\partial \tau_{ij}}{\partial x_j} \quad (2)$$

$$\frac{\partial \overline{\theta}}{\partial t} + \frac{\partial}{\partial x_j}(\overline{u}_j \overline{\theta}) = \frac{1}{\mathrm{Re}\,\mathrm{Pr}} \frac{\partial^2 \overline{\theta}}{\partial x_j \partial x_j} + \frac{\partial q_j}{\partial x_j} \quad (3)$$

where the over bars refer to the grid-filtered quantities of LES, while τ_{ij} and q_j are the stress and heat flux at the sub-grid scale (SGS), respectively. The parts that need to be modeled in the LES: τ_{ij} is usually modeled as being proportional to the velocity gradient for which currently, either a constant value is used or it is determined by considering conditions such as the flow or wall [76].

Although compressible flow is sometimes considered in the LES of film cooling [77], the internal cooling passage is simulated as incompressible flow by using the above equations. Although there is a problem with the convection term becoming unstable when simulating film cooling, which is an external flow [78], this does not occur when analyzing a ribbed channel, which is an internal flow. In addition, periodicity is often imposed in simulations of fully developed flows in ribbed channels. When developing a flow that is not sensitive to inflow, there is no inflow generation issue appearing during film cooling [78].

Commercial packages such as Fluent have been mainly used for LES of film cooling [78], but as summarized in Table 1, in-house codes have been mainly used for LES studies of ribbed channels. The Reynolds number was lower than that of actual gas turbines in the early 2000s, but recently LES was performed in the range of 10,000 to 100,000, which is a typical value for gas turbines. Blockage ratio (=e/H) and pitch ratio (=p/e) are mostly 0.1 and 10, respectively. The most common rib angle is 90°, and some LES for angled ribs are also being performed.

Ribbed channel flow has other considerations not applicable to film cooling. First, to impose cyclic boundary conditions at the inlet and outlet in a fully developed flow, the streamwise pressure drop and heat transfer must be corrected. This is solved by decouplin the mean pressure and temperature gradient, as is shown in the following equations [79

$$P(\mathbf{x}, t) = -\beta x + p(\mathbf{x}, t) \quad (4$$

$$T(\mathbf{x}, t) = \gamma x + \theta(\mathbf{x}, t), \quad (5$$

where β and γ are the mean streamwise gradients of the pressure and temperature, th determination of which satisfies the conservation of global momentum and energy, respe tively [79].

Second, there is a sharp edge in the rib that can cause wiggle in the data near th upstream edge [36] or it may become difficult to construct a body-fitted grid. Since LE data are rougher than RANS data, this numerical instability can be solved by increasing th grid resolution: it takes around 1 million grids to solve one periodic box and approximatel 10 million grids to solve one passage in a ribbed channel [20,21].

Th author of [38] compared quasi- direct numerical simulation (DNS) data performe in the same grid system with LES data by adopting the dynamic SGS model [80] in ribbed duct, which showed a difference of 10% to 15% in heat transfer prediction. Th authors of [64] also showed that LES, course-grid LES, and detached eddy simulatio (DES) similarly predicted local heat transfer. Most of the research teams in Table 1 adopte the dynamic SGS model, albeit the SGS turbulent viscosity value obtained using scal similarity may be unstable. To solve this problem, a method of averaging while trackin flow particles [81] was proposed, while local averaging [82] has often been used in ribbe channel studies.

Unlike using RANS, LES consistently predicts the flow and heat transfer characteristic of ribbed channels [20,21]. Looking at the time-averaged streamlines presented in Figure 2b the two LES data examples predict the location of the reattachment point similarly to th particle image velocimetry (PIV) data [11]. When using RANS, although there are sligh differences depending on the model, the reattachment point is generally predicted to appea slightly downstream from the experimental value.

Figure 3 shows a comparison of the results for local heat transfer in the channe wall between ribs predicted by using RANS and LES. It has been reported that local hea transfer predictions using RANS differ by up to 250% depending on the model [20,21], a demonstrated in Figure 3a. The k-ε shear stress transport (SST) and Spalart–Allmaras (SA models greatly underpredict heat transfer, while the other models predict the local hea transfer distribution differently.

Figure 3b shows three comparisons of LES data with experimentally obtained result The three sets of experimental data were obtained using TLC [47], naphthalene sublima tion [85], or laser holography [86]. In all three, the first peak appears at $3 < x/e < 4$ whil the second peak appears before $x/e = 9$. The LES results obtained by the three differen groups provided predicted local peaks that converge within a much narrower range tha the RANS data in Figure 3a.

The difference between the LES and RANS data for ribbed channel flow is mor evident in the turbulence data in Figure 4 than in the time-averaged velocity field i Figure 2b. With LES, the turbulent kinetic energy (TKE) appears high in the shear laye from the top surface of the rib to the downstream, which is not predicted by the k-ω SST model at all. Although the v^2-f model predicts high TKE of the shear layer to some exten it predicts the location where the maximum TKE occurs differently than when using LES

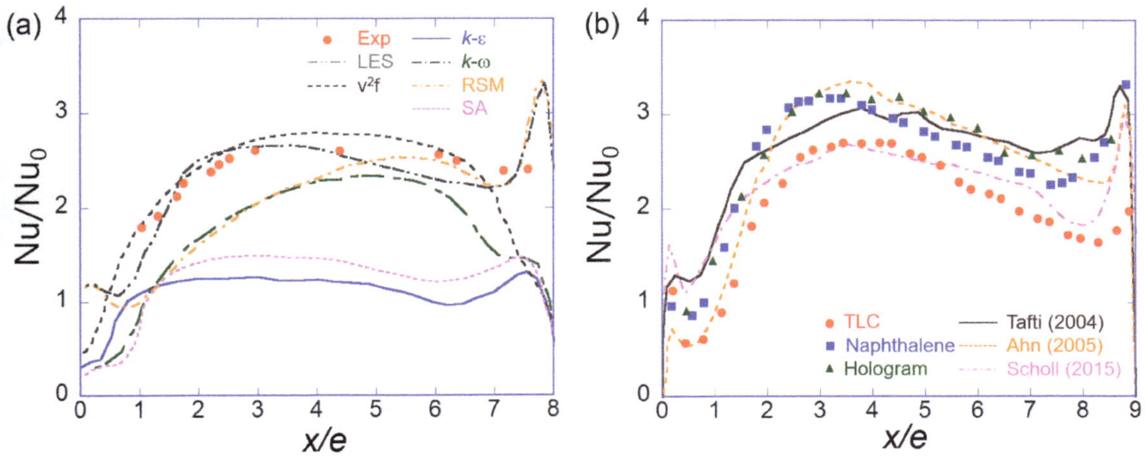

Figure 3. The local Nusselt number distribution on the channel wall between the ribs; (**a**) RANS data [12] (v^2f from [83]) compared to experiment [84]; (**b**) Three experiments [47,85,86] and three LES data [38,47,57]. SA: Spalart–Allmaras.

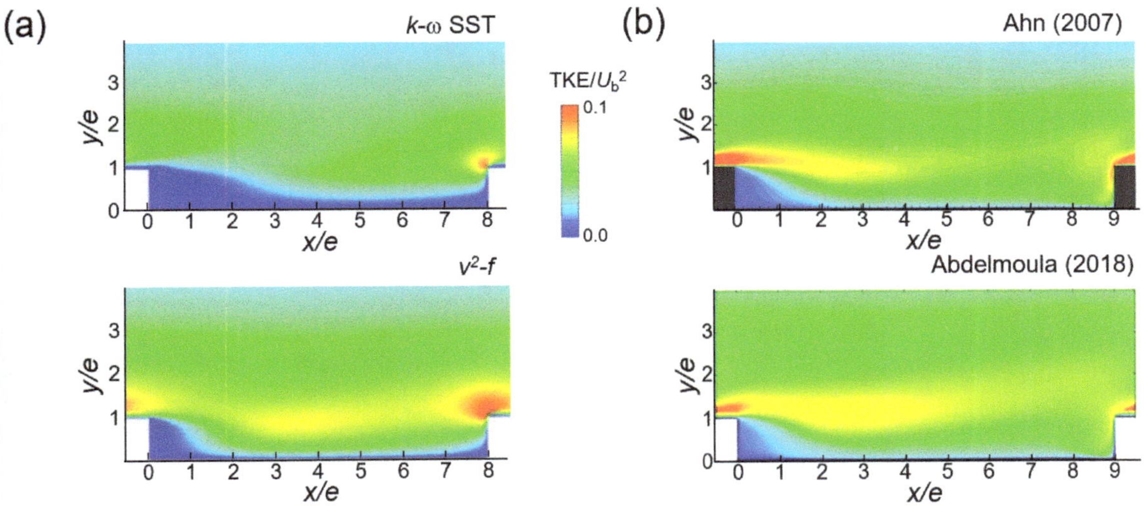

Figure 4. Turbulent kinetic energy (TKE) contours; (**a**) RANS data [12]; (**b**) LES data [48,72].

3. Instantaneous Flow and Thermal Fields

The local heat transfer in Figure 3b can be explained by the flow field in Figure 2b [86]. The first peak appearing at $3 < x/e < 4$ can be accounted for by reattachment of the separated flow; the reattachment point in Figure 2b is around $4 < x/e < 5$, and there is a slight difference in location. The second peak occurring before $x/e = 9$ can be explained by the influence of a corner vortex. However, although one appears in front of the rib in the RANS data in Figure 2b, the second peak is not clear except for the Reynolds stress model (RSM) of the heat transfer data in Figure 2a.

The above issues can be explained in terms of the instantaneous flow and temperature fields obtained using LES. When comparing the vector plot and temperature contour in Figure 5a, it can be seen that shear layer vortices are generated downstream of the rib while the cold main flow is introduced so that the heat transfer peak occurs upstream of the reattachment line. The second peak can be explained by looking at the flow field and temperature field near the channel wall shown in Figure 5b. Around z/e = 4.5 before x/e = 9, the velocity vector spreads to both sides along the rib as the reattached flow collides with the rib, and the blue streak spreads to both sides in the temperature field at that location. This impinging flow creates a second heat transfer peak.

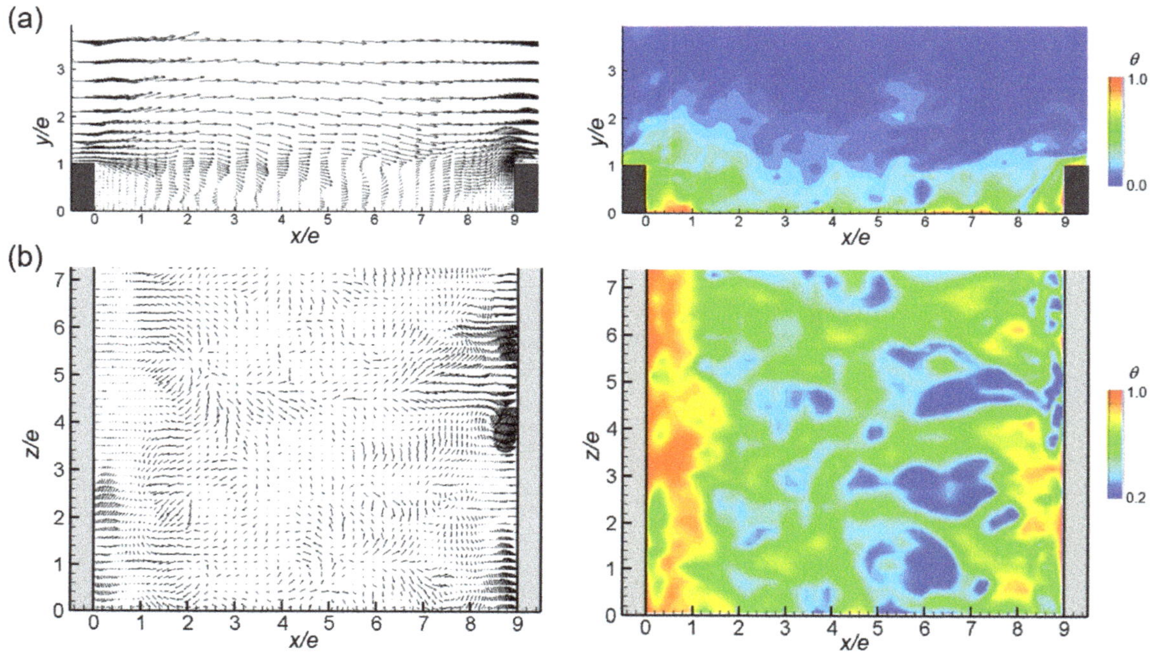

Figure 5. Instantaneous flow and thermal fields: (**a**) in the xy plane; (**b**) in the xz plane at y/e = 0.06.

Instantaneous flow field and temperature field information is limited to a specific moment and can be selective, so statistical verification is also being performed. The authors of [55] performed proper orthogonal decomposition (POD), thereby showing that 80 modes exist. Meanwhile, [75] performed dynamic mode decomposition (DMD), thereby showing that there are dominant modes in the downstream and inter-rib ribs and that minor modes are also involved in heat transfer.

4. Heat Transfer on the Rib

The heat transfer promotion of the rib turbulator peaks in front of the reattachment point and weakens downstream, so the ribs are installed periodically in the flow direction [1]. The installation spacing (p) of the ribs is usually around 10 times their height (e). In this case, if the cross-section of the rib is square, its surface area accounts for 25% of the heat transfer area. Since the heat transfer rate of the rib is expected to be higher than that of the channel wall, information on the heat transfer through the rib is very important for the design of the cooling passage. However, most of the heat transfer results for the ribbed channel were determined at the channel wall only [18].

The authors of [18] could not identify the details of the heat transfer on the surface of the rib by using thermocouple data, and since access from the front and back of the rib for optical measurement apparatus such as TLC or using infrared cameras is difficult, measurement data are rare [3,15]. The authors of [86] measured the heat transfer data on the rib surface by obtaining laser holography images from the side of the test section, while [87] measured the heat transfer distribution on the rib surface by securing optical access from the high blockage rib. The heat transfer distribution on the rib surface can be obtained using CFD, and LES provides more reliable data than RANS.

Figure 6 shows a comparison of the heat transfer on the rib obtained using LES and obtained experimentally. Compared with the channel wall in Figure 3b, the deviation between the data is large. Among the three surfaces of the rib, the front one has the largest Nusselt number ($0 < s/e < 1$). Although the LES results predict a sharp heat transfer peak at the upstream edge, this is not clear from the experimental data.

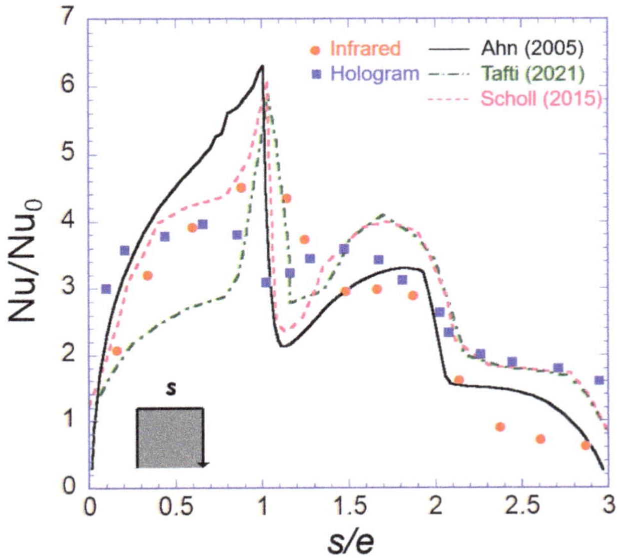

Figure 6. Local heat transfer distribution on the rib (2 experiments [86,87] and 3 LES data [44,47,57]).

At the top surface ($1 < s/e < 2$), the flow detached from the upstream edge becomes reattached (Figure 2b), resulting in a heat transfer peak at $1.5 < s/e < 2$, which is commonly predicted by using the three sets of LES results and the experimental data in [87]. Meanwhile, only the results in [86] with measurements obtained by using a laser hologram do not follow this trend. Heat transfer at the rear surface ($2 < s/e < 3$), which is the weakest among the three surfaces, decreases from the edge to the corner. This trend is unanimous when comparing the five datasets in Figure 6.

5. Conjugate Heat Transfer

The rib installed in the cooling passage of the gas turbine also serves as a fin to widen the heat transfer area. To obtain accurate measurements of the heat transfer performance, the fin performance while considering the thermal conduction from the rib must be obtained [88,89]. To perform this, it is necessary to find the temperature inside the rib. Since it is difficult to plant a sensor inside the rib in experiments, it is obtained by measuring the temperature distribution on the surface and performing a separate conduction analysis using this as a boundary condition [89].

In CFD, conjugate heat transfer simulation including thermal conduction must be performed. The authors of [88] analyzed the conjugate heat transfer of the ribbed channel using RANS, but did not consider the conduction at the channel wall. When trying to analyze the conjugate heat transfer for a ribbed channel with LES, a problem arises in that the time scales of conduction and convection are significantly different. The authors of [58] solved this by exchanging time-averaged boundary conditions intermittently while applying separate solvers for solids and fluids. The authors of [10,44] performed a fully coupled analysis by applying the immersed boundary method.

Figure 7a shows a comparison of the temperature distribution after conjugate heat transfer analysis with that of pure convection. Since the thermal conductivity of gas turbine materials is around 600 times that of cooling air [10,89], the solid interior is close isothermal and most of the temperature change occurs inside the rib. Therefore, the heigh of the rib [10] is better than the thickness of the channel wall for use as the characterist length in the Biot number calculation [90] until the thermal conductivity decreases, at whic point the latter becomes more appropriate [70].

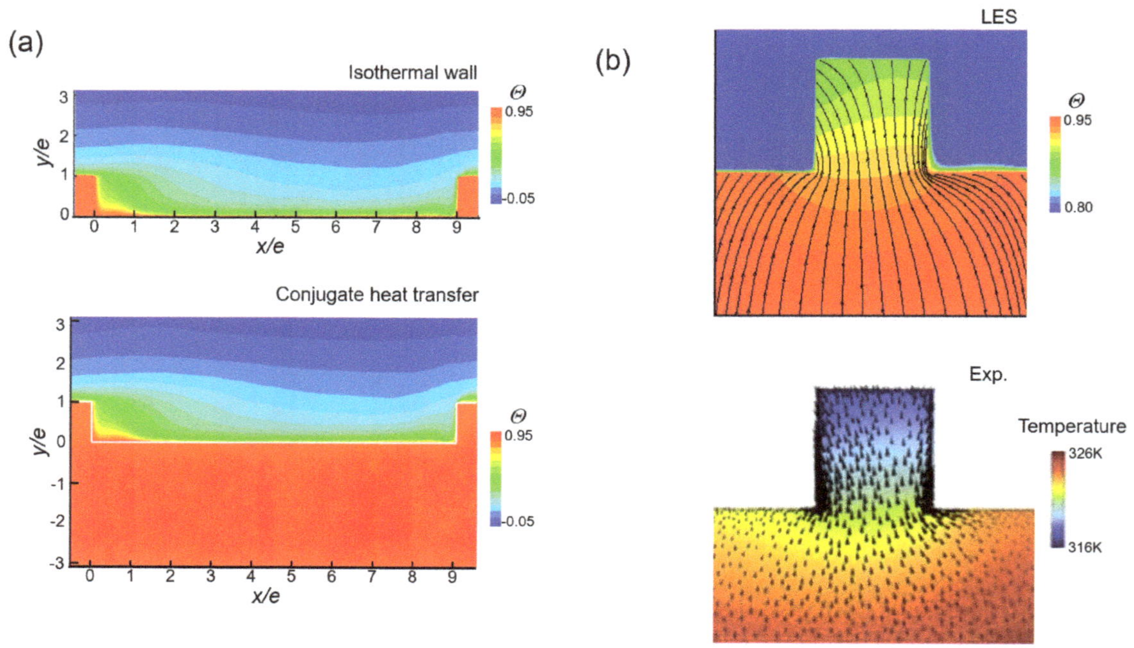

Figure 7. Thermal fields in a ribbed channel with a conducting wall [10]: (**a**) temperature contours i the *xy* plane; (**b**) the temperature around the rib.

Figure 7b depicts a comparison of the temperature distribution inside the rib obtaine by LES with experimentally obtained results; the temperature distribution and heat flu vector obtained by using LES and experimentally are in good agreement. Moreover, it ca be seen that the channel wall should be included in the analysis because of the effect of th heat conduction from the rib penetrating the channel wall.

Figure 8a shows a comparison of the heat transfer data in a conducting channel wa obtained experimentally and via LES. At the channel wall, the effect of conduction is no significant, so the conjugate heat transfer data appears similar to the iso-flux data. In th experiment, the conjugate data [89] (blue squares) occur sooner after the rib ($0 < x/e < 2$ and in front of the rib ($8 < x/e < 9$) than the iso-flux data (red circles) [87]. However, thi trend cannot clearly be seen in the LES results. Although this tendency is evident in [10 the blockage ratio is different, so additional investigations are needed.

Figure 8b illustrates a comparison of the heat transfer on the rib surface. Compare with the iso-flux data (red circles), the conjugate data (blue square) show a decrease i heat transfer at the front surface ($0 < s/e < 1$) and the rear surface ($2 < s/e < 3$). Althoug a decrease in heat transfer at the rear surface was predicted by [44], a change at the fron surface was not clearly observed in the LES data. The authors of [44,57,58] performed LES under the same conditions and shapes as the experiments performed by [87,89], so th difference in the heat transfer at the front surface seems to be due to the difference in th time scales for conduction and convection.

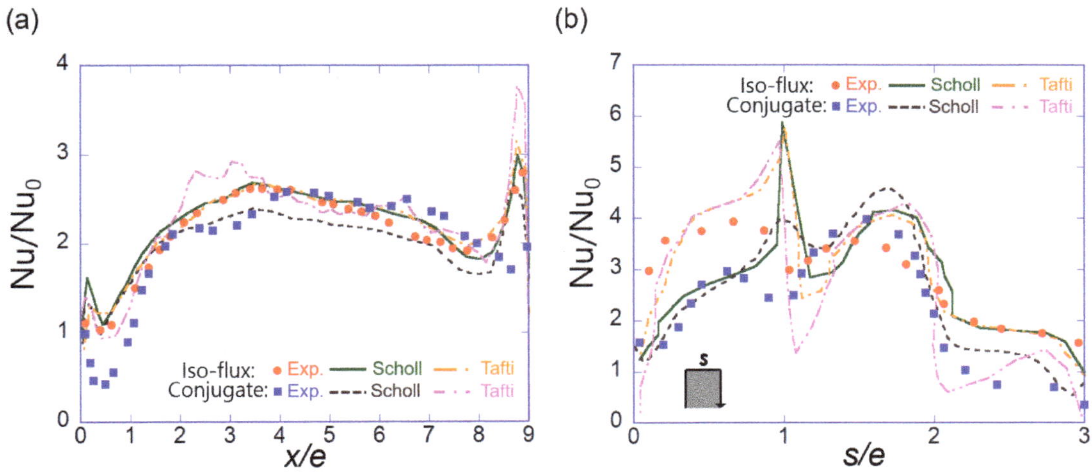

Figure 8. Local heat transfer distributions: (**a**) on the channel wall; (**b**) on the rib. The experimental data [87,89] are represented by symbols and the LES data [44,57,58] by lines.

6. The Effects of Rotating the Gas Turbine Blade

Rotating the gas turbine blade affects the heat transfer in the cooling channel [3,4], and since it is difficult to measure this experimentally, CFD can be usefully employed instead. However, the effects of rotation are not well-represented in RANS [36,48], so LES was identified as being more suitable for examining this phenomenon. In LES, the centrifugal force or rotating buoyancy effect of using a high rotation speed can be implemented more easily than can be achieved experimentally [42,50].

As shown in Figure 9a, to test the heat transfer of the rotating ribbed duct, the duct and the sensor including the camera must be rotated together and the balance must be well-matched to obtain a high rotation number [3]. The experimental results in Figure 9b show that the heat transfer characteristics change by rotation. However, rotational buoyancy caused by rotation with centrifugal force is difficult to observe experimentally.

The local heat transfer distribution of the rotating ribbed channel obtained by LES is presented in Figure 9c,d. The rotation effect not well-predicted by using RANS can be by using LES. Heat transfer is promoted in the trailing wall when LES is used with an Ro value of 0.3 (Figure 9c) [48,50], which could not be obtained experimentally by [91]. The results from [48,50] agree well in Figure 9c even though the channel aspect ratio is different (see Table 1).

Although heat transfer can be seen to decrease at the leading wall in Figure 9d, LES underpredicts the amount of heat transfer reduction for $Ro = 0.1$. The data in [50] show a significantly different distribution from other LES or experimentally obtained data. The heat transfer at the leading wall appears to be more sensitive to the channel aspect ratio than that at the trailing wall (see Table 1). When replacing the test duct in the rotation rig in Figure 9a, the sensor must be installed and the balance must be re-adjusted. Thus, LES is a useful tool for studying the rotation effect according to the duct aspect ratio.

Rotating buoyancy, which is difficult to experimentally implement close to that in a real gas turbine, can be imposed by using LES. Around 1 million grids are required to analyze one cycle in the flow direction of a ribbed channel using LES, while tens of millions of grids are required to analyze a two-pass duct [20,21]. Due to recent improvements in hardware performance, data on a two-pass duct can be obtained via LES (Figure 10) [43].

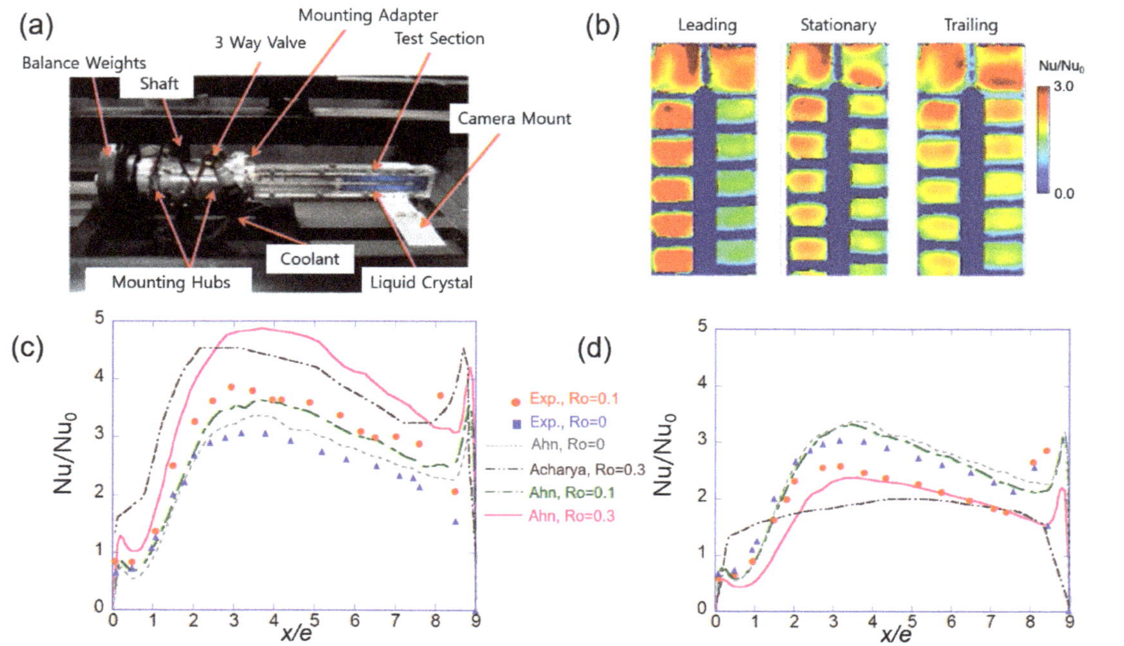

Figure 9. Heat transfer in the rotating ribbed channel: (**a**) A typical experimental setup [3]; (**b**) local heat transfer distributions [3]; (**c**) the local Nusselt number on the trailing wall; (**d**) the local Nusselt number on the leading wall. *Ro*: rotation number (LES data [48,50] compared to experiment [91]).

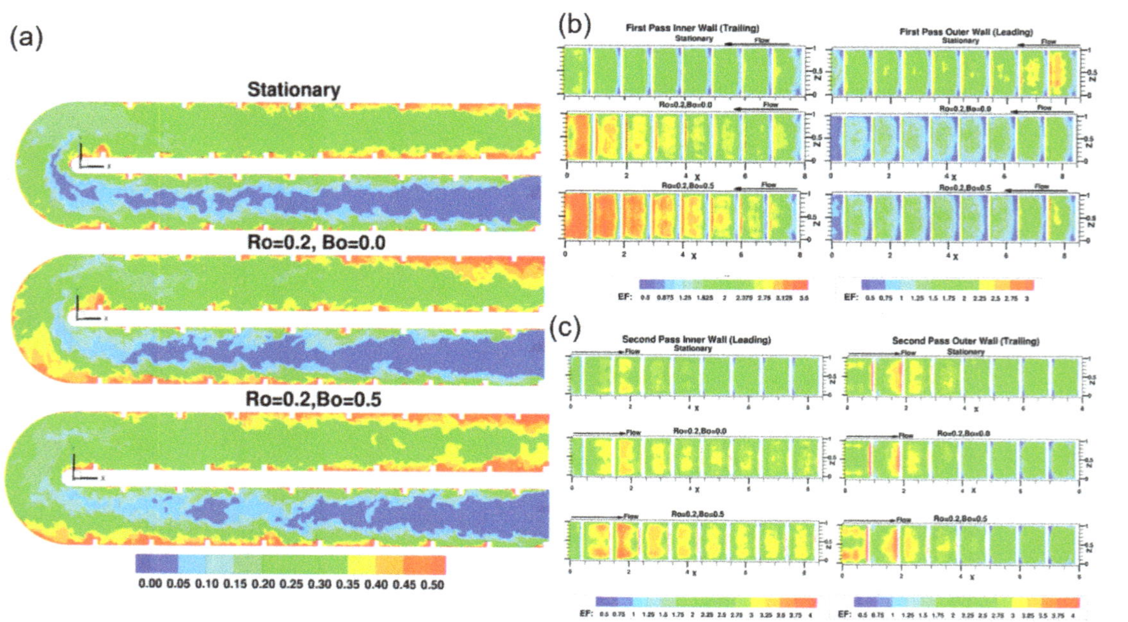

Figure 10. Large eddy simulation of heat transfer in a two-pass rotating ribbed duct [43]: (**a**) thermal fields in the channel; (**b**) heat transfer on the channel wall during the first pass; (**c**) heat transfer on the channel wall during the second pass. *Bo*: buoyancy number.

In Figure 10a, it can be seen that the effect of buoyancy number appears downstream in the first pass where the coolant is heated in the second pass. In the heat transfer distribution of the first pass shown in Figure 10b, the entry effect appears to be common and both the rotation effect and the buoyancy effect occur. The effect of a sharp 180° turn occurs during the heat transfer in the second pass (Figure 9c). Many studies have been conducted on shapes to secure flow uniformity during sharp 180° turns [92], for which LES is a useful tool that will be widely used in such studies.

7. Geometrical Shapes for Performance Improvement

Although a rib turbulator installed in the internal cooling passage of a gas turbine promotes heat transfer, it also causes additional pressure loss. In addition, heat transfer becomes locally non-uniform, especially in the low heat transfer region behind the rib. To improve this problem, various rib shapes have been proposed [1], as shown in Figure 11. Since the gas turbine blade has an airfoil-shaped cross-section, those in the cooling passage near the leading edge and trailing edge are semicircular or triangular, respectively (see Figure 11). In most studies on gas turbine internal flow channels, the researchers have dealt with the effects of the geometric parameters on heat transfer and pressure drop [1].

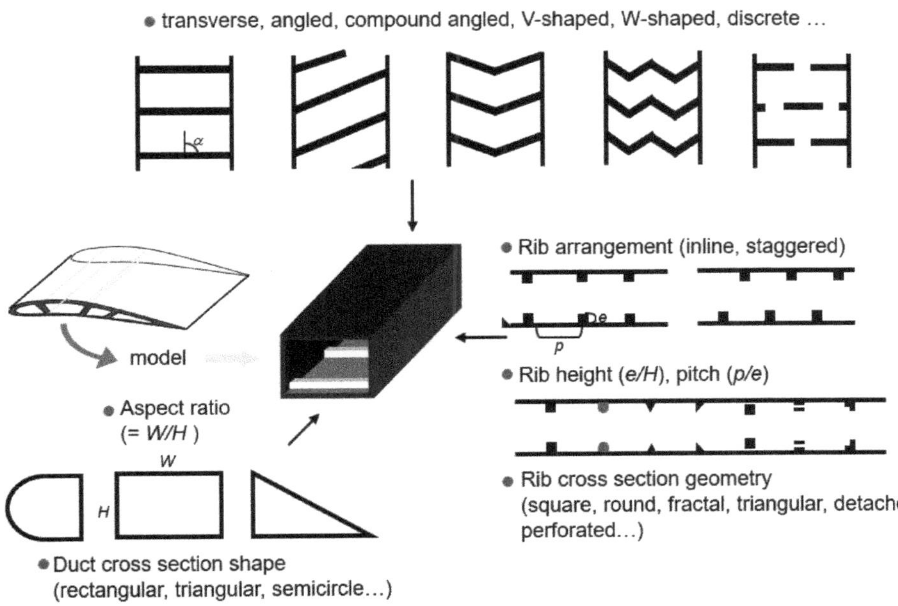

Figure 11. Geometrical shapes and variables affecting heat transfer in the internal cooling passage of a gas turbine.

The shape of ribs, especially ribs angled as V and W shapes, has been actively studied over the past 40 years, with articles on the subject being published recently [93]. When the rib is at an angle to the main flow, the secondary flow is generated in the duct to promote heat transfer. Compared with using transverse ribs, the latter can be relatively well-predicted by using RANS and the results are in good agreement with the experimentally obtained ones. The authors of [94] performed RANS under the same conditions as used in the experiment in [95] and obtained a local heat transfer distribution profile similar to the experimentally obtained one (Figure 12).

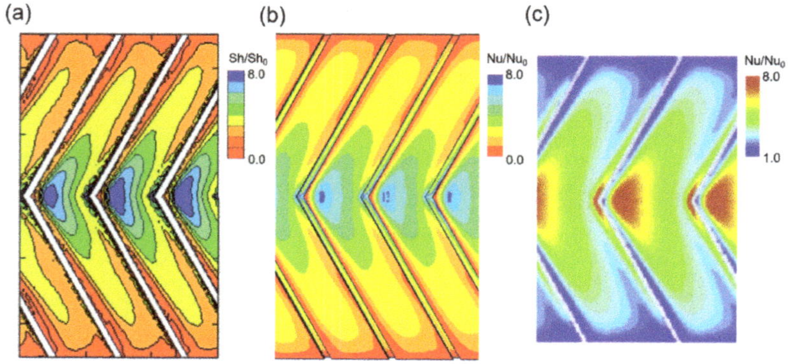

Figure 12. Local heat transfer distribution on a channel wall with V-shaped ribs: (**a**) experiment data [95]; (**b**) RANS data with RSM [94]; (**c**) RANS data with the *k*-model [96].

Comparing Figure 12a,b demonstrates that heat transfer is underpredicted compare with the experiment and the reattachment point is pushed slightly downstream, so obtai ing better results with LES remains possible. The *k*-ε model, which does not predict hea transfer in the transverse rib well, also provides a heat transfer distribution close to th experimentally obtained one when using V-shaped ribs, as shown in Figure 12c. Whe performing LES on a model using angled ribs, the Reynolds number in [26] is rather lo whereas that in [55] is over 10,000. Meanwhile, although [73] performed DES for V-shape ribs and obtained results in good agreement with the experimentally obtained ones, the could not provide instantaneous large flow structures. Moreover, [64,65] performed LE but did not include heat transfer.

The most studied topic for rib geometries is their cross-sectional shape [97,98]. Figure 13 shows a comparison of the streamlines obtained in channels with semicircular ribs an detached ribs by using LES. The semicircular rib shows a flow structure similar to that c the square rib, and the heat transfer distribution comparison shown in Figure 13b is als similar [47]. The detached rib promotes heat transfer in the low heat transfer region behin the rib [99], albeit the heat transfer promotion effect decreases near the reattachment poir (Figure 13b) [49].

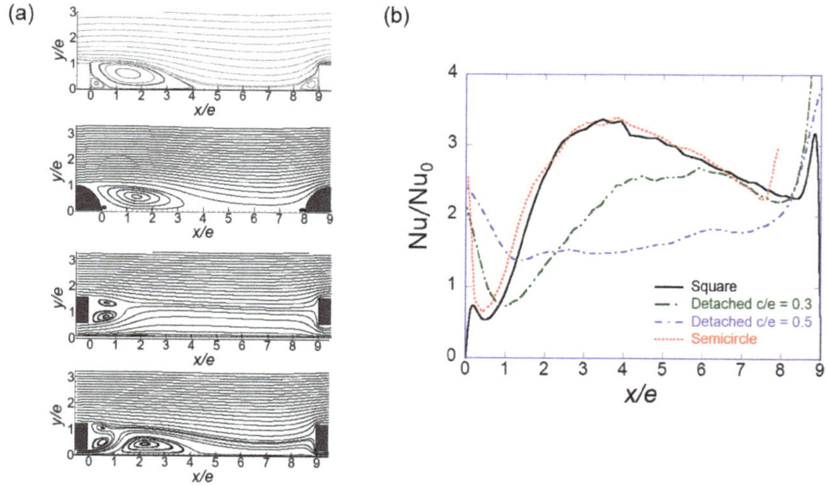

Figure 13. Effects of rib cross-section geometry on flow and heat transfer in a ribbed channel [47,49 (**a**) time-averaged streamlines; (**b**) local heat transfer distributions on the channel wall.

The authors of [97] performed RANS on 16 different rib cross-sectional shapes and presented data with the best heat transfer performance for boot-type ribs with the downstream side shaved off. The authors of [45] recently performed LES on the forward step rib and the backward step rib, and the backward step rib obtained better heat transfer performance. The authors of [55] performed LES on 10 rib cross-sections and obtained good heat transfer performance in the shape where the downstream side of the rib was shaved as in [97].

8. Conclusions

Herein, 50 studies published over the past 20 years on the flow and heat transfer in a ribbed channel by simulating the flow path inside a gas turbine using LES are reviewed. The main findings can be summarized as follows.

In ribbed channel flow, instability in the convection term or inflow generation problem does not appear, so LES can be performed more easily than with film cooling. The effect of the SGS model on the heat transfer results is not significant, and the LES results obtained by several groups are very consistent, unlike those obtained using RANS. Including sharp corners in the geometrical shape of the ribs causes grid configuration and stability issues. When considering the thermal conduction by the rib, there is a time scale difference between flow and conduction. An immersed boundary method that supports conjugate heat transfer has been used as a solution to these numerical problems.

LES can be used to not only accurately predict heat transfer in ribbed channels but also provide instantaneous flow and temperature field and turbulence statistics that using RANS cannot. Based on this information, it is possible to explain the local heat transfer distribution process and identify the mechanism that promotes heat transfer. In addition, LES provides local heat transfer distribution or conjugate heat transfer data for the ribs, for which there is a lack of experimentally obtained data. High rotation and buoyancy effects, which are difficult to measure experimentally, can also be simulated by using LES.

9. Future Directions

Improvements in computing hardware have enabled LES results to be obtained for two-pass ducts, and using LES to simulate the entire internal cooling passage is expected to be possible soon. Compared with the shape of the rib, not much data on duct geometry using LES have been published. It is anticipated that using LES for a triangular duct [100] appearing near the trailing edge or semicircular duct near the leading edge will be available soon. Problems with experimental high-resolution optical measurement methods can be overcome using LES, which thereby can make a useful contribution.

In addition to the duct geometry, heat transfer promoters such as dimples and pin fins are used near the trailing edge. LES results on using a dimpled channel have been published since 2008 [101,102], showing that LES can be used to predict local heat transfer much better than RANS. LES results on using pin fins have recently begun to be published [103], while research including conjugate heat transfer will be needed in the future. Besides ribs, impinging jets and vortices are also widely used [104], so LES models including these should provide useful data.

Funding: This work was supported by an Agency for Defense Development Grant funded by the Korean Government (UD220004JD).

Data Availability Statement: Data are contained within this article.

Conflicts of Interest: The author declares no conflict of interest.

Nomenclature

d	thickness of the channel wall [m]
D_h	hydraulic diameter of the channel [m]
e	rib height [m]
f	friction factor
h	heat transfer coefficient [W/m^{2}K]
H	channel height [m]
k_f	thermal conductivity of fluid [W/mK]
Nu	Nusselt number (=$h\ D_h/k_f$)
p	rib-to-rib pitch [m]
Pr	Prandtl number (=ν/α)
Re	bulk Reynolds number (=$U_b\ D_h/\nu$)
Ro	rotation number (=$\omega\ D_h/U_b$)
q	heat transfer rate [W]
t	time [s]
T	temperature [K]
T_b	bulk temperature [K]
T_w	wall temperature [K]
U_b	bulk velocity [m/s]
W	channel width [m]
Greek symbols	
α	thermal diffusivity [m^2/s]
ν	kinematic viscosity [m^2/s]
θ	dimensionless temperature (=$(T - T_b)/(T_w - T_b)$)
Θ	time-averaged dimensionless temperature
Subscripts	
0	fully developed value in a smooth pipe

References

1. Han, J.C. Advanced cooling in gas turbines 2016 Max Jakob memorial award paper. *J. Heat Transf.* **2018**, *140*, 113001. [CrossRef]
2. Bunker, R.S. Gas turbine heat transfer: Ten remaining hot gas path challenges. *J. Turbomach.* **2017**, *129*, 193–201. [CrossRef]
3. Ekkad, S.V.; Singh, P. Detailed heat transfer measurements for rotating turbulent flows in gas turbine systems. *Energies* **2021**, *14*, 39. [CrossRef]
4. Yeranee, K.; Yu, R.A.O. A review of recent studies on rotating internal cooling for gas turbine blades. *Chin. J. Aeronaut.* **2021**, *34*, 85–113. [CrossRef]
5. Ligrani, P. Heat transfer augmentation technologies for internal cooling of turbine components of gas turbine engines. *Int. J. Rotating Mach.* **2013**, *2013*, 275653. [CrossRef]
6. Du, W.; Luo, L.; Jiao, Y.; Wang, S.; Li, X.; Sunden, B. Heat transfer in the trailing region of gas turbines—A state-of-the-art review. *Appl. Therm. Eng.* **2021**, *199*, 117614. [CrossRef]
7. Chang, S.W.; Wu, P.-S.; Wan, T.-Y.; Cai, W.-L. A review of cooling studies on gas turbine rotor blades with rotation. *Inventions* **2023**, *8*, 21. [CrossRef]
8. Takeishi, K. Evolution of turbine cooled vanes and blades applied for large industrial gas turbines and its trend toward carbon neutrality. *Energies* **2022**, *15*, 8935. [CrossRef]
9. Alam, T.; Saini, R.P.; Saini, J.S. Heat and flow characteristics of air heater ducts provided with turbulators—A review. *Renew. Sustain. Energy Rev.* **2014**, *31*, 289–304. [CrossRef]
10. Ahn, J.; Song, J.C.; Lee, J.S. Fully coupled large eddy simulation of conjugate heat transfer in a ribbed channel with a 0.1 blockage ratio. *Energies* **2021**, *14*, 2096. [CrossRef]
11. Casarsa, L.; Arts, T. Experimental investigation of the aerothermal performance of a high blockage rib-roughened cooling channel. *J. Turbomach.* **2005**, *127*, 580–588. [CrossRef]
12. Keshmiri, A.; Osman, K.; Benhamadouche, S.; Shokri, N. Assessment of advanced RANS models against large eddy simulation and experimental data in the investigation of ribbed passages with passive heat transfer. *Numer. Heat Transf. B* **2016**, *69*, 96–110. [CrossRef]
13. Sharma, S.K.; Kalamkar, V.R. Computational Fluid Dynamics approach in thermo-hydraulic analysis of flow in ducts with rib roughened walls—A review. *Renew. Sustain. Energy Rev.* **2016**, *55*, 756–788. [CrossRef]
14. Nidhul, K.; Yadav, A.K.; Anish, S.; Kumar, S. Critical review of ribbed solar air heater and performance evaluation of various V-rib configuration. *Renew. Sustain. Energy Rev.* **2021**, *142*, 110871. [CrossRef]

Ekkad, S.V.; Singh, P. Liquid crystal thermography in gas turbine heat transfer: A review on measurement techniques and recent investigations. *Crystals* **2021**, *11*, 1332. [CrossRef]

Jang, H.N.; Park, J.S.; Kwak, J.S. Experimental study on heat transfer characteristics in a ribbed channel with dimples, semi-spherical protrusions, or oval protrusions. *Appl. Therm. Eng.* **2018**, *131*, 734–742. [CrossRef]

Unnikrishnan, U.; Yang, V. A review of cooling technologies for high temperature rotating components in gas turbine. *Propuls. Power Res.* **2022**, *11*, 293–310. [CrossRef]

Taslim, M.E.; Wadsworth, C.M. An experimental investigation of the rib surface-averaged heat transfer coefficient in a rib-roughened square passage. *J. Turbomach.* **1997**, *119*, 381–389. [CrossRef]

Kumar, S.; Amano, R.S. An investigation in the numerical approach to solve the heat transfer phenomenon in gas turbine. *J. Energy Resour. Technol.* **2021**, *143*, 080805. [CrossRef]

Tyacke, J.C.; Tucker, P.G. Future use of large eddy simulation in aero-engines. *J. Turbomach.* **2015**, *137*, 081005. [CrossRef]

Tyacke, J.; Vadlamani, N.R.; Trojak, W.; Watson, R.; Ma, Y.; Tucker, P.G. Turbomachinery simulation challenges and the future. *Prog. Aerosp. Sci.* **2019**, *110*, 100554. [CrossRef]

Han, J.C.; Dutta, S. Recent developments in turbine blade internal cooling. *Ann. N. Y. Acad. Sci. USA* **2001**, *934*, 162–178. [CrossRef] [PubMed]

Yadav, A.S.; Agrawal, A.; Sharma, A.; Gupta, A. Revisiting the effect of ribs on performance of solar air heater using CFD approach. *Mater. Today Proc.* **2022**, *60*, 240–252. [CrossRef]

Iacovides, H.; Launder, B.E. Computational fluid dynamics applied to internal gas-turbine blade cooling: A review. *Int. J. Heat Fluid Flow* **1995**, *16*, 454–470. [CrossRef]

Murata, A.; Mochizuki, S. Large eddy simulation with a dynamic subgrid-scale model of turbulent heat transfer in an orthogonally rotating rectangular duct with transverse rib turbulators. *Int. J. Heat Mass Transf.* **2000**, *43*, 1243–1259. [CrossRef]

Murata, A.; Mochizuki, S. Comparison between laminar and turbulent heat transfer in a stationary square duct with transverse or angled rib turbulators. *Int. J. Heat Mass Transf.* **2001**, *44*, 1127–1141. [CrossRef]

Murata, A.; Mochizuki, S. Large eddy simulation of turbulent heat transfer in an orthogonally rotating square duct with angled rib turbulators. *J. Heat Transf.* **2001**, *123*, 858–867. [CrossRef]

Murata, A.; Mochizuki, S. Effect of centrifugal buoyancy on turbulent heat transfer in an orthogonally rotating square duct with transverse or angled rib turbulators. *Int. J. Heat Mass Transf.* **2001**, *44*, 2739–2750. [CrossRef]

Murata, A.; Mochizuki, S. Effect of cross-sectional aspect ratio on turbulent heat transfer in an orthogonally rotating rectangular duct with angled rib turbulators. *Int. J. Heat Mass Transf.* **2003**, *46*, 3119–3133. [CrossRef]

Murata, A.; Mochizuki, S. Centrifugal buoyancy effect on turbulent heat transfer in a rotating two-pass smooth square channel with sharp 180-deg turns. *Int. J. Heat Mass Transf.* **2004**, *47*, 3215–3231. [CrossRef]

Murata, A.; Mochizuki, S. Aiding and opposing contributions of centrifugal buoyancy on turbulent heat transfer in a two-pass transverse-or angled-rib-roughened channel with sharp 180 turns. *Int. J. Heat Mass Transf.* **2004**, *47*, 3721–3743. [CrossRef]

Murata, A.; Mochizuki, S. Effects of centrifugal buoyancy and Reynolds number on turbulent heat transfer in a two-pass angled-rib-roughened channel with sharp 180 turns investigated by using large eddy simulation. *Int. J. Rotating Mach.* **2008**, *2008*, 764720. [CrossRef]

Watanabe, K.; Takahashi, T. LES analysis and measurement of fully developed transverse ribbed channel flows and heat transfer. *Trans. Jpn. Soc. Mech. Eng. B* **2002**, *68*, 240–245. [CrossRef]

Watanabe, K.; Takahashi, T. LES and measurement of heat transfer and flow in rectangular channels with crossed angled ribs. *Trans. Jpn. Soc. Mech. Eng. B* **2005**, *17*, 1459–1464. [CrossRef]

Cui, J.; Patel, V.C.; Lin, C.L. Large-eddy simulation of turbulent flow in a channel with rib roughness. *Int. J. Heat Fluid Flow* **2003**, *24*, 372–388. [CrossRef]

Lee, J.S.; Meng, N.; Pletcher, R.H.; Liu, Y. Numerical study of the effects of rotation on heat transfer in channels with and without ribs. *Int. J. Heat Mass Transfer* **2004**, *47*, 4673–4684. [CrossRef]

Abdel-Wahab, S.; Tafti, D.K. Large eddy simulation of flow and heat transfer in a 90 deg ribbed duct with rotation: Effect of Coriolis and centrifugal buoyancy forces. *J. Turbomach.* **2004**, *126*, 627–636. [CrossRef]

Tafti, D.K. Evaluating the role of subgrid stress modeling in a ribbed duct for the internal cooling of turbine blades. *Int. J. Heat Fluid Flow* **2005**, *26*, 92–104. [CrossRef]

Sewall, E.A.; Tafti, D.K.; Graham, A.B.; Thole, K.A. Experimental validation of large eddy simulations of flow and heat transfer in a stationary ribbed duct. *Int. J. Heat Fluid Flow* **2006**, *27*, 243–258. [CrossRef]

Viswanathan, A.K.; Tafti, D.K. Detached eddy simulation of turbulent flow and heat transfer in a two-pass internal cooling duct. *Int. J. Heat Fluid Flow* **2006**, *27*, 1–20. [CrossRef]

Viswanathan, A.K.; Tafti, D.K. Detached eddy simulation of flow and heat transfer in fully developed rotating internal cooling channel with normal ribs. *Int. J. Heat Fluid Flow* **2006**, *27*, 351–370. [CrossRef]

Sewall, E.A.; Tafti, D.K. Large eddy simulation of flow and heat transfer in the developing flow region of a rotating gas turbine blade internal cooling duct with Coriolis and buoyancy forces. *J. Turbomach.* **2008**, *130*, 011005. [CrossRef]

Tafti, D.; Dowd, C.; Tan, X. High Reynold number LES of a rotating two-pass ribbed duct. *Aerospace* **2018**, *5*, 124. [CrossRef]

44. Oh, T.K.; Tafti, D.K.; Nagendra, K. Fully coupled large eddy simulation-conjugate heat transfer analysis of a ribbed coolin passage using the immersed boundary method. *J. Turbomach.* **2021**, *143*, 041012. [CrossRef]
45. Sreekesh, K.; Tafti, D.K.; Vengadesan, S. Large-eddy simulation investigation of modified rib shapes on heat transfer in a ribbe duct. *J. Heat Transf.* **2021**, *143*, 112101. [CrossRef]
46. Sreekesh, K.; Tafti, D.K.; Vengadesan, S. The combined effect of Coriolis and centrifugal buoyancy forces on internal cooling turbine blades with modified ribs using large eddy simulation (LES). *Int. J. Therm. Sci.* **2022**, *182*, 107797. [CrossRef]
47. Ahn, J.; Choi, H.; Lee, J.S. Large eddy simulation of flow and heat transfer in a channel roughened by square or semicircle ribs. *Turbomach.* **2005**, *127*, 263–269. [CrossRef]
48. Ahn, J.; Choi, H.; Lee, J.S. Large eddy simulation of flow and heat transfer in a rotating ribbed channel. *Int. J. Heat Mass Tran* **2007**, *50*, 4937–4947. [CrossRef]
49. Ahn, J.; Lee, J.S. Large eddy simulation of flow and heat transfer in a channel with a detached rib array. *Int. J. Heat Mass Tran* **2010**, *53*, 445–452. [CrossRef]
50. Tyagi, M.; Acharya, S. Large eddy simulations of flow and heat transfer in rotating ribbed duct flows. *J. Heat Transf.* **2005**, *12* 486–498. [CrossRef]
51. Saha, A.K.; Acharya, S. Flow and heat transfer in an internally ribbed duct with rotation: An assessment of large eddy simulatior and unsteady Reynolds-averaged Navier-Stokes simulations. *J. Turbomach.* **2005**, *127*, 306–320. [CrossRef]
52. Saha, A.; Acharya, S. Turbulent heat transfer in ribbed coolant passages of different aspect ratios: Parametric effects. *J. Heat Tran* **2007**, *129*, 449–463. [CrossRef]
53. Liu, Y.; Tucker, P.G.; Iacono, G.L. Comparison of zonal RANS and LES for a non-isothermal ribbed channel flow. *Int. J. Heat Flu Flow* **2006**, *27*, 391–401. [CrossRef]
54. Tyacke, J.; Tucker, P.G. Large eddy simulation of turbine internal cooling ducts. *Comput. Fluids* **2015**, *114*, 130–140. [CrossRef]
55. Tyacke, J.C.; Dai, Y.; Tucker, P.G. Impact of rib shape on heat transfer using LES. *Appl. Math. Model.* **2021**, *97*, 244–267. [CrossRe
56. Lohász, M.M.; Rambaud, P.; Benocci, C. Flow features in a fully developed ribbed duct flow as a result of MILES. *Flow Turb Combust.* **2006**, *77*, 59–76. [CrossRef]
57. Scholl, S.; Verstraete, T.; Torres-Garcia, J.; Duchaine, F.; Gicquel, L.Y.M. Influence of the thermal boundary conditions on the he transfer of a rib-roughened cooling channel using LES. *Proc. Inst. Mech. Eng. A* **2015**, *229*, 498–507. [CrossRef]
58. Scholl, S.; Verstraete, T.; Duchaine, F.; Gicquel, L. Conjugate heat transfer of a rib-roughened internal turbine blade coolin channel using large eddy simulation. *Int. J. Heat Fluid Flow* **2016**, *61*, 650–664. [CrossRef]
59. Ramgadia, A.G.; Saha, A.K. Large eddy simulation of turbulent flow and heat transfer in a ribbed coolant passage. *J. Appl. Mat* **2012**, *2012*, 246313. [CrossRef]
60. Borello, D.; Salvagni, A.; Hanjalić, K. Effects of rotation on flow in an asymmetric rib-roughened duct: LES study. *Int. J. Heat Flu Flow* **2015**, *55*, 104–119. [CrossRef]
61. Salvagni, A.; Borello, D.; Rispoli, F.; Hanjalić, K. Large-eddy simulations of heat transfer in asymmetric rib-roughened duct Effects of rotation. *Int. J. Heat Fluid Flow* **2017**, *68*, 373–385. [CrossRef]
62. Labbé, O. Large-eddy-simulation of flow and heat transfer in a ribbed duct. *Comput. Fluids* **2013**, *76*, 23–32. [CrossRef]
63. Jiang, Z.; Xiao, Z.; Shi, Y.; Chen, S. Constrained large-eddy simulation of turbulent flow and heat transfer in a stationary ribbe duct. *Int. J. Numer. Methods Heat Fluid Flow* **2016**, *26*, 1069–1091. [CrossRef]
64. Fang, X.; Yang, Z.; Wang, B.C.; Tachie, M.F.; Bergstrom, D.J. Highly-disturbed turbulent flow in a square channel with V-shape ribs on one wall. *Int. J. Heat Fluid Flow* **2015**, *56*, 182–197. [CrossRef]
65. Fang, X.; Yang, Z.; Wang, B.C.; Tachie, M.F.; Bergstrom, D.J. Large-eddy simulation of turbulent flow and structures in a squar duct roughened with perpendicular and V-shaped ribs. *Phys. Fluids* **2017**, *29*, 065110. [CrossRef]
66. Mahmoodi-Jezeh, S.V.; Wang, B.C. Direct numerical simulation of turbulent heat transfer in a square duct with transverse rib mounted on one wall. *Int. J. Heat Fluid Flow* **2021**, *89*, 108782. [CrossRef]
67. Matsubara, K.; Ohta, H.; Miura, T. Entrance region heat transfer in a channel with a ribbed wall. *J. Heat Transf.* **2016**, *138*, 12200 [CrossRef]
68. Matsubara, K.; Ohta, H.; Ishino, T. Direct simulation of inlet region heat transfer in a channel with repeated ribs under iso-therma wall heating condition. *Int. J. Therm. Sci.* **2020**, *154*, 106408. [CrossRef]
69. Song, J.C.; Ahn, J.; Lee, J.S. An immersed-boundary method for conjugate heat transfer analysis. *J. Mech. Sci. Technol.* **2017**, *3* 2287–2294. [CrossRef]
70. Ahn, J.; Song, J.C.; Lee, J.S. Dependence of conjugate heat transfer in ribbed channel on thermal conductivity of channel wall: A LES study. *Energies* **2021**, *14*, 5698. [CrossRef]
71. Ahn, J.; Song, J.C.; Lee, J.S. Large eddy simulation of conjugate heat transfer in a ribbed channel: Reynolds number effect. *Processe* **2022**, *10*, 1928. [CrossRef]
72. Abdelmoula, A.; Younis, B.A.; Spring, S.; Weigand, B. Large-eddy simulations of heated flows in ribbed channels with spanwis rotation. *Numer. Heat Transf. A* **2018**, *74*, 895–916. [CrossRef]
73. Ruck, S.; Arbeiter, F. Detached eddy simulation of turbulent flow and heat transfer in cooling channels roughened by variousl shaped ribs on one wall. *Int. J. Heat Mass Transf.* **2018**, *118*, 388–401. [CrossRef]

74. Duchaine, F.; Gicquel, L.; Grosnickel, T.; Koupper, C. Large-eddy simulation of the flow developing in static and rotating ribbed channels. *J. Turbomach.* **2020**, *142*, 041003. [CrossRef]
75. Perrot, A.; Gicquel, L.; Duchaine, F.; Odier, N.; Dombard, J.; Grosnickel, T. Unsteady analysis of heat transfer coefficient distribution in a static ribbed channel for an established flow. *J. Turbomach.* **2021**, *143*, 121004. [CrossRef]
76. Zhiyin, Y. Large-eddy simulation: Past, present and the future. *Chin. J. Aeronaut.* **2015**, *28*, 11–24. [CrossRef]
77. Oliver, T.A.; Bogard, D.G.; Moser, R.D. Large eddy simulation of compressible, shaped-hole film cooling. *Int. J. Heat Mass Transf.* **2019**, *140*, 498–517. [CrossRef]
78. Ahn, J. Large eddy simulation of film cooling: A review. *Energies* **2022**, *15*, 8876. [CrossRef]
79. Patankar, S.V.; Liu, C.H.; Sparrow, E.M. Fully developed flow and heat transfer in ducts having streamwise-periodic variations of cross-sectional area. *J. Heat Transf.* **1977**, *99*, 180–186. [CrossRef]
80. Germano, M.; Piomelli, P.; Moin, P.; Cabot, W.H. A dynamic sub-grid scale eddy viscosity model. *Phys. Fluids* **1991**, *A3*, 1760–1765. [CrossRef]
81. Anderson, R.; Meneveau, C. Effects of the similarity model in finite-difference LES of isotropic turbulence using a Lagrangian dynamic mixed model. *Flow Turbul. Combust.* **1999**, *62*, 201–225. [CrossRef]
82. Stoll, R.; Porté-Agel, F. Large-eddy simulation of the stable atmospheric boundary layer using dynamic models with different averaging schemes. *Bound. Layer Meteorol.* **2008**, *126*, 1–28. [CrossRef]
83. Ooi, A.; Iaccarino, G.; Durbin, P.A.; Behnia, M. Reynolds averaged simulation of flow and heat transfer in ribbed ducts. *Int. J. Heat Fluid Flow* **2002**, *23*, 750–757. [CrossRef]
84. Rau, G.; Çakan, M.; Moeller, D.; Arts, T. The effect of periodic ribs on the local aerodynamic and heat transfer performance of a straight cooling channel. *J. Turbomach.* **1998**, *120*, 368–375. [CrossRef]
85. Cho, H.H.; Wu, S.J.; Kwon, H.J. Local heat/mass transfer measurements in a rectangular duct with discrete ribs. *J. Turbomach.* **2000**, *122*, 579–586. [CrossRef]
86. Liou, T.M.; Hwang, J.J. Effect of ridge shapes on turbulent heat transfer and friction in a rectangular channel. *Int. J. Heat Mass Transf.* **1993**, *36*, 931–940. [CrossRef]
87. Cukurel, B.; Arts, T. Local heat transfer dependency on thermal boundary condition in ribbed cooling channel geometries. *J. Heat Transfer* **2013**, *135*, 101001. [CrossRef]
88. Iaccarino, G.; Ooi, A.; Durbin, P.A.; Behnia, M. Conjugate heat transfer predictions in two-dimensional ribbed passages. *Int. J. Heat Fluid Flow* **2002**, *23*, 340–345. [CrossRef]
89. Cukurel, B.; Arts, T.; Selcan, C. Conjugate heat transfer characterization in cooling channels. *J. Therm. Sci.* **2012**, *21*, 286–294. [CrossRef]
90. Jung, E.Y.; Chung, H.; Choi, S.M.; Woo, T.-K.; Cho, H.H. Conjugate heat transfer on full-coverage film cooling with array jet impingement with various Biot numbers. *Exp. Thermal Fluid Sci.* **2017**, *83*, 1–8. [CrossRef]
91. Kim, K.M.; Kim, Y.Y.; Lee, D.H.; Rhee, D.H.; Cho, H.H. Influence of duct aspect ratio on heat/mass transfer in coolant passages with rotation. *Int. J. Heat Fluid Flow* **2007**, *28*, 357–373. [CrossRef]
92. Schüler, M.; Zehnder, F.; Weigand, B.; von Wolfersdorf, J.; Neumann, S.O. The effect of turning vanes on pressure loss and heat transfer of a ribbed rectangular two-pass internal cooling channel. *J. Turbomach.* **2011**, *133*, 021017. [CrossRef]
93. Krishnaswamy, K.; Sivan, S. Improvement in thermal hydraulic performance by using continuous V and W-Shaped rib turbulators in gas turbine blade cooling application. *Case Stud. Therm. Eng.* **2021**, *24*, 100857. [CrossRef]
94. Ahn, J.; Kim, M.S.; Jang, S. Effects of cross-sectional aspect ratio of V-shaped ribs and blockage ratio on heat transfer in a channel at a low Reynolds number. *J. Mech. Sci. Technol.* **2018**, *32*, 5465–5473. [CrossRef]
95. Lee, D.H.; Rhee, D.H.; Kim, K.M.; Cho, H.H.; Moon, H.K. Detailed measurement of heat/mass transfer with continuous and multiple V–shaped ribs in rectangular channel. *Energy* **2009**, *24*, 1770–1778. [CrossRef]
96. Jin, D.; Zuo, J.; Quan, S.; Xu, S.; Gao, H. Thermohydraulic performance of solar air heater with staggered multiple V-shaped ribs on the absorber plate. *Energy* **2017**, *127*, 68–77. [CrossRef]
97. Moon, M.A.; Park, M.J.; Kim, K.Y. Evaluation of heat transfer performances of various rib shapes. *Int. J. Heat Mass Transf.* **2014**, *71*, 275–284. [CrossRef]
98. Togun, H.; Hamidatou, S.; Mohammed, H.I.; Abed, A.M.; Hasan, H.A.; Homod, R.Z.; Al-Fatlawi, A.W.; Al-Thamir, M.; Abdulrazzaq, T. Numerical Simulation on Heat Transfer Augmentation by Using Innovative Hybrid Ribs in a Forward-Facing Contracting Channel. *Symmetry* **2023**, *15*, 690. [CrossRef]
99. Liou, T.; Wang, W.; Chang, Y. Holographic interferometry study of spatially periodic heat transfer in a channel with ribs detached from one wall. *J. Heat Transfer.* **1995**, *117*, 32–39. [CrossRef]
100. Kim, S.; Suh, S.; Baek, S.; Hwang, W. The effect of single-sided ribs on heat transfer and pressure drop within a trailing edge internal channel of a gas turbine blade. *J. Thermal Sci. Eng. Appl.* **2022**, *14*, 081005. [CrossRef]
101. Elyyan, M.A.; Tafti, D.K. Large eddy simulation investigation of flow and heat transfer in a channel with dimples and protrusions. *J. Turbomach.* **2008**, *130*, 041016. [CrossRef]
102. Lee, Y.O.; Ahn, J.; Kim, J.; Lee, J.S. Effect of dimple arrangements on the turbulent heat transfer in a dimpled channel. *J. Enhanced Heat Transf.* **2012**, *19*, 359–367. [CrossRef]

103. Hao, Z.; Gorlé, C. Large eddy simulations of forced heat convection in a pin-fin array with a priori examination of an eddy-viscosity turbulence model. *Int. J. Heat Fluid Flow* **2019**, *77*, 73–83. [CrossRef]
104. Fan, X.; Li, L.; Zou, J.; Zhou, Y. Cooling methods for gas turbine blade leading edge: Comparative study on impingement cooling, vortex cooling and double vortex cooling. *Int. Commun. Heat Mass Transf.* **2019**, *100*, 133–145. [CrossRef]

energies

MDPI

rticle

.arge Eddy Simulation of Forced Convection around Wavy :ylinders with Different Axes

lin-Ki Kim [1], Chin-Hyuk Chang [2], Seok-Hyun Nam [1] and Hyun-Sik Yoon [1,*]

[1] Department of Naval Architecture and Ocean Engineering, Pusan National University, 2, Busandaehak-ro 63beon-gil, Gumjeong-Gu, Busan 46241, Republic of Korea; minkiiekim@gmail.com (M.-K.K.); nsh219@naver.com (S.-H.N.)

[2] School of Mechanical Engineering, Pusan National University, 2, Busandaehak-ro 63beon-gil, Geumjeong-gu, Busan 46241, Republic of Korea; chinhyuk@nate.com

* Correspondence: lesmodel@pusan.ac.kr

Abstract: Four types of undulated cylinders with streamwise undulation, transverse undulation, in-phase undulation and antiphase undulation are employed to investigate the undulation-axis effect on the structure of heat transfer around wavy cylinders. The flows around these undulated cylinders are numerically simulated by large eddy simulation at $Re = 3000$. The force coefficients and Nusselt numbers of the cylinders with transverse undulation and in-phase undulation are significantly influenced by wavelength and wave amplitude. On the other hand, the cylinders with streamwise undulation and antiphase undulation show a very weak dependence of the force coefficients and Nusselt numbers on the combinations of wavelength and wave amplitude. It is noted that the cylinder with antiphase undulation, under certain wavy conditions, provides about the same Nusselt number as the smooth cylinder, even though the force coefficients are considerably decreased. The thermal characteristics, according to the combination of wavy geometric parameters, are supported by the surface distribution of the Nusselt numbers. In addition, the isothermal distribution, which depends on the wake flow, explains the variation in the Nusselt numbers. The present results suggest that a proper modification of geometry can improve both heat transfer and aerodynamic performances.

Keywords: undulation axis; undulated cylinder; heat transfer

tation: Kim, M.-K.; Chang, C.-H.; am, S.-H.; Yoon, H.-S. Large Eddy nulation of Forced Convection ound Wavy Cylinders with fferent Axes. *Energies* **2024**, *17*, 894. tps://doi.org/10.3390/en17040894

cademic Editor: Artur Bartosik

eceived: 26 December 2023
evised: 13 February 2024
ccepted: 13 February 2024
ıblished: 14 February 2024

1. Introduction

The flow around bluff bodies is observed daily in natural phenomena and industrial applications. For this reason, in order to identify the flow characteristics, many researchers have studied different geometrical cross-sections in various applications, for instance, electric and heat transfer equipment, and aerodynamics and hydrodynamics of land or offshore architectural structures. Among them, several studies on wavy cylinders have been recently implemented and published to closely investigate the effect of rectified cylinder geometry on thermal characteristics. From a global perspective, two types of cylinder geometries are mainly taken into account for a detailed description and grouping hereafter, one for wavy cylinders and the other for wavy elliptic cylinders.

For wavy cylinders, Ahn et al. [1] investigated the effect of undulation on fluid flow and forced convection heat transfer around wavy cylinders with different wavelengths, such as $\pi/2$, $\pi/3$ and $\pi/4$, with a fixed amplitude of 0.1 at a Reynolds number and a Prandtl number of 300 and 0.71, respectively. The fluid dynamics and heat transfer around these undulating cylinders are influenced by both the position along the spanwise direction and the wavelength. For a wavy cylinder with a half wavelength ($\lambda/2$), the averaged Nusselt number over time and the entire surface are greater than those of a smooth cylinder. Conversely, for cylinders with wavelengths of $\lambda/4$ and $\lambda/3$, the averaged Nusselt number is smaller compared to that of a smooth cylinder. While Ahn et al. [1] performed their study at a Reynolds number of 300, Kim and Yoon [2] extended the investigation at a higher

Reynolds number of 3000 in the subcritical regime. Kim and Yoon [2] also investigated th effect of wavelength (λ/Dm) on the flow and thermal field. They considered a wide rang of wavelengths from 1.136 to 6.06 at a Reynolds number of 3000 in the subcritical regim and a Prandtl number of 0.7. They showed that the changes in the mean Nusselt numbe are correlated with the force coefficients. They found that the critical wavelength occurs a the transition of the point where the maximum Nusselt number is located, shifting fror the node to the saddle.

Moon et al. [3] studied the effect of asymmetric wavy (ASW) perturbation on force convection heat transfer as a passive technique to control the force coefficients by mear of a large eddy simulation at a Reynolds number of 3000. The ASW cylinder exhibit the lowest mean and fluctuation in the time- and total surface-averaged Nusselt numbe compared to the smooth (CY) and symmetric wavy (SW) cylinders. The Nusselt numbe averaged over time and locally in the spanwise direction for the SW cylinder, exhibit region-dependent characteristics, with one region remaining invariant and another regio showing an increase. The ASW cylinder shows increasing and decreasing performance a the short and long wavelengths, respectively.

Yoon et al. [4] adopted the shape of a double-wavy (DW) cylinder as a geometrica perturbation to control the fluid flow and heat transfer at a Reynolds number and a Prand number of 3000 and 0.7, respectively. Among the different geometrical perturbation such as CY and SW, the ASW of a DW cylinder achieves the smallest mean drag and li fluctuation in terms of not only the force coefficient but also the Nusselt number. The DV cylinder yields a reduced spanwise local Nusselt number across the span when compare to the SW and ASW cylinders. The wake alteration exhibited by the DW cylinder allow the attenuation in heat transfer.

Yoon et al. [5] performed an initial research study on the forced convection hea transfer around a helically twisted elliptic (HTE) cylinder influenced by the structure an design of a daffodil stem. They also investigated the influence of Reynolds number on th system of the laminar flow by means of a numerical simulation at the range of Reynold numbers of $60 \leq Re \leq 150$ and a Prandtl number (*Pr*) of 0.7. The HTE cylinder has muc lower drag and lift fluctuation than a smooth cylinder. The variation in Nusselt numbe along the spanwise direction was discerned through an examination of the flow structure and the distribution of isotherms.

Yoon and Moon [6] conducted a numerical evaluation of the performance of a variabl pitch helically twisted elliptic (VPHTE) cylinder at a Reynolds number of 3000. A com prehensive investigation, including a parametric study on the variable pitch ratio an simulations of both smooth and HTE cylinders for a comparative analysis, was conducte The obtained results affirm the efficacy of the VPHTE disturbance as a passive flow contro strategy for achieving drag reduction and suppressing lift fluctuations, which is in line wit previous research findings. The VPHTE cylinder presents a relatively smaller value of tota surface-averaged Nusselt number than the smooth and HTE cylinders, with a stabilize time series.

There have been more published studies on wavy elliptic cylinders than wavy cylin ders. Kim and Yoon [7] studied the forced convection heat transfer around a biomimeti elliptic cylinder inspired by a harbor seal vibrissa (HSV) by utilizing the technique of larg eddy simulation at a Reynolds number and a Prandtl number of 500 and 0.7, respectivel The temporal trends of the surface-averaged Nusselt number demonstrated that the HSV contributed to the stable heat transfer behavior by significantly suppressing its fluctuatior whose characteristics aligned with the HSV's distinctive ability to suppress lift fluctuatior

Kim and Yoon [8] employed numerical methods to explore the influence of Reynold number (Re) on the fluid flow and heat transfer characteristics of a biomimetic elliptica cylinder inspired by an HSV at Reynolds numbers ranging from 50 to 500 and a Prandt number of 0.7. The findings confirm the effect of the distinctive geometry of the HSV in th low Reynolds number regime, demonstrating a reduction in drag and a suppression of lif fluctuation. The root-mean-square (RMS) amplitude of the fluctuating lift is exceedingly small due to the unsteady behavior of the HSV within this Reynolds number range.

Yoon et al. [9] numerically investigated the influence of geometrical characteristics of an HSV on forced convection heat transfer at a Reynolds number of 500. Seven types of HSV geometries were defined to utilize and combine the minor-axis undulation, the major-axis undulation, the both-axis undulation and the offset angle. The inclusion of HSV features in all modified geometries results in a reduction in the time-averaged drag coefficient when compared to the mean drag coefficient of an elliptic cylinder. The authors concluded that it can be cautiously inferred that among the geometrical features of the HSV, the undulation along the major axis is a crucial factor influencing the forced convection heat transfer.

Yoon et al. [10] studied the effect of wavelengths of a wavy elliptic cylinder with different aspect ratios on forced convection heat transfer. They defined four types of wake thermal structures, such as quasi-2D unsteady and steady structures, which are formed by short wavelengths, and complex 3D unsteady and steady structures, which are originated from long wavelengths. The surface distribution of the mean Nusselt number varied based on the characteristics of the thermal structures. In summary, the authors suggested that a streamlined shape plays a dominant role in stabilizing the forced convection as the aspect ratio (AR) decreases, irrespective of the wavelength.

From a review of previous studies, it is found that there is no dedicated study that takes into account the specific types of undulated geometry disturbance to find the effect of heat transfer characteristics and the correlation between undulation geometries. Thus, four types of undulated cylinders with streamwise undulation (SU), transverse undulation (TU), in-phase undulation (IPU) and antiphase undulation (APU) are considered in this study to classify the undulation-axis effect on the wake of flow and the reduction in force coefficients.

The main objective of this study is to confirm whether a geometrical modification of wavy cylinders can well explained the structure of heat transfer around these wavy cylinders. Particularly, it is noted that the motivation for conducting this study is to clarify the thermal and heat transfer characteristics of the proposed undulated cylinders and to identify the correlation among the undulated cylinders with SU, TU, IPU and APU, even though the wavelength (λ) and amplitude (a) of each type are randomly selected and applied, with the combination of λ/D_m = 2.273 and a/D_m = 0.091 being considered as Case 1, while the combination of λ/D_m = 6.06 and a/D_m = 0.152 is considered as Case 2.

2. Numerical Details

2.1. Geometry Definitions

The shapes of smooth cylinders (CY) with in-phase undulation (IPU), transverse undulation (TU), streamwise undulation (SU) and antiphase undulation (APU) are depicted in Figure 1. The local diameter (D_z) of a cylinder with IPU is defined as follows:

$$D_z = D_m + 2a\cos\left(\frac{2\pi z}{\lambda}\right) \quad (1)$$

where a and λ are the amplitude and wavelength of the waviness, respectively; D_m is the mean diameter of the wavy cylinder, which is the mean value of the maximum and minimum local diameters (D_{max} and D_{min}); and D_m is the same as the diameter of a smooth cylinder.

The locations of D_{max}, D_m and D_{min} are at the node, the middle and the saddle, respectively, as shown in Figure 2. As mentioned above, the combination of λ/D_m = 2.273 and a/D_m = 0.091 is considered Case 1, while the combination of λ/D_m = 6.06 and a/D_m = 0.152 is considered Case 2.

Unlike the IPU cylinder, the TU, SU and APU cylinders have local axis lengths in the streamwise and transverse directions. The geometry definitions of the TU, SU and APU cylinders are shown in Figure 2b–d, respectively. The local lengths in the streamwise direction ($L_{z,s}$) and transverse direction ($L_{z,t}$) of the TU, SU and APU cylinders are defined in Equations (2)–(4), respectively, as follows:

$$L_{z,s} = D_m,\ L_{z,t} = D_m + 2a\cos\left(\frac{2\pi z}{\lambda}\right) \tag{2}$$

$$L_{z,s} = D_m + 2a\cos\left(\frac{2\pi z}{\lambda}\right),\ L_{z,t} = D_m \tag{3}$$

$$L_{z,s} = D_m + 2a\cos\left(\frac{2\pi z}{\lambda}\right),\ L_{z,t} = D_m - 2a\cos\left(\frac{2\pi z}{\lambda}\right) \tag{4}$$

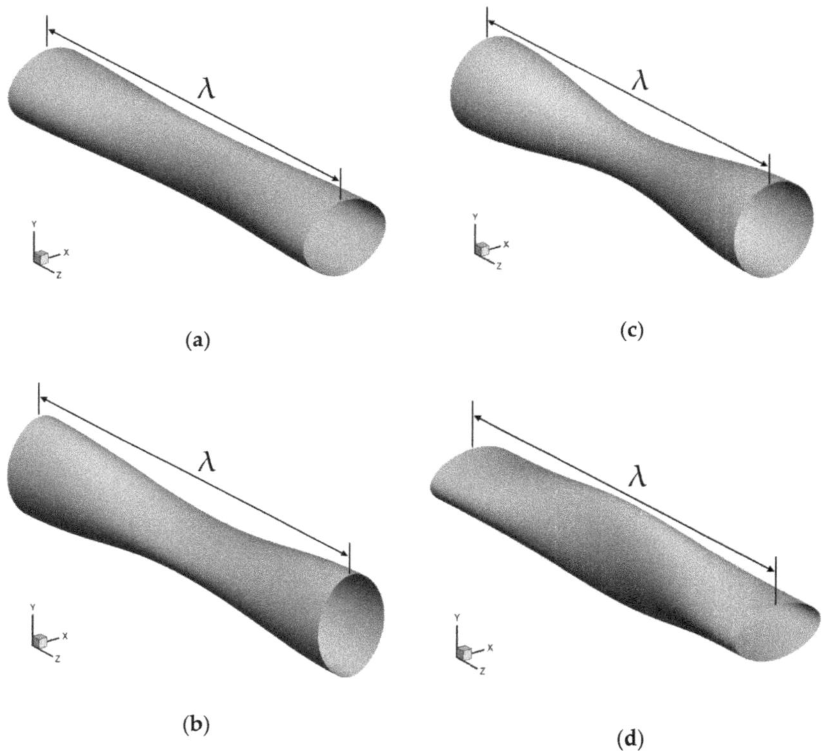

Figure 1. The geometries of cylinders with (**a**) SU, (**b**) TU, (**c**) IPU and (**d**) APU.

The TU cylinder has an undulation in the transverse direction along the center of the cylinder. This wavy shape is similar to the MI cylinder among the HSV-shaped cylinders presented in Yoon et al. [9,10]. The SU cylinder has an undulation in the streamwise direction along the center of the cylinder. This wavy shape is similar to the MA cylinder among the HSV-shaped cylinders presented in Yoon et al. [9,10]. The APU cylinder has undulations in both the streamwise and transverse directions like the IPU cylinder. However, it has an antiphase undulation in the streamwise and transverse directions. This wavy shape is similar to the BA cylinder among the HSV-shaped cylinders presented in Yoon et al. [9,10]. All undulation cases have the same mean diameter of D_m, though these three cylinders have different lengths in the streamwise and transverse directions according to the spanwise direction.

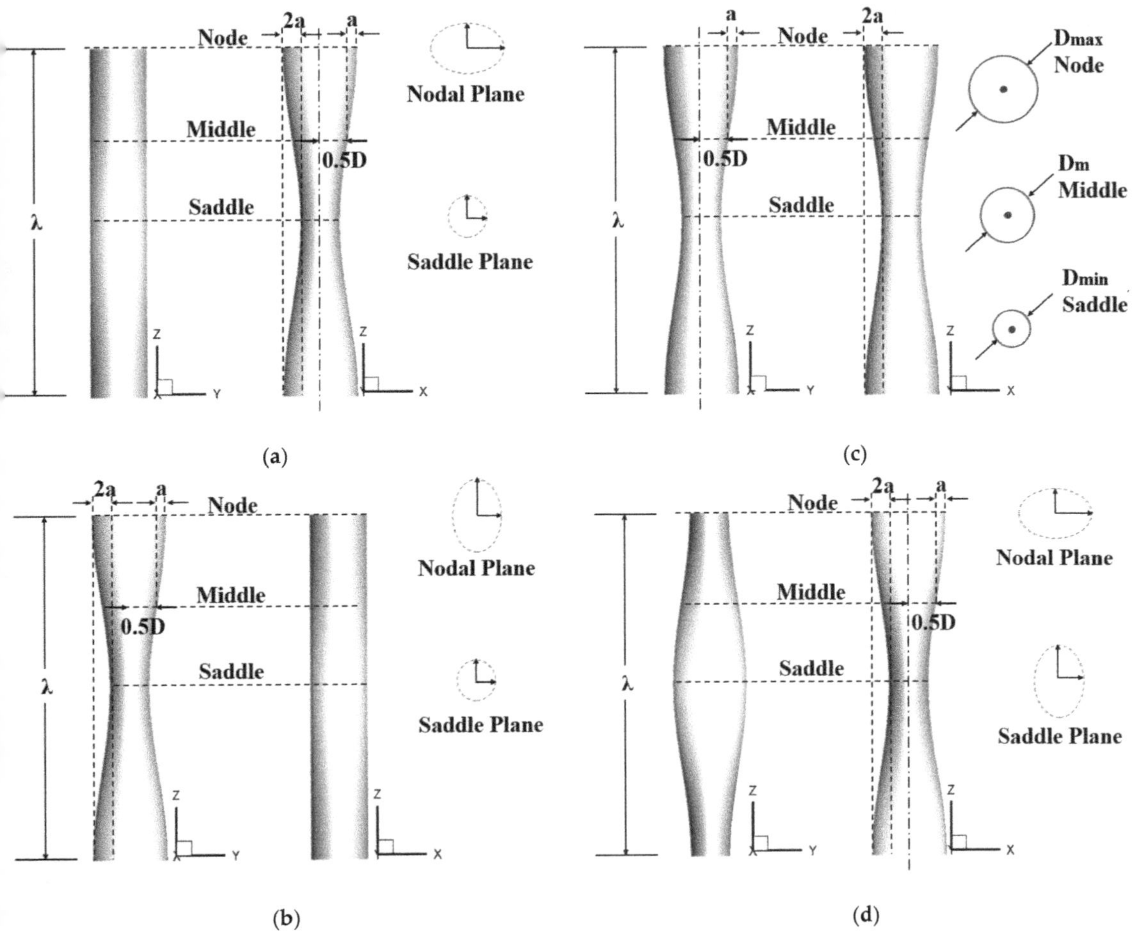

Figure 2. Definition of geometries for cylinders with (**a**) SU, (**b**) TU, (**c**) IPU and (**d**) APU.

2.2. Governing Equations and Numerical Methods

The Navier–Stokes, continuity and energy equations were considered to simulate the unsteady three-dimensional incompressible turbulent flow and thermal fields around the wavy cylinders. The filtered governing equations were used in the large eddy simulation (LES), and their expressions are as follows:

$$\frac{\partial \tilde{u}_i}{\partial t} + \frac{\partial \tilde{u}_i \tilde{u}_j}{\partial x_j} = -\frac{\partial \tilde{p}}{\partial x_i} + \frac{1}{Re}\frac{\partial^2 \tilde{u}_i}{\partial x_j \partial x_j} - \frac{\partial \tau_{ij}}{\partial x_j} \tag{5}$$

$$\frac{\partial \tilde{u}_i}{\partial x_i} = 0 \tag{6}$$

$$\frac{\partial \tilde{T}}{\partial t} + \frac{\partial \tilde{u}_j}{\partial x_j} = \frac{1}{RePr}\frac{\partial^2 \tilde{T}}{\partial x_j^2} - \frac{\partial q_j}{\partial x_j} \tag{7}$$

where u_i is the corresponding velocity component; p is the pressure; Re is the Reynolds number; and τ_{ij} is the subgrid-scale stress tensor ($\tau_{ij} = \widetilde{u_i u_j} - \tilde{u}_i\, \tilde{u}_j$), resulting in an effect of the subgrid scales on the resolved scales. T is the temperature, Pr is the Prandtl number, and q_i is the subgrid flux ($q_i = \widetilde{Tu_j} - \tilde{T}\, \tilde{u}_j$). An overbar denotes a large-scale (or resolved) flow obtained from grid filtering.

The dynamic Smagorinsky subgrid-scale (SGS) model by Germano et al. [11] wa used to represent the effects of unresolved small-scale fluid motions. Further details c the SGS model are available from Yoon et al. [12], Jung and Yoon [13], Yoon et al. [14 Bohgi et al. [15], Ouro et al. [16], Silva et al. [17], Hubbard et al. [18] and Shi et al. [19]. Jun and Yoon [13] successfully used this model to simulate the flow around a helical twiste elliptic cylinder at the same sub-critical Reynolds number of 3000. This study uses the sam LES method as Yoon et al. [12].

All variables were non-dimensionalized according to the free-stream velocity U_∞ the cylinder surface temperature T_S, and the mean diameter $D_m = \frac{D_{max}+D_{min}}{2}$, where D_{ma} and D_{min} represent the maximum and minimum local diameters of the wavy cylinde The non-dimensionalization results in two dimensionless parameters: $Re = U_\infty D_m/v$ an $Pr = C_p\mu/k$, where C_p is the specific heat, and μ and k are the dynamic viscosity an thermal conductivity, respectively. The Prandtl number is $Pr = 0.7$, which corresponds t air, and the Reynolds number is $Re = 3000$.

For spatial discretization, a second-order central difference scheme based on the finit volume method (FVM) was applied to the structured grids under consideration. A secon order implicit scheme was employed for temporal discretization of the governing equation The second-order Adams–Bashforth scheme and the Crank–Nicolson scheme were used fc the convection and diffusion terms, respectively. We derived the governing equations usin a generalized curvilinear coordinate system. Detailed information on the implementatio can be found in the STAR CCM+ manuals [20].

2.3. Computational Domain and Boundary Conditions

Figure 3 presents the domain sizes and boundary conditions imposed in each directio The outflow boundary is at a distance of 16 D_m from the cylinder. The spanwise domai size varies according to the wavelength. We determined the spanwise domain size based o Yoon et al. [21], Lam and Lin [22] and Lin et al. [23] for smaller and larger wavelengths c $\lambda/D_m \leq 2.818$ and $\lambda/D_m \geq 3.79$, respectively. Lam and Lin [22] and Lin et al. [23] explaine the determination of the spanwise domain size corresponding to the wavelength.

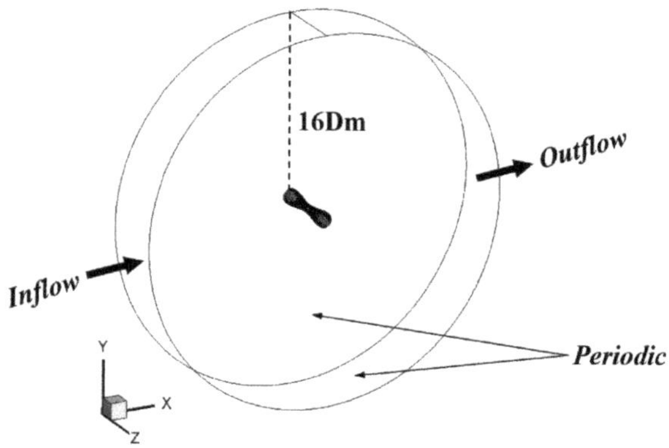

Figure 3. Schematic of the computational domain and boundary conditions.

A fluid flow was uniformly imposed on the outflow [inflow] region along the x-axis a the inlet boundary, with a constant temperature of T_∞ and a convective boundary conditio A no-slip condition was applied at the wall boundary with a constant surface temperature o T_s, and a Neumann-type boundary condition was used for pressure. A periodic boundar condition was imposed in the spanwise direction.

2.4. Grid Dependence Test and Validation

To verify the presented numerical method and grid system, a validation test of ou results and a grid dependence test were carried out. Comparisons of the time-average drag coefficient ($\overline{C_D}$), the root mean square (RMS) of lift fluctuation ($C_{L,RMS}$), the Strouhal number (St) and the time- and total surface-averaged Nusselt number ($\ll\overline{Nu}\gg$) with previous data were performed for the smooth cylinder, as shown in Table 1.

Table 1. Validation of the present numerical methods.

Case			$\overline{C_D}$	$C_{L,RMS}$	St	$\ll\overline{Nu}\gg$
Present study	Numerical	Coarse	0.9725	0.0850	0.20	26.61
		Medium	0.9756	0.0836	0.20	26.71
		Fine	0.9757	0.0837	0.20	26.71
Lam and Lin [22]	Numerical		1.09	0.177	0.21	-
Norberg [24]	Experimental		0.98–1.03	-	0.210–0.213	-
Norberg [25]	Summarized		-	0.05, 0.07	0.21	-
Lu et al. [26]	Experimental		1.02	N/A	N/A	-
Yoon et al. [27]	Numerical		1.023	0.1648	0.21	-
Zukauskas and Ziuhzda [28]	Analytical		-	-	-	27.94
Sanitjai and Goldstein [29]	Analytical		-	-	-	26.98
Hilpert [30]	Analytical		-	-	-	25.66

The drag and lift coefficients and the Strouhal number are defined as follows:

$$C_D = \frac{2F_D}{\rho U_\infty^2 A} \tag{8}$$

$$C_L = \frac{2F_L}{\rho U_\infty^2 A} \tag{9}$$

$$St = \frac{f_s D_m}{U_\infty} \tag{10}$$

where F_D and F_L are the total drag and lift forces, respectively; ρ is the fluid density; U_∞ is the free-stream velocity; A is the projected area of the cylinder facing the flow direction; and f_s is the frequency of vortex shedding. The Nusselt number is the ratio of convective heat transfer to conductive heat transfer:

$$Nu = \frac{hD_m}{k} \tag{11}$$

where h is the convective heat transfer coefficient, and k is the thermal conductivity. Once the velocity and temperature fields are obtained, the local Nusselt number, the spanwise local surface-averaged Nusselt number, the total surface-averaged Nusselt number, the time-averaged local Nusselt number, the time- and spanwise local surface-averaged Nusselt number, and the time- and total surface-averaged Nusselt number are defined as follows:

$$\begin{gathered} Nu = \left.\frac{\partial T}{\partial n}\right|_{wall},\ \langle Nu\rangle = \frac{1}{W_z}\int_0^{W_z} Nu dS,\ \ll Nu\gg = \frac{1}{L}\int_0^{L}\langle Nu\rangle dz \\ \overline{Nu} = \frac{1}{t_p}\int_0^{t_p} Nu dt,\ \langle\overline{Nu}\rangle = \frac{1}{t_p}\int_0^{t_p}\langle Nu\rangle dt,\ \ll\overline{Nu}\gg = \frac{1}{t_p}\int_0^{t_p}\ll Nu\gg dt \end{gathered} \tag{12}$$

where n is the normal direction to the walls, W_z is the spanwise local surface area, L is the spanwise arc length, and t_p is the period of time integration. The results agree well with those of previous research. The force coefficients and flow structures of the wavy cylinders at different wavelengths were also compared with previous results for validation. Three different grid systems were considered for the grid dependence test: coarse (5.5 million grids), medium (7.5 million grids) and fine (9.5 million grids). The values of the force coefficients and the Nusselt numbers for the three different grid systems show small variations that are not significant, so the medium grid system is used to discuss the results.

Figure 4 shows a typical grid system. The grid distribution is non-uniform in the x–y plane and uniform along the z direction. The distance between the cylinder surface and the nearest grid points is confined to $y^+ = \rho u_\tau \Delta y / \mu \approx 1$, where $u_\tau = \sqrt{\frac{\tau_w}{\rho}}$ is the friction velocity and τ_w is the wall shear stress. Grid stretching was used to increase the grid resolutions near the surface and in the wake region. There are 250 grids around the cylinder circumference, which corresponds to 61,752 cells in the x–y plane. The number of uniform grid layers along the z direction ranges from 65 to 121, depending on the wavelength of the sinusoidal cylinder.

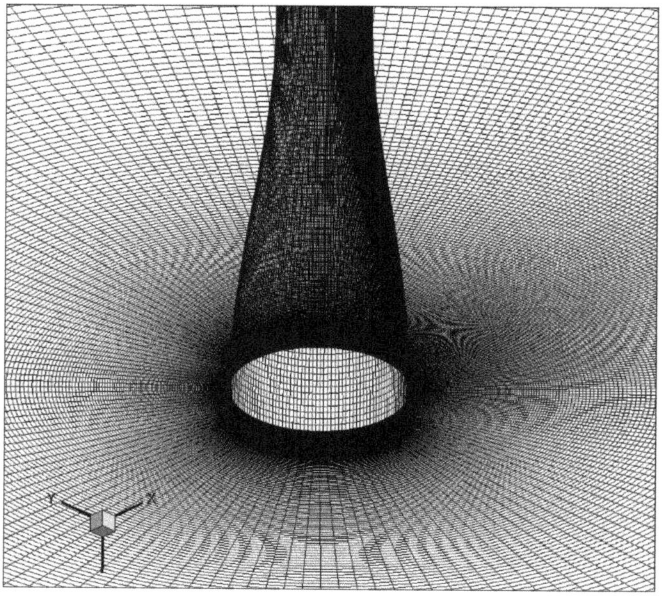

Figure 4. The grid distribution of a wavy cylinder.

3. Results and Discussion

Figure 5a,b show the time histories of C_D and C_Lf that correspond to Case 1 and Case 2, respectively. For both cases, a smooth cylinder was included to investigate (a) C_D and C_L for Case 1 and (b) C_D and C_L for Case 2 as an objective of the comparison basis. In addition, for a quantitative comparison of the force coefficients, the reduction rates of $\overline{C_D}$ and $C_{L,RMS}$ of the undulated cylinders compared to those of a smooth cylinder are described in Figure 5c. The reduction rates of $\overline{C_D}$ and $C_{L,RMS}$ are defined as $\Delta\overline{C_D}/\overline{C}_{D,CY}$ and $\Delta C_{L,RMS}/C_{L,RMS,CY}$, respectively, where $\Delta\overline{C_D} = \overline{C}_{D,CY} - \overline{C_D}$, $\Delta C_{L,RMS} = C_{L,RMS,CY} - C_{L,RMS}$, and the subscript of CY stands for a smooth cylinder.

For C_D in Case 1, the APU cylinder provides the smallest value among the different undulated cylinders, as shown in Figure 5a. The APU cylinder presents the largest reduction rate of $\overline{C_D}$ at 14.8%, as shown in Figure 5c. Otherwise, the TU cylinder achieves the largest value of C_D, resulting in the smallest reduction rate of 7.6% for $\overline{C_D}$. The values of C_D for the SU and IPU cylinders are in between those of the APU and TU cylinders, as shown in Figure 5a. The SU cylinder shows the second lowest value of C_D, leading to a 13.5% reduction rate of $\overline{C_D}$, as shown in Figure 5c.

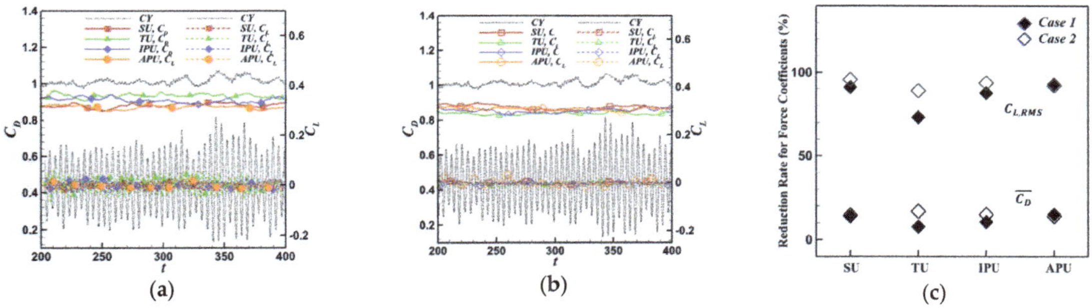

Figure 5. Time histories of (**a**) C_D and C_L for Case 1 and (**b**) C_D and C_L for Case 2, and (**c**) reduction rates of the drag and lift coefficients for Case 1 and Case 2.

For C_L in Case 1, the APU cylinder also provides the smallest value among the various types of undulated cylinders, as described in Figure 5a. Therefore, the APU cylinder presents the largest reduction rate of $C_{L,RMS}$ at 92.9%, as shown in Figure 5c. On the other hand, the TU cylinder also represents the largest value of C_L, resulting in the smallest reduction rate of 72.9% for $C_{L,RMS}$. The values of C_L for the SU and IPU cylinders are in between those of the APU and TU cylinders, as shown in Figure 5a. The SU cylinder's C_L is ranked as the second lowest value, resulting in a 91% reduction rate of $C_{L,RMS}$, as shown in Figure 5c.

For C_D in Case 2, the TU cylinder exhibits the smallest value among the different undulated cylinders, as illustrated in Figure 5b, leading to the largest reduction rate of $\overline{C_D}$ at 16.8%, as shown in Figure 5c, whereas the APU cylinder shows the largest value of C_D, resulting in the smallest reduction rate of 13.9% for $\overline{C_D}$. The values of C_D for the SU and TU cylinders are in the middle of those of the TU and APU cylinders, as shown in Figure 5b. The IPU cylinder shows the second lowest value of C_D, leading to a 14.9% reduction rate of $\overline{C_D}$, as shown in Figure 5c.

For C_L in Case 2, the SU cylinder provides the smallest value among the different undulated cylinders, as described in Figure 5b. Thus, the SU cylinder shows the largest reduction rate of $C_{L,RMS}$ at 95.6%, as shown in Figure 5c. By contrast, the TU cylinder presents the largest value of C_L, resulting in the smallest reduction rate of 89.1% for $C_{L,RMS}$. The values of C_L for the IPU and APU cylinders are in the middle of those of the SU and TU cylinders, as shown in Figure 5b. The APU cylinder's C_L is ranked as the second lowest value, leading to a 92% reduction rate of $C_{L,RMS}$, as shown in Figure 5c.

The TU and IPU cylinders are significantly influenced by the wavelength and wave amplitude, with the reduction rates of force coefficients being the smallest in Case 1 and the drag reduction being the largest in Case 2. On the other hand, the SU and APU cylinders in this study are not strongly affected by the considered combinations of wavelength and wave amplitude.

Figure 6 shows the 3D vortical structures as visualized through the utilization of the methodology proposed by Zhou et al. [31]. Apart from the types of undulation, a smooth cylinder is also included as (a) CY in Figure 6 for comparison. The vortical structures are identified by the negative of λ_2, which is the second largest eigenvalue of $S_{ij}^2 + \Omega_{ij}^2$, where S_{ij}^2 and Ω_{ij}^2 are the strain-rate and rotation-rate tensors, respectively. To analyze the impact of flow structures on the thermal field, temperature contours were overlaid onto the vortical structures.

It is clearly depicted in Figure 6a–e that the CY, SU, TU, IPU and APU cylinders in Case 1 show a clear visualization of Kármán vortices in the wake region. On the other hand, the corresponding undulated cylinders in Case 2 cause a further delay in the vortex roll-up and a greater suppression of vortex shedding than those in Case 1, as shown in Figure 6a–e. Therefore, in general, the SU, TU, IPU cylinders in Case 2 contribute to a greater reduction in C_D and C_L fluctuations than those in Case 1.

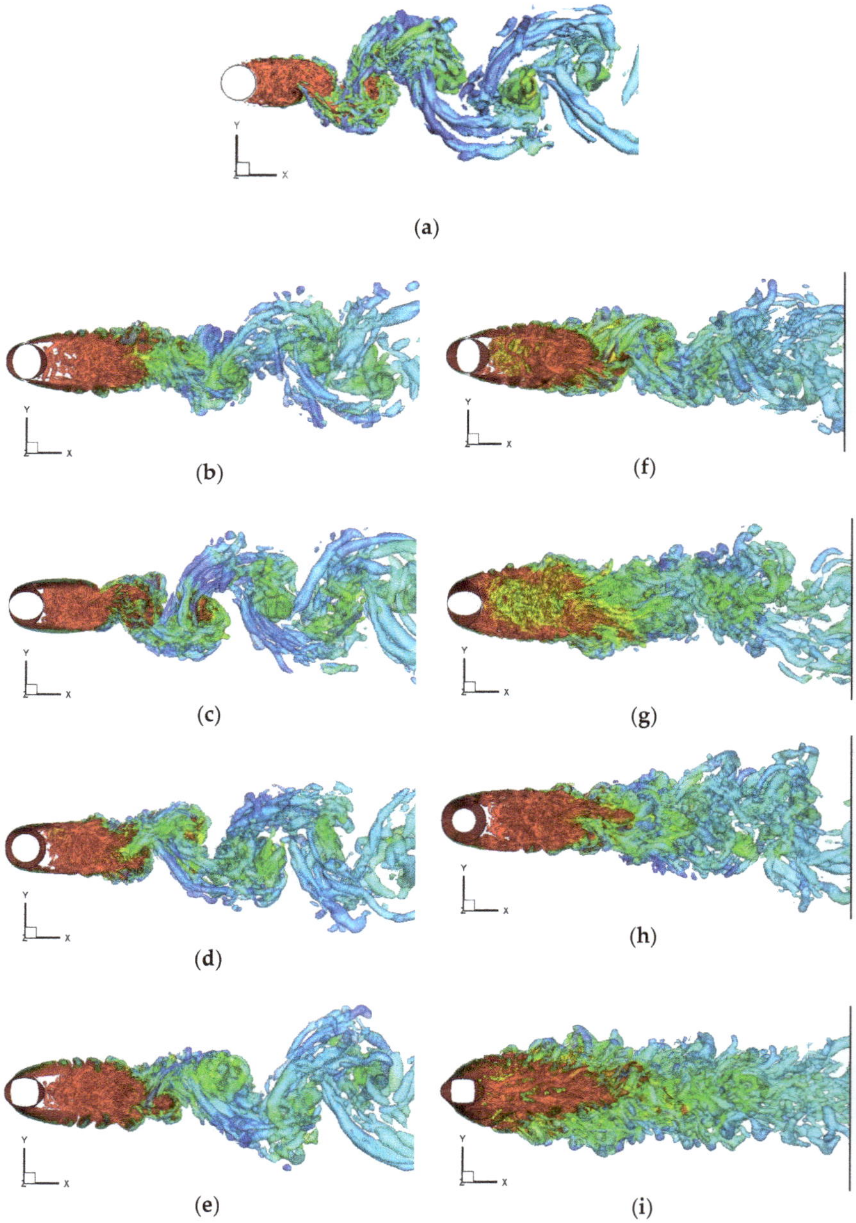

Figure 6. Side views of temperature contours on the vortical structures for Case 1 (1st column) an Case 2 (2nd column): (**a**) CY, (**b**,**f**) SU, (**c**,**g**) TU, (**d**,**h**) IPU and (**e**,**i**) APU.

More specifically, for Case 1, the vortical structure of the TU cylinder in Figure 6 shows that vortex shedding occurs relatively early in the near-wake region, which i strongly associated with and is supported by the highest values of C_D and C_L, as shown i Figure 5a, corresponding to the lowest values for the reduction rates of C_D and C_L show in Figure 5c.

The vortical structure of the APU cylinder in Figure 6e shows that the shear layer i further elongated and vortex shedding is more delayed, compared to the other undulate cylinders. These features of the wake structure are associated with the smallest values c force coefficients, which result in the largest reduction rates of $\overline{C_D}$ and $C_{L,RMS}$, as already observed in Figure 5a,c, respectively.

In addition, the TU cylinder has the smallest zero-vorticity area, but the APU cylinder presents the largest area of zero vorticity, as shown in Figure 6c,e, respectively, which also support the smallest and largest reduction rates of force coefficients of the TU and APU cylinders seen in Figure 5a,c.

For Case 2, the SU cylinder forms the earliest rolling-up and vortex shedding among the undulated cylinders, as shown in Figure 6f, which supports the largest value of C_L seen in Figure 5b. The TU cylinder in Figure 6g shows a weaker shear-layer elongation among the undulated cylinders in Case 2, which supports the largest value of C_D seen in Figure 5b.

The APU cylinder in Figure 6i presents the bluffiest near wake and an almost disappearance of vortex shedding, which support the lowest values of C_D and C_L seen in Figure 5b. From the perspective of vorticity near each cylinder, the SU cylinder presents a wide area of zero vorticity, as shown in Figure 6f, which supports the reduction rates of $\overline{C_D}$ and $C_{L,RMS}$ seen in Figure 5b,c.

Figure 7 presents the Cp contours in the x-z plane for all the undulated cylinders. A wide and close Cp distribution along a cylinder represents high values of C_D and C_L, for instance, (c) the TU cylinder in Case 1 and the (f) SU and (i) APU cylinders in Case 2 in Figure 7, which is confirmed by (a) the C_D and C_L values for Case 1 and (b) the C_D and C_L values for Case 2 seen in Figure 5. A narrow and far Cp distribution along a cylinder shows low values of C_D and C_L, for example, (e) the APU cylinder in Case 1 and (g) the TU cylinder in Case 2 in Figure 7, which is also well matched with (a) the C_D and C_L values for Case 1 and (b) the C_D and C_L values for Case 2 seen in Figure 5.

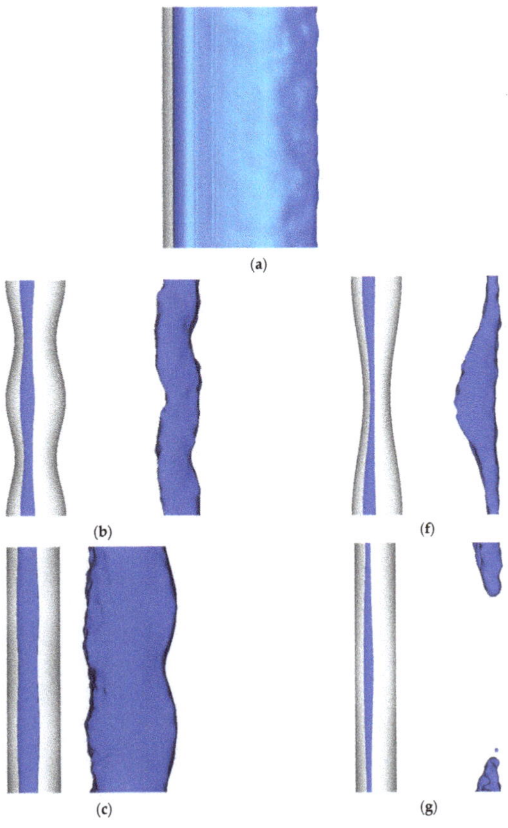

Figure 7. *Cont.*

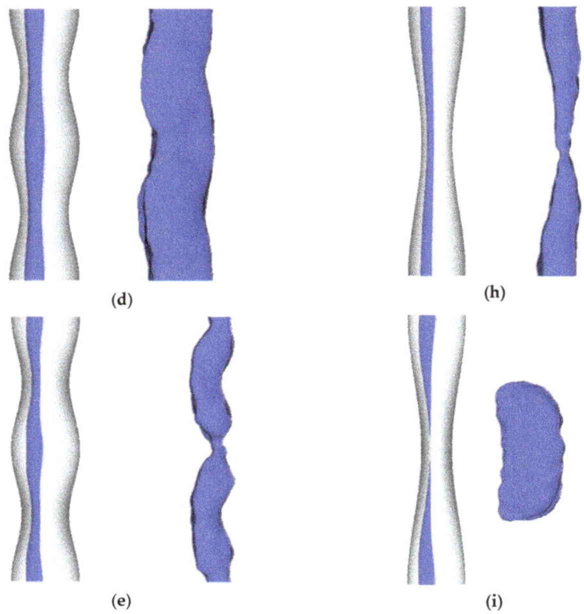

Figure 7. Cp contours in the x-z plane: (**a**) CY, (**b**) SU, (**c**) TU, (**d**) IPU and (**e**) APU for Case 1, and (**f**) SU, (**g**) TU, (**h**) IPU and (**i**) APU for Case 2.

Figure 8 shows the time histories of the total surface-averaged Nusselt number for Case 1 and Case 2, respectively. It is evident in Figure 8a,b that for Case 1 and Case 2, the values of $\langle\langle Nu \rangle\rangle$ for the undulated cylinders are smaller than those for the smooth cylinder.

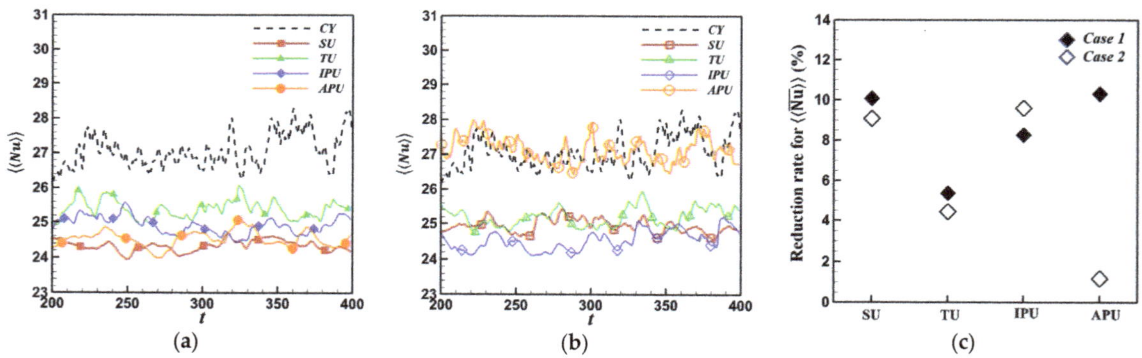

Figure 8. Time histories of surface-averaged Nusselt number for (**a**) Case 1 and (**b**) Case 2, and (**c**) reduction rate of time- and surface-averaged Nusselt number.

Among the undulated cylinders in Case 1, the TU cylinder has the largest value and the IPU cylinder has the next highest value of $\langle\langle Nu \rangle\rangle$. The SU and APU cylinders reveal almost the same values of $\langle\langle Nu \rangle\rangle$ that are in between those of the TU and IPU cylinders.

For Case 2 in Figure 8b, the APU cylinder presents a plot of $\langle\langle Nu \rangle\rangle$ with the highest values over the entire time. The $\langle\langle Nu \rangle\rangle$s of the TU, SU and IPU cylinders are ranked next in order, but there is a big discrepancy between the APU cylinder and the other undulated cylinders.

The reduction rates of the time- and total surface-averaged Nusselt number ($\langle\langle \overline{Nu} \rangle\rangle$) for the undulated cylinders in Case 1 and Case 2 are presented in Figure 8c. The reduction rate of $\ll\overline{Nu}\gg$ is defined as $\Delta\ll\overline{Nu}\gg/\ll\overline{Nu}\gg)_{CY}$, where $\Delta\ll\overline{Nu}\gg = \ll\overline{Nu}\gg_{CY} - \ll\overline{Nu}\gg$, and the subscript of CY stands for a smooth cylinder. For Case 1, the APU and SU cylinders accomplish a considerable reduction in $\ll\overline{Nu}\gg$ in comparison with the smooth cylinder, as described in Figure 8c, leading to approximately 10.3% and 10.1% attenuation in $\ll\overline{Nu}\gg$ compared to the smooth cylinder, respectively. The IPU cylinder is ranked next, with an 8.3% reduction rate for $\ll\overline{Nu}\gg$. The TU cylinder shows the lowest value of 5.4% in the reduction rate among the cylinders in Case 1.

For Case 2, the TU cylinder still gives a low reduction rate of 4.4%. The $\ll\overline{Nu}\gg$s of the IPU and SU cylinders show the highest reduction rate of 9.6% and the second highest rate of 9.1%, respectively, which are comparable to the reductions rates for Case 1. Otherwise, the APU cylinder in Case 2 presents a considerably different $\ll\overline{Nu}\gg$ to that in Case 1, resulting in the lowest reduction rate of 1.2 for $\ll\overline{Nu}\gg$. It is noted that the APU cylinder in Case 2 provides about the same $\ll\overline{Nu}\gg$ compared to the smooth cylinder. Specifically, there is almost no attenuation in the heat transfer performance in comparison with the smooth cylinder.

In general, a disturbance of geometry in forced convection establishes a connection between the Nusselt number and force coefficients. This correlation prompts a reduction in $\overline{C_D}$ and $C_{L,RMS}$, which occurs concurrently with a decrease in $\ll\overline{Nu}\gg$. In recent studies, geometric disturbances, such as through the use of a helically twisted elliptic cylinder [5], a wavy cylinder [2], an asymmetric wavy cylinder [3], or a double-wavy cylinder [4], have been examined for their impact on forced convection. These studies have indicated that a stable near wake is distinguished by a prolonged vortex formation length, a low vorticity, and a regime of zero vorticity. Consequently, an altered wake pattern caused by geometric disturbances plays a role in diminishing the mixing in the near wake, ultimately leading to a reduction in heat transfer.

However, the APU cylinder under the condition of Case 2 gives an inconsistent pattern with regard to the force coefficients and Nusselt number. Regardless of whether it is Case 1 or Case 2, the APU cylinder presents a significant reduction in $\overline{C_D}$ and $C_{L,RMS}$. But the APU cylinder reveals almost no reduction in $\ll\overline{Nu}\gg$. It means that a proper modification of geometry can achieve a reduction in force coefficients by maintaining the heat transfer performance.

Figure 9a,b present the time- and spanwise local surface-averaged Nusselt number ($\langle\overline{Nu}\rangle$) along the spanwise direction for the undulated cylinders in Case 1 and Case 2, respectively. In addition, the ratios of the amplitude of $\langle\overline{Nu}\rangle$ to $\ll\overline{Nu}\gg$ ($\langle\overline{Nu}\rangle_A/\ll\overline{Nu}\gg$) for the undulated cylinders are shown in Figure 9c, where the amplitude of $\langle\overline{Nu}\rangle$ is defined as $\langle\overline{Nu}\rangle_A = (\langle\overline{Nu}\rangle_{Max} - \langle\overline{Nu}\rangle_{Min})/2$. The $\langle\overline{Nu}\rangle$ for the smooth cylinder remains consistent along the spanwise direction due to the statistically homogeneous condition imposed by its two-dimensional geometry.

For Case 1, the values of $\langle\overline{Nu}\rangle$ along the spanwise direction are smaller than those of the smooth cylinder, as shown in Figure 9a, which supports that the values of $\ll\overline{Nu}\gg$ for all undulated cylinders in Case 1 are smaller than those of the smooth cylinder, as already shown in Figure 8a,c. In addition, the undulated cylinders in Case 1 reveal a much weaker spanwise dependence of $\langle\overline{Nu}\rangle$ on the spanwise direction than those in Case 2. Thus, the values of $\langle\overline{Nu}\rangle_A$ for all undulated cylinders in Case 1 are smaller than those of the corresponding cylinders in Case 2, as shown in Figure 9c.

For Case 2, the undulated cylinders provide about the same spanwise variation in $\langle\overline{Nu}\rangle$. Specifically, as the cross-sectional area's spanwise location shifts from $z/\lambda = 0$ to 0.5, $\langle\overline{Nu}\rangle$ experiences an increase, reaching its peak near $z/\lambda = 0.5$. As the position of the cross-sectional area progresses from $z/\lambda = 0.5$ to 1, there is a subsequent decrease in $\langle\overline{Nu}\rangle$.

In contrast to Case 1, the undulated cylinders with TU, IPU and APU in Case 2 give locally larger values of $\langle\overline{Nu}\rangle$ than the smooth cylinder, as shown in Figure 9b. Particularly, the APU cylinder forms a wider spanwise range that exhibits larger values of $\langle\overline{Nu}\rangle$ than the smooth cylinder, as shown in Figure 9b, and the ratio of the amplitude is also the smallest, as shown in Figure 9c. As a result, the APU cylinder in Case 2 has the smallest reduction rate of $\ll\overline{Nu}\gg$ among the undulated cylinders considered in this study, a already observed in Figure 8c.

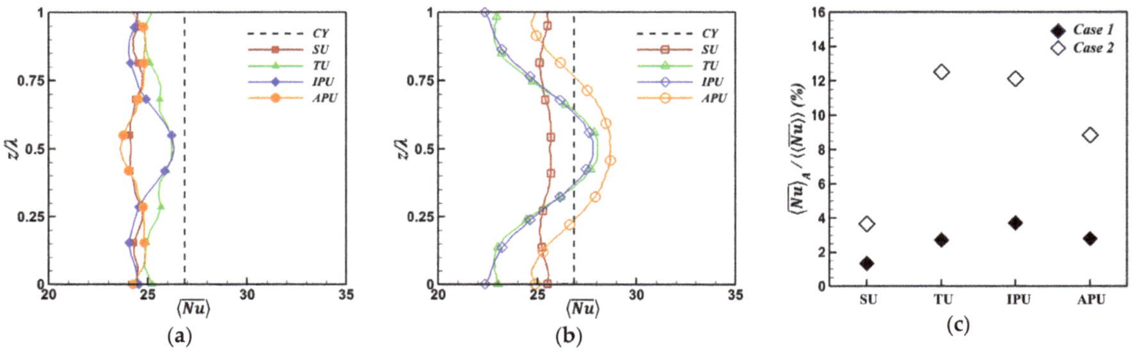

Figure 9. The time- and spanwise local surface-averaged Nusselt number ($\langle\overline{Nu}\rangle$) for (**a**) Case 1 an (**b**) Case 2 along the spanwise direction for undulated cylinders, and (**c**) the amplitude ratio of th time- and spanwise local surface-averaged Nusselt number to the time- and total surface-average Nusselt number.

Figure 10 illustrates the distribution of the time-averaged local Nusselt number ($\overline{N}$ across the surface from various perspectives for the different undulated cylinders in Case and Case 2. In addition to the various undulation types, a smooth cylinder is also presente in Figure 10a as CY for the purpose of comparison. This visualization aligns with th spanwise variation in $\langle\overline{Nu}\rangle$ depicted in Figure 9a,b. For more information, it is noted tha the distribution of the time-averaged local Nusselt number is denoted as 2λ in Figure 1 whereas the spanwise local Nusselt number is depicted as λ in Figure 9.

For Case 1, Figure 10a–e show the front, top and rear views of the 3D contours of $\overline{N}$ for the CY, SU, TU, IPU and APU cylinders.

From the front view, the SU and APU cylinders form a local maximum of $\langle\overline{Nu}\rangle$ nea the nodal position along the streamwise direction, as shown in Figure 10b,e, respectivel In contrast, the TU and IPU cylinders form a local maximum of $\langle\overline{Nu}\rangle$ near the saddle alon the transverse direction, as shown in Figure 9c,d for Case 1, respectively. Particularly, th front view of the SU cylinder exposes a diminished three-dimensional reliance of $\overline{Nu}$ o both the spanwise and transverse directions in comparison to other undulated cylinder which contributes to the weakest variation in $\langle\overline{Nu}\rangle$ seen in Figure 9a,c.

From the top view, the SU cylinder shows that the sinusoidal profiles exhibit maximu and minimum deflection points at the nodes and saddles, respectively, as shown in th upstream side in the left half of the top view. Alternatively, for the downstream side, th local maximum and minimum of $\overline{Nu}$ emerge near the nodes and saddles, respectively, a shown in Figure 10b. The TU cylinder shows a nearly 2D distribution of $\overline{Nu}$, except for th region near the rear stagnation point, as shown in the middle column of Figure 10c.

The top view of the upstream side for the IPU cylinder depicts that the surfac distribution of $\overline{Nu}$ showcases a 3D wavy formation, as illustrated in Figure 10d. As th spanwise location of the cross-sectional area shifts from z/Dm = 0 to 0.5, the Nu of the IP cylinder is larger and forms a denser distribution, which clearly explains the larger valu of $\langle Nu\rangle$ near the saddle around z/Dm = 0.5, as depicted in Figure 9a. The APU cylind forms a more apparent wavy pattern of $\overline{Nu}$ compared to the IPU cylinder, not only in th upstream side but also in the downstream side, as illustrated in Figure 10e.

For Case 2 in Figure 10f–i, the configurations of $\overline{Nu}$ contours for the undulate cylinders are very similar with those for the corresponding undulated cylinders in Case However, the distributions of $\overline{Nu}$ in Case 2 are more 3D-dependent than those in Case regardless of the undulation type. In addition, the ranges of $\overline{Nu}$ in Case 2 are wider tha those of $\overline{Nu}$ in Case 1. In particular, the distributions of $\overline{Nu}$ on the top and rear surfac are more locally dependent, leading to the clear identification of the local maximum and minimum positions, as shown in Figure 10f–i. Therefore, these strong 3D distributions of $\overline{Nu}$ in Case 2 cause a larger spanwise variation in $\langle\overline{Nu}\rangle$, as seen in Figure 9b,c.

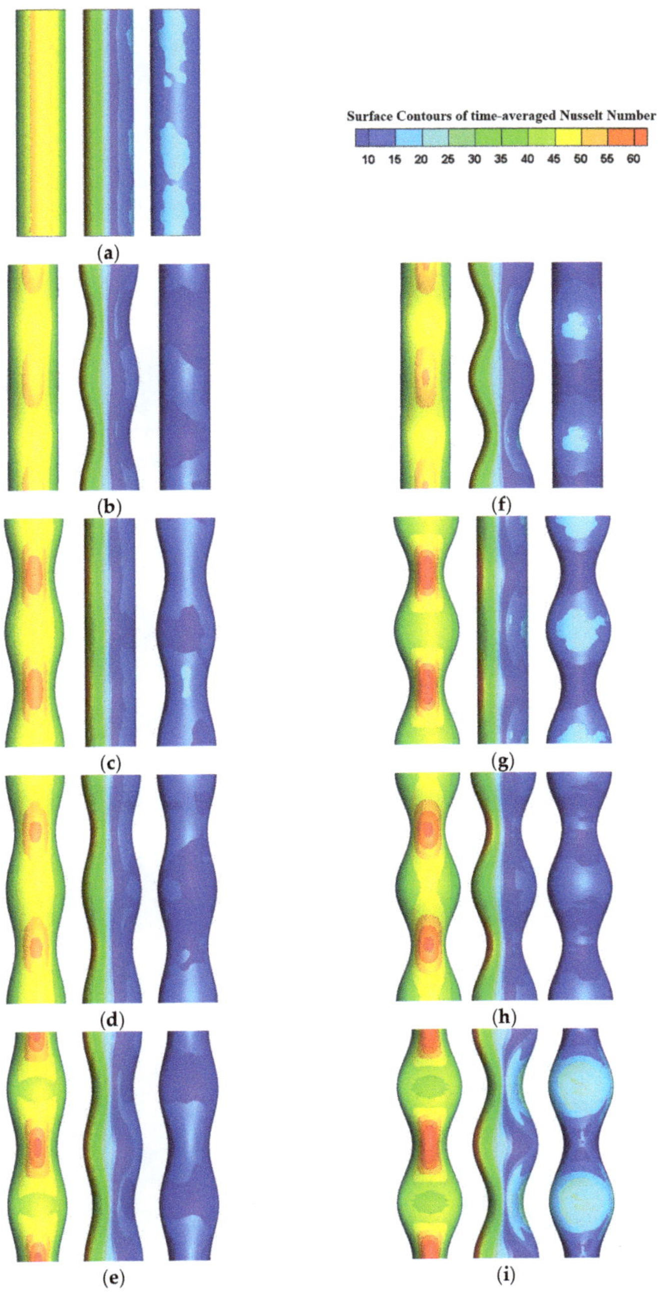

Figure 10. Front (left column), top (middle column) and rear (right column) views of the surface contours of time-averaged Nusselt number: (**a**) CY, (**b**,**f**) SU, (**c**,**g**) TU, (**d**,**h**) IPU and (**e**,**i**) APU.

To identify the downstream and rear distributions of $\overline{Nu}$ observed in Figure 10, the mean isotherms in the x-z plane at y = 0.0 and in the x-y plane at the nodes and saddles for the four different undulated cylinders are presented in Figure 11. In addition to the various undulated cylinders, a smooth cylinder is also featured as (a) CY in Figure 9 to serve as a basis for comparison. In general, the undulated cylinders in Case 1, as shown in Figure 11b–e, form a weaker spanwise-dependent thermal boundary layer than those in Case 2, as shown in Figure 11f–i.

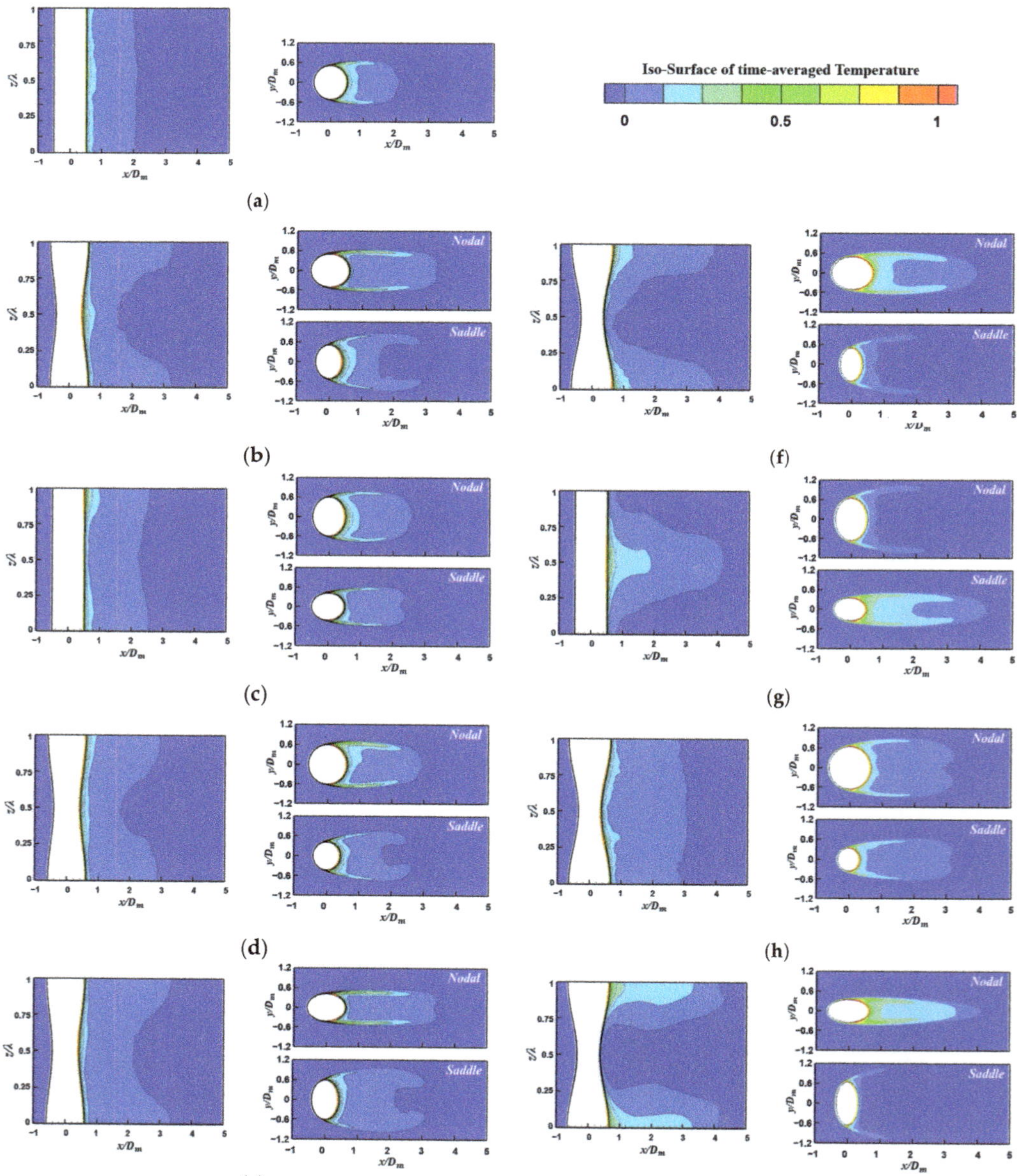

Figure 11. Time-averaged isotherms in the x-z plane at y = 0.0 (left column) and in the x-y plane at the nodal plane (right column, upper) and saddle plane (right column, lower): **(a)** CY, **(b)** SU, **(c)** TU, **(d)** IPU and **(e)** APU for Case 1, and **(f)** SU, **(g)** TU, **(h)** IPU and **(i)** APU for Case 2.

In Case 1, the SU, IPU and APU cylinders form denser and coarser isotherms near the nodes and saddles, respectively, as shown in Figure 11b,d,e, respectively. Therefore, thinner and thicker thermal boundaries appear near the nodes and saddles on the rear surface in the x-z plan. These local distributions of the isotherms in the x-z plane are consistent with the isothermal distributions in the x-y plane. The isotherms near the rear surface in the nodal plane for these undulated cylinders are much denser than those in the saddle plane, as shown in Figure 11b,d,e, which explains the alternate appearance of locally large and small values of $\overline{Nu}$ near the nodes and saddles, as already depicted in Figure 10b,d,e.

By contrast, the TU cylinder in Case 1 forms an opposite pattern of isothermal distribution when compared to the SU, IPU and APU cylinders. Specifically, the TU cylinder forms coarser and denser isotherms near the nodes and saddles, respectively, as shown in Figure 11c, which explains the alternate appearance of locally small and large values of $\overline{Nu}$ near the nodes and saddles, as already shown in Figure 10c.

In Case 2, the undulated cylinders form isotherm patterns opposite to those of the corresponding undulated cylinders in Case 1. Thus, the SU, IPU and APU cylinders form denser and coarser isotherms near the saddles and nodes in the x-z and x-y planes, as shown in Figure 11f,h,i, respectively. Successively, the TU cylinder forms denser and coarser isotherms near the nodes and saddles, respectively, as shown in Figure 11g.

Particularly, the APU cylinder in Case 2 causes a much wider region with a very thin isothermal boundary layer compared to the other undulated cylinders, which results in a wider region containing locally large values of $\overline{Nu}$ on the real surface, as shown in Figure 10i. As a result, the $\langle\overline{Nu}\rangle$ values of the APU cylinder become much larger through the wide region surrounding the saddle than those of the smooth cylinder, as already depicted in Figure 9b showing the spanwise variation in the local surface-averaged Nusselt number.

These strong 3D isothermal distributions for the APU cylinder in Case 2 can be associated with the wake flow structures. Thus, the distributions of the spanwise vorticity in the x-y plane at the nodes and the saddles are presented in Figure 12. The spanwise vorticities in the nodal plane are constrained to a very narrow width, as shown in Figure 12a. The upper and lower shear layers are very close to each other and are further elongated downstream. The corresponding streamlines within the upper and lower shear layers reveal almost no reverse flow in the nodal plane, as shown in Figure 12b. Therefore, the isotherms near the nodes are very coarse due to the weak mixing effect, as already observed in Figure 11i.

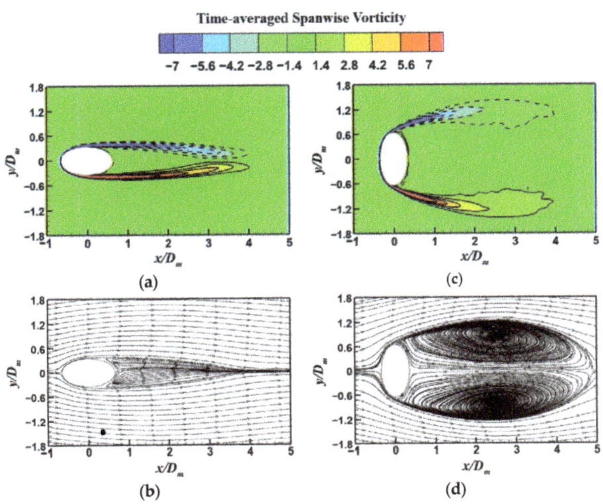

Figure 12. Time-averaged spanwise vorticity and streamlines for the APU cylinder in Case 2 in the x-y plane: (**a**,**b**) at the nodes and (**c**,**d**) at the saddles.

Otherwise, a large spanwise vortex pair occurs in the saddle plane, as shown i Figure 12c,d. Thus, short and wide spanwise vorticities are formed in the wake. Thes strong revere flows contribute to the mixing effect, which causes the very dense isotherma distribution and corresponding thin thermal boundary layer near the saddle that ar previously seen Figure 11i. As a result, the APU cylinder presents larger $\langle\overline{Nu}\rangle$ values nea the saddle than the smooth cylinder, as already shown in Figure 9b. Consequently, the AP cylinder in Case 2 provides about the same $\ll\overline{Nu}\gg$ values as those of the smooth cylinde as depicted in Figure 8 showing the time- and total surface-averaged Nusselt number.

4. Conclusions

The present study focused on an assessment of the influence of undulation on the hea transfer characteristics around wavy cylinders at a Reynolds number of 3000 by using LE The presented numerical methods were well validated by comparing previous results an a grid dependence test was successively conducted. Four types of wavy cylinders wit SU, TU, IPU and APU were taken into account to confirm the heat transfer characteristic together with two different cases of wavelength and amplitude combinations. In additio for the purpose of comparison, a smooth cylinder was considered.

From the different aspects of wavelength and amplitude, for the TU and IPU cylinder there is an obvious discrepancy between Cases 1 and 2 in the values of C_D and C_L. Fo the SU and APU cylinders, however, there is little discrepancy between Case 1 and Cas 2. Therefore, it is concluded that the TU and IPU cylinders are significantly influence by wavelength and wave amplitude, with the reduction rate of force coefficients bein the smallest in Case 1 and the drag reduction being the largest in Case 2. On the othe hand, the SU and APU cylinders in this study are not strongly affected by the considere combinations of wavelength and wave amplitude.

For the Nusselt number, it is observed that the largest and smallest reduction rate occur for the APU and TU cylinders in Case 1, respectively. But, in Case 2, the IP and APU cylinders provide the largest and smallest reduction rates. The spanwise loca surface-averaged Nusselt numbers for the undulated cylinders in Case 1 are smaller tha those of the smooth cylinder along the spanwise direction. However, the undulate cylinders with TU, IPU and APU in Case 2 present locally larger values of $\langle\overline{Nu}\rangle$ than th smooth cylinder. In general, Case 2 causes a stronger spanwise dependence of $\langle\overline{Nu}\rangle$ tha Case 1. These characteristics, according to the combination of the parameters of the wav geometry, are supported by the surface distribution of the Nusselt numbers. In additio the isothermal distribution, which depends on the wake flow, explains the variation in th Nusselt numbers.

It is noted that the APU cylinder in Case 2 provides about the same $\ll\overline{Nu}\gg$ value as those of the smooth cylinder. Specifically, there is almost no attenuation in the hea transfer performance in comparison with the smooth cylinder. However, the APU cylinde under the condition of Case 2 gives an inconsistent pattern of force coefficients and Nusse number. Regardless of whether it is Case 1 or Case 2, the APU cylinder presents a significa reduction in $\overline{C_D}$ and $C_{L,RMS}$. But the APU cylinder reveals almost no reduction in $\ll\overline{Nu}\gg$ It means that a proper modification of geometry can achieve a reduction in force coefficient by maintaining the heat transfer performance.

Based on the forementioned geometrical characteristics of undulated cylinders and th correlation of heat transfer performance of each wavy cylinder, it is feasible to predict th structure of heat transfer from the modification of wavy cylinder geometry. It is recognize that how we modify the geometry of a wavy cylinder to obtain what we desire affects th structure characteristics of heat transfer accordingly. Furthermore, it enables us to predic the expected thermal performance of heat transfer structure through the generalization c the geometry of wavy cylinders without any time-consuming and costly simulation.

It is expected that an extended investigation based on this study is to be carried ou in the near future in order to clarify the relationship between undulated cylinders tha results from the superposition and combination of these undulated cylinders from the flo structure and heat transfer point of view. Specifically, a wide range of wavelength and wave amplitude will be considered to find the more general dependence of heat transfer characteristics on wavy geometric parameters.

Author Contributions: Conceptualization, C.-H.C.; writing—original draft preparation, M.-K.K.; writing—review and editing, H.-S.Y.; supervision, H.-S.Y.; software, H.-S.Y.; validation, S.-H.N. All authors have read and agreed to the published version of the manuscript.

Funding: This work was supported by the National Research Foundation of Korea (NRF) funded by the Korea government (MSIT) through grant no. NRF-2019R1A2C1009081.

Institutional Review Board Statement: Not applicable.

Informed Consent Statement: Not applicable.

Data Availability Statement: The data presented in this study are available on request from the corresponding author.

Conflicts of Interest: The authors declare no conflicts of interest.

Nomenclature

Re	Reynolds number
λ	Wavelength
D_m	Mean diameter of a cylinder
Pr	Prandtl number
a	Wave amplitude
D_z	Local diameter
D_{max}	Maximum local diameter
D_{min}	Minimum local diameter
$L_{z,s}$	Local length in the streamwise direction
$L_{z,t}$	Local length in the transverse direction
t	Time
x_i	Cartesian coordinates
u_i, u_j	Velocity components
p	Pressure
T	Temperature
τ_{ij}	Subgrid-scale stress tensor
q_i	Subgrid flux
T_S	Cylinder surface temperature
T	Temperature
C_p	Specific heat capacity
k	Thermal conductivity
C_D	Drag coefficient
C_L	Lift coefficient
St	Strouhal number
F_D	Total drag force
F_L	Total lift force
U_∞	Free-stream velocity
A	Projected area of a cylinder
f_s	Frequency of vortex shedding
Nu	Nusselt number
$\langle Nu \rangle$	Spanwise local surface-averaged Nusselt number
$\ll Nu \gg$	Total surface-averaged Nusselt number
$\overline{Nu}$	Time-averaged local Nusselt number
$\langle \overline{Nu} \rangle$	Time- and spanwise local surface-averaged Nusselt number
$\ll \overline{Nu} \gg$	Time- and total surface-averaged Nusselt number
n	Normal direction to the walls

W_z	Spanwise local surface area
L	Spanwise arc length
t_p	Period of time integration
y^+	Wall scale
u_τ	Friction velocity
τ_w	Wall shear stress
λ_2	Second largest eigenvalue
S_{ij}^2	Strain-rate tensor
Ω_{ij}^2	Rotation-rate tensor
Greek symbols	
ρ	Density
μ	Dynamic viscosity
ν	Kinematic viscosity
Superscripts	
$-$	Time-averaged quantity
$\sim$	Grid filtering
Abbreviations	
ASW	Asymmetric wavy
CY	Smooth
SW	Symmetric wavy
DW	Double wavy
HTE	Helically twisted elliptic
VPHTE	Variable-pitch helically twisted elliptic
HSV	Harbor seal vibrissa
RMS	Root mean square
SU	Streamwise undulation
TU	Transverse undulation
IPU	In-phase undulation
APU	Antiphase undulation
SGS	Smagorinsky subgrid scale
FVM	Finite volume method

References

1. Ahn, C.T.; Yoon, H.S.; Ha, M.Y.; Lee, H.G. Waviness effect of a wavy circular cylinder on the heat transfer at a Reynolds number of 300. *Int. J. Heat Mass Transf.* **2009**, *52*, 345–354. [CrossRef]
2. Kim, M.I.; Yoon, H.S. Large eddy simulation of forced convection heat transfer from wavy cylinders with different wavelengths. *Int. J. Heat Mass Transf.* **2018**, *127*, 683–700. [CrossRef]
3. Moon, J.; Yoon, H.S.; Kim, H.J.; Kim, M.I. Forced convection heat transfer from an asymmetric wavy cylinder at a subcritical Reynolds number. *Int. J. Heat Mass Transf.* **2019**, *129*, 707–720. [CrossRef]
4. Yoon, H.S.; Moon, J.; Kim, M.I. Effect of a double wavy geometric disturbance on forced convection heat transfer at a subcritical Reynolds number. *Int. J. Heat Mass Transf.* **2019**, *141*, 861–875. [CrossRef]
5. Yoon, H.S.; Kim, H.J.; Wei, D.J. Forced convection heat transfer from the helically twisted elliptic cylinder inspired by a daffodil stem. *Int. J. Heat Mass Transf.* **2018**, *119*, 105–116. [CrossRef]
6. Yoon, H.S.; Moon, J. Effect of variable pitch on forced convection heat transfer around a helically twisted elliptic cylinder. *Int. J. Heat Mass Transf.* **2021**, *173*, 121205. [CrossRef]
7. Kim, H.J.; Yoon, H.S. Forced convection heat transfer from the biomimetic cylinder inspired by a harbor seal vibrissa. *Int. J. Heat Mass Transf.* **2018**, *117*, 548–558. [CrossRef]
8. Kim, H.J.; Yoon, H.S. Reynolds number effect on the fluid flow and heat transfer around a harbor seal vibrissa shaped cylinder. *Int. J. Heat Mass Transf.* **2018**, *126*, 618–638. [CrossRef]
9. Yoon, H.S.; Nam, S.H.; Kim, M.I. Effect of the geometric features of the harbor seal vibrissa based biomimetic cylinder on the forced convection heat transfer. *Int. J. Heat Mass Transf.* **2020**, *159*, 120086. [CrossRef]
10. Yoon, H.S.; Nam, S.H.; Kim, M.I. Laminar forced convection heat transfer around wavy elliptic cylinders with different aspect ratios. *Int. J. Heat Mass Transf.* **2022**, *194*, 123038. [CrossRef]
11. Germano, M.; Piomelli, U.; Moin, P.; Cabot, W.H. A dynamic subgrid-scale eddy viscosity model. *Phys. Fluids* **1991**, *3*, 1760–1765. [CrossRef]

2. Yoon, H.S.; Balachandar, S.; Ha, M.Y. Large eddy simulation of flow in an unbaffled stirred tank for different Reynolds numbers. *Phys. Fluids* **2009**, *21*, 8. [CrossRef]
3. Jung, J.H.; Yoon, H.S. Large eddy simulation of flow over a twisted cylinder at a subcritical Reynolds number. *J. Fluid Mech.* **2014**, *759*, 579–611. [CrossRef]
4. Yoon, H.S.; Balachandar, S.; Ha, M.Y. Large eddy simulation of passive scalar transport in a stirred tank for different diffusivities. *Int. J. Heat Mass Transf.* **2015**, *91*, 885–897. [CrossRef]
5. Boghi, A.; Di Venuta, I.; Gori, F. Passive scalar diffusion in the near field region of turbulent rectangular submerged free jets. *Int. J. Heat Mass Transf.* **2017**, *112*, 1017–1031. [CrossRef]
6. Ouro, P.; Fraga, B.; Viti, N.; Angeloudis, A.; Stoesser, T.; Gualtieri, C. Instantaneous transport of a passive scalar in a turbulent separated flow. *Environ. Fluid Mech.* **2018**, *18*, 487–513. [CrossRef] [PubMed]
7. Silva, M.I.; De Bortoli, A.L. Improving the mixing of a turbulent jet diffusion flame of methane via linear quadratic regulator control. *Phys. Fluids* **2021**, *33*, 11. [CrossRef]
8. Hubbard, J.A.; Cheng, M.-D.; Domino, S.P. Mixing in Low Reynolds Number Reacting Impinging Jets in Crossflow. *J Fluids Eng.* **2023**, *145*, 051303. [CrossRef]
9. Shi, J.; Zhu, D.; Cai, Z.; Cao, X. Non-hydrostatic simulation of two-layer density stratified flows over a sill. *Ocean Eng.* **2024**, *291*, 116432. [CrossRef]
0. *Siemens Digital Industries Software*, Simcenter STAR-CCM+, version 16.04; Siemens: Munich, Germany, 2021.
1. Yoon, H.S.; Nam, S.H.; Kim, M.I. Effect of the geometric features of the harbor seal vibrissa based biomimetic cylinder on the flow over a cylinder. *Ocean Eng.* **2020**, *218*, 108150. [CrossRef]
2. Lam, K.; Lin, Y.F. Large eddy simulation of flow around wavy cylinders at a subcritical Reynolds number. *Int. J. Heat Fluid Flow* **2008**, *29*, 1071–1088. [CrossRef]
3. Lin, Y.F.; Bai, H.L.; Alam, M.M.; Zhang, W.G.; Lam, K. Effects of large spanwise wavelength on the wake of a sinusoidal wavy cylinder. *J. Fluids Struct.* **2016**, *61*, 392–409. [CrossRef]
4. Norberg, C. *Effects of Reynolds Number and a Low-Intensity Freestream Turbulence on the Flow Around a Circular Cylinder*; Chalmers University Technological Publications: Göteborg, Sweden, 1987; Volume 87, pp. 1–55.
5. Norberg, C. Fluctuating lift on a circular cylinder: Review and new measurements. *J. Fluids Struct.* **2003**, *17*, 57–96. [CrossRef]
6. Lu, X.; Dalton, C.; Zhang, J. Application of large eddy simulation to flow past a circular cylinder. *J. Offshore Mech. Arctic Eng.* **1997**, *119*, 219–225. [CrossRef]
7. Yoon, H.S.; Shin, H.; Kim, H. Asymmetric disturbance effect on the flow over a wavy cylinder at a subcritical Reynolds number. *Phys. Fluids.* **2017**, *29*, 9. [CrossRef]
8. Zukauskas, A. *Advances in Heat Transfer*; Academic Press: New York, NY, USA, 1972; pp. 93–160.
9. Sanitjai, S.; Goldstein, R.J. Forced convection heat transfer from a circular cylinder in crossflow to air and liquids. *Int. J. Heat Mass Transf.* **2004**, *47*, 4795–4805. [CrossRef]
0. Hilpert, R. Warmeabgabe von geheizen Drahten und Rohren. *Forsch. Geb. Ingenieurwes.* **1933**, *4*, 220.
1. Zhou, J.; Adrian, R.J.; Balachandar, S.; Kendall, T. Mechanisms for generating coherent packets of hairpin vortices in channel flow. *J. Fluid Mech.* **1999**, *387*, 353–396. [CrossRef]

energies

MDPI

Article

Calibration of a Near-Wall Differential Reynolds Stress Model Using the Updated Direct Numerical Simulation Data and Its Assessment

Lev Usov *, Alexei Troshin, Kirill Anisimov and Vladimir Sabelnikov *

Central Aerohydrodynamic Institute (TsAGI), 1 Zhukovsky Str., 140180 Zhukovsky, Russia; troshin.ai@phystech.edu (A.T.)
* Correspondence: usov.la@phystech.edu (L.U.); sabelnikov@free.fr (V.S.)

Abstract: In the article, a differential Reynolds stress model is recalibrated using turbulent channe flow direct numerical simulation data in the range of friction Reynolds numbers 550–5200. Th calibration aims to produce a RANS sublayer model for use within the hybrid RANS/LES framewor The model is designed to capture the average field of a thin near-wall part of a boundary layer a accurately as possible. An *a posteriori* procedure is employed in which one-dimensional channe flow calculations are performed for all variations of the model coefficients at each stage of th optimization procedure. The coefficients are initialized with their original values and then optimize by minimizing the appropriately chosen norm. An improved representation of the mean velocit profile and peak Reynolds stress values is demonstrated. Both models—baseline and recalibrated– are implemented in an in-house CFD code, and several simulations, including a channel flow, flat plate boundary layer and a boundary layer separation from a rounded step, are performe The latter benchmark flow is also simulated in hybrid RANS/LES mode. The updated model i compared to the original one, demonstrating improvements over the baseline model in the cases it wa designed for.

Keywords: differential Reynolds stress model; channel flow; boundary layer; near-wall turbulenc RANS; LES; hybrid method

Citation: Usov, L.; Troshin, A.; Anisimov, K.; Sabelnikov, V. Calibration of a Near-Wall Differential Reynolds Stress Model Using the Updated Direct Numerical Simulation Data and Its Assessment. *Energies* **2023**, *16*, 6826. https://doi.org/10.3390/en16196826

Academic Editors: Marco Marengo and Artur Bartosik

Received: 31 August 2023
Revised: 22 September 2023
Accepted: 25 September 2023
Published: 26 September 2023

1. Introduction

Differential Reynolds stress models (DRSM), which emerged in the middle of the 20t century [1,2], currently represent a very promising approach to turbulence modeling [3 The main reasons for this are the recognition of restrictions imposed by the eddy viscosit models and DRSMs' potential for improved prediction of flow separation, which is ofte relevant in practical applications [4].

The first DRSM to acquire widespread popularity was the Launder–Reece–Rodi (LRR model [5,6]. The LRR model assumes a common approximation of the redistribution terr which is linear concerning the Reynolds stress tensor. A more general approximation tha is quadratic with respect to the Reynolds stress tensor was suggested by Lumley [7], and it particular case simplification was proposed in the Speziale–Sarkar–Gatski (SSG) model [8 Combining these two models and transitioning from the equation for the dissipation rat of turbulence kinetic energy ε to an equation for the turbulence frequency ω led to th creation of the SSG/LRR-ω model [9,10]. It became popular due to adopting concept from a successful eddy viscosity model SST [11], a relatively simple model formulatio and a reasonable level of accuracy in practical external aerodynamic applications [12]. I is worth noting that although the SSG/LRR-ω model suggests integrating the equation to wall and setting the no-slip boundary condition (with a demand that the first nea wall cell height correspond to $y_1^+ \leq 1$ in law-of-the-wall variables), this model cannot b considered accurate for detailed description of the near-wall portion of the boundary laye (such models are called "low-Reynolds-number models"). The model does not contain near-wall corrections to the redistribution term, and the dissipation rate tensor is assumed to be isotropic. This leads to substantial deviations of the Reynolds stress $R_{ij} = \overline{u_i u_j}$ (the overbar denotes a Reynolds average and u_i is a fluctuating part of the i-th velocity vector component) and ε profiles from reference data.

Although the development of low-Reynolds-number DRSMs is a relatively niche research area, several prominent papers have been published on this topic. A large amount of work in this direction was put in by the Manchester University group led by B.E. Launder. In a span of two decades (1978–1998), they described several models in papers [13–16] and presented the final results of their research in the paper by T.J. Craft [17] in 1998. A group led by K. Hanjalić has been developing their low-Reynolds-number model since 1994; they proposed two models [18,19] and summarized their findings in the work of Jakirlić and Hanjalić [20], which was devoted to their JHh model of 2002. Subsequently, the model [20] became a basis for more practice-oriented low-Reynolds-number models JHh-v1, 2, 3 [10,21,22], in which efforts were made to empirically improve the prediction of separated flows and to account for the large-scale unsteadiness. The JHh model formulation was also used in the JH-ω^h [23] and Jakirlić & Maduta [24] models, in which the ω-equation is utilized, leading to improved computational stability. Another direction of low-Reynolds-number DRSMs is being developed by the group of G.A. Gerolymos (Pierre and Marie Curie University, Paris) [25,26]; their latest developed model, GLVY [27], was published in 2014.

The creation of hybrid Reynolds-averaged Navier–Stokes/Large Eddy Simulation (RANS/LES) methods is of particular interest in which a RANS turbulence model is only used in a thin near-wall layer [28]. In contrast, the outer part of the boundary layer and freestream turbulent regions are modeled within the LES framework, meaning that the large-scale turbulence is resolved. Only the unresolved small-scale turbulence is described by a subgrid-scale model. The RANS region being constricted to a thin sublayer (of about 0.1δ typically, where δ is the boundary layer thickness) facilitates the calibration process. Notably, outer flow conditions (e.g., the presence of an adverse pressure gradient) have less influence on the mean characteristics of this sublayer than on the outer layer [29]. This allows us to perform the calibration using only basic flows, such as the developed channel flow, on which high-quality modern data from direct numerical simulation (DNS) is available. There are currently several popular hybrid RANS/LES methods, among which are Improved Delayed Detached Eddy Simulation (IDDES) [28], Partially Averaged Navier–Stokes (PANS) [30], Partially Integrated Transport Modeling (PITM) [31], and Stress-Blended Eddy Simulation (SBES) [32]. Most of them are based on eddy viscosity turbulence models, which simplifies their formulation and ensures computational robustness. DRSM-based hybrid RANS/LES methods are still at an early stage of their development.

The paper modifies one of the low-Reynolds-number DRSM models to create a finely tuned RANS sublayer model for hybrid RANS/LES simulations. To the best of the authors' knowledge, no studies in the literature are devoted to this problem. The papers on DRSMs cited above focus on developing "general purpose" models. Such models were calibrated against a wide range of benchmark flows, inevitably making their performance in a near-wall layer the result of compromises. Furthermore, the calibration of these models mostly relied on early low-Reynolds-number DNS results of the 1990s. In contrast, in the current paper, the calibration is performed using only one type of flow, namely the turbulent channel flow, and the reference data are more recent, with higher Reynolds number values.

The Jakirlić & Maduta model [24] was taken as a baseline model. Aside from the improved computational stability mentioned above, additional advantages of this model over the other models include using a relatively simple linear tensor model of the redistribution term and the extra attention given to the dissipation rate tensor modeling. The models developed by the Manchester University team and the Gerolymos team have a much more complex formulation and may be investigated in future work if necessary.

The paper is organized as follows. In Section 2, the baseline Jakirlić & Maduta DRSM [24] is formulated, the coefficient calibration technique is described, and the calibration results are presented. Section 3 compares the modified model to the baseline one in the pure RANS regime using three benchmark flows. Two of them do not include strong pressure gradients, and the modified model is shown to perform better than the baseline. The third flow features a separation from a smooth surface, and the modified model fails to replicate the mean flow field. However, as shown in Section 4, the same model performs remarkably well in the same flow when applied as part of a hybrid RANS/LES method. The results are discussed in Section 5.

2. Model Formulation and Coefficient Calibration

2.1. Baseline Model

Differential Reynolds stress models contain six equations for the independent stress tensor components. With density ρ and the kinematic viscosity ν assumed constant, these equations can be written as follows:

$$\frac{\partial R_{ij}}{\partial t} + \frac{\partial}{\partial x_k}\left[R_{ij}U_k + \underbrace{\overline{u_i u_j u_k}}_{T^{(v)}_{ijk}} + \underbrace{\left(\overline{p u_i}\delta_{jk} + \overline{p u_j}\delta_{ik}\right)/\rho}_{T^{(p)}_{ijk}} - \frac{\nu}{2}\frac{\partial R_{ij}}{\partial x_k}\right] = \\ = \underbrace{-\overline{u_i u_k}\frac{\partial U_j}{\partial x_k} - \overline{u_j u_k}\frac{\partial U_i}{\partial x_k}}_{P_{ij}} + \underbrace{\overline{\frac{p}{\rho}\left(\frac{\partial u_i}{\partial x_j} + \frac{\partial u_j}{\partial x_i}\right)}}_{\Phi_{ij}} - \underbrace{\left(2\nu\overline{\frac{\partial u_i}{\partial x_k}\frac{\partial u_j}{\partial x_k}} - \frac{\nu}{2}\frac{\partial^2 R_{ij}}{\partial x_k \partial x_k}\right)}_{\varepsilon^h_{ij}}, \tag{1}$$

where U_i is the i-th mean velocity component, p is the pressure fluctuation. Hereinafter, a summation of the repeated indices is assumed. These equations include the production term P_{ij} (does not require closure in DRSMs), the redistribution term Φ_{ij}, responsible for the redistribution of energy between velocity fluctuations in different directions, the "homogeneous" dissipation rate tensor ε^h_{ij} [20], turbulent transport by fluctuating velocity $T^{(v)}_{ijk}$ and turbulent transport by fluctuating pressure $T^{(p)}_{ijk}$ (the sum of these fluxes will be denoted T_{ijk}). Let us additionally define the turbulence kinetic energy $k = R_{ii}/2$, its production and "homogeneous" dissipation rates $P = P_{ii}/2$ and $\varepsilon^h = \varepsilon^h_{ii}/2$ correspondingly, the turbulent Reynolds number $Re_t = k^2/\left(\nu\varepsilon^h\right)$, the stress-anisotropy tensor $a_{ij} = R_{ij}/k - 2\delta_{ij}/3$ and its second and third invariants $A_2 = a_{ij}a_{ji}$ and $A_3 = a_{ij}a_{jk}a_{ki}$, and Lumley's flatness parameter $A = 1 - \frac{9}{8}(A_2 - A_3)$. Similarly, one may define an anisotropy tensor for the dissipation rate $e_{ij} = \varepsilon^h_{ij}/\varepsilon^h - 2\delta_{ij}/3$, its invariants $E_2 = e_{ij}e_{ji}$ and $E_3 = e_{ij}e_{jk}e_{ki}$, and the parameter $E = 1 - \frac{9}{8}(E_2 - E_3)$.

The Jakirlić & Maduta model (denoted as "conventional 'steady' RSM model" in [24]) aims to reproduce the term-by-term Reynolds stress balance in the near-wall region. The model was developed from the JHh model [20], based on the research conducted in the 1990s to early 2000s and calibrated using the experimental data and DNS results from this timeframe. Inevitably, this means the data they used was at low Reynolds numbers, typically at $Re_\tau = u_\tau h/\nu$ from 180 to 590 where $u_\tau = \sqrt{\tau_w/\rho}$ is the friction velocity, $\tau_w = \mu|\nabla U_{wall}|$ is the wall shear stress, $\mu = \rho\nu$ is the dynamic viscosity and h is the channel half-height or the boundary layer thickness.

The Jakirlić & Maduta model [24] uses the following closures:

$$T_{ijk} = -\frac{\nu_t}{\sigma_R}\frac{\partial R_{ij}}{\partial x_k}, \tag{2}$$

where $\nu_t = 0.144A\sqrt{k}\max\{10\nu^{3/4}/(\varepsilon^h)^{1/4}, k^{3/2}/\varepsilon^h\}$ is the turbulence eddy viscosity;

$$\begin{aligned}\Phi_{ij} = & \underbrace{-C_1\varepsilon^h a_{ij}}_{\Phi_{ij,1}} + \underbrace{C_1^w f_w \frac{\varepsilon^h}{k}\left(R_{km}n_k n_m \delta_{ij} - \frac{3}{2}R_{ik}n_k n_j - \frac{3}{2}R_{jk}n_k n_i\right)}_{\Phi_{ij,1}^w} \\ & \underbrace{-C_2\left(P_{ij} - \frac{2}{3}P\delta_{ij}\right)}_{\Phi_{ij,2}} + \underbrace{C_2^w f_w\left(\Phi_{km,2}n_k n_m \delta_{ij} - \frac{3}{2}\Phi_{ik,2}n_k n_j - \frac{3}{2}\Phi_{jk,2}n_k n_i\right)}_{\Phi_{ij,2}^w},\end{aligned} \tag{3}$$

where empirical coefficients and functions $C_1 = C_{11}AF^{1/4}f + C_{12}\sqrt{A}E^2$, $F = \min\{C_F, A_2\}$, $f = \min\{(Re_t/Re_{t0})^{3/2}, 1\}$, $C_1^w = \max\{1 - C_{11}^w AF^{1/4}f, C_{12}^w\}$, $f_w = \min\{k^{3/2}/(C_l\varepsilon^h d_w), C_{fw}\}$, $C_2 = C_{21}\sqrt{A}$, $C_2^w = \min\{A, C_{21}^w\}$ are used, d_w is the distance to the nearest wall, n_k is the k-th component of the normal vector $\mathbf{n}$ (note that Φ_{ij} is separated into several parts which correspond to different physical processes: $\Phi_{ij,1}$ is the "slow" term which represents a return of the turbulence to an isotropic state in the absence of mean velocity gradients, $\Phi_{ij,2}$ is the "rapid" term which represents the effect of mean velocity gradients on the Reynolds stresses, and $\Phi_{ij,1}^w$ and $\Phi_{ij,2}^w$ are near-wall corrections to $\Phi_{ij,1}$ and $\Phi_{ij,2}$ correspondingly);

$$\varepsilon_{ij}^h = \left(f_s\frac{R_{ij}}{k} + (1 - f_s)\frac{2}{3}\delta_{ij}\right)\varepsilon^h. \tag{4}$$

Equation (4) can be rewritten as $e_{ij} = f_s a_{ij}$, which reveals that the dissipation-anisotropy is linearly related to the stress-anisotropy through the empirical function f_s. This function in the considered model equals $1 - \sqrt{A}E^2$. It should be noted that a dependency of f_s on E makes Equation (4) implicit. Therefore, it should be solved iteratively. In the Jakirlić & Maduta model [24], the value of ε^h is calculated as $\varepsilon^h = k\omega^h$, where ω^h is the characteristic turbulence frequency (another interpretation is the inverse turbulence time scale) obtained from the following equation:

$$\frac{\partial\omega^h}{\partial t} + \frac{\partial}{\partial x_k}\left[\omega^h U_k - \left(\frac{\nu}{2} + \frac{\nu_t}{\sigma_\omega}\right)\frac{\partial\omega^h}{\partial x_k}\right] = C_{\omega1}\frac{\omega^h}{k}P - C_{\omega2}\left(\omega^h\right)^2 + \frac{1}{k}P_{\omega3} + S_l + \frac{2}{k}\left(0.55\frac{\nu}{2} + C_{cr2}\frac{\nu_t}{\sigma_\omega}\right)\frac{\partial k}{\partial x_k}\frac{\partial\omega^h}{\partial x_k}. \tag{5}$$

In the above, $P_{\omega3} = 2C_{\omega3}\nu\nu_t(\partial^2 U_i/\partial x_j\partial x_k)(\partial^2 U_i/\partial x_j\partial x_k)$ represents the additional production of ω^h due to the second velocity derivatives, $S_l = \max\{((\nabla L_t\cdot\mathbf{n}/C_l)^2 - 1)(\nabla L_t\cdot\mathbf{n}/C_l)^2, 0\}A(\omega^h)^2$ represents a correction which eliminates the streamline back-bending in the reattachment region, $L_t = k^{3/2}/\varepsilon^h$ is the turbulence length scale.

The expressions above contain the following empirical constants:

$$\begin{array}{llll} C_{11} = 2.5, & C_{12} = 1.0, & C_F = 0.6, & Re_{t0} = 150, \\ C_{11}^w = 1.75, & C_{12}^w = 0.3, & C_l = 2.5, & C_{fw} = 1.4, \\ C_{21} = 0.8, & C_{21}^w = 0.3, & C_{\omega1} = 0.44, & C_{\omega2} = 0.8, \\ \sigma_R = 1.1, & \sigma_\omega = 1.1, & C_{\omega3} = 1.0, & C_{cr2} = 0.275. \end{array} \tag{6}$$

2.2. Wall Boundary Conditions

The boundary conditions at the wall are prescribed the following way. The no-slip condition demands that all the Reynolds stress tensor components be zero at the wall: $(R_{ij})_{wall} = 0$. It is known that ω^h asymptotically tends to $\omega^{h+} \cong 1/(y^+)^2$ as $y^+ \to 0$ [33], where $\omega^{h+} = \omega^h\nu/u_\tau^2$ and $y^+ = yu_\tau/\nu$. This asymptotic behavior may be used to specify ω^h in one or (as recommended in [34]) several off-wall grid points. The current paper analyzed the behavior of $\omega^{h+}(y^+)$ in DNS data of channel flow at $550 \le Re_\tau \le 2000$ [35].

Using the least squares method, a more accurate approximation for the asymptotic behavic was found:

$$\omega^{h+} = \frac{1}{(y^+)^2} + 0.0257 + (0.0454y^+)^2 - (0.102y^+)^4 + (0.116y^+)^6 - (0.110y^+)^8. \quad (7$$

This equation corresponds well to the DNS data for $0 < y^+ < 6$, as seen in Figure This boundary condition was used in all following simulations, ω^h was prescribed at th first three off-wall grid points.

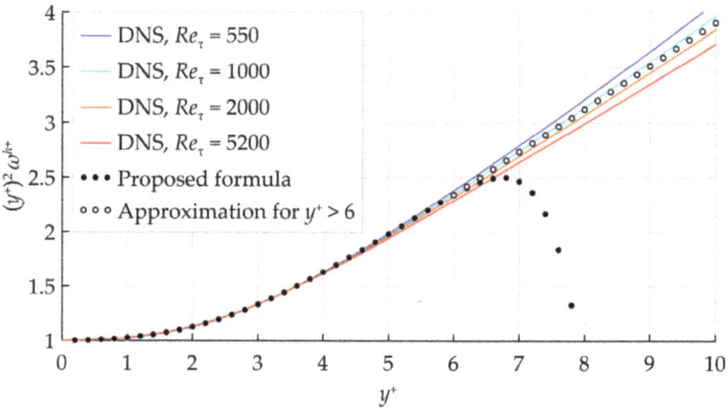

Figure 1. Near-wall ω^{h+} distributions extracted from the channel flow DNS data [35] and the approximation. ω^{h+} is multiplied by $(y^+)^2$ for convenience.

Note that Equation (7) is designed for use only within the viscous sublayer. If th value of y^+ exceeds 6, some other form of expression for ω^{h+} should be adopted, e.g $\omega^{h+} = 0.39/y^+$, which is shown in Figure 1 with open symbols. In the simulation presented below, however, y^+ was always lower than 6 at the points where ω^{h+} wa prescribed, which allowed to use the Equation (7) without any extensions.

2.3. Model Recalibration Method

A recalibration of the Jakirlić & Maduta model [24] was done by an *a posteriori* procedure, to improve the correspondence of the developed channel flow simulation results to the DNS data [35]. In contrast to an *a priori* calibration, in which DNS data is substituted into the model relations, during the *a posteriori* calibration, the model is used at each step of the procedure for the chosen flow simulation, which allows the estimate of its accuracy and ensures computational stability.

The *a posteriori* model calibration was conducted the following way. Sixteen model coefficients were varied. Their original values are stated above (see (6)). Let us denote them as $C_i, i = 1, \ldots, 16$. Sets of coefficients can be viewed as points in an N-dimensional space. An iterative procedure was run; at a preliminary iteration, the model with its original coefficient values (6) was used in 4 simulations of a one-dimensional channel flow at various Reynolds numbers (see Appendix A). A supplementary code was written for this. All of the following iterations were similar; let us consider steps at the k-th iteration ($k \in \mathbb{N}$), at the beginning of which the coefficients correspond to the point $C^{(k)} = \left\{ C_i^{(k)} \,\middle|\, i = 1, \ldots, 16 \right\}$

First, the norm of the difference between the solution from the previous iteration and the DNS data is calculated through the following formula:

$$N_k = \left[\sum_{r=1}^{N_{Re_\tau}} \sum_{i=1}^{5} w_i \cdot \frac{\sum_{n=0}^{N(r)} \left(\tilde{D}^{k-1}_{i,r,n} - D_{i,r,n}\right)^2}{\sum_{r=1}^{N_{Re_\tau}} \sum_{n=0}^{N(r)} \left(D_{i,r,n}\right)^2} \right]^{1/2}, \quad (8$$

where $D_{i,r,n}$ and $\tilde{D}{}^{k-1}_{i,r,n}$ are the DNS data and the solution with the model at the $(k-1)$-th iteration correspondingly. The subscript indices are introduced as follows: $i = 1, \ldots, 5$ is the flow parameter (U, R_{xx}, R_{yy}, R_{zz}, R_{xy}), $r = 1, \ldots, 4$ is the case number ($Re_\tau \approx 550$, 1000, 2000, 5200), n is the number of a point in space ($n = 0$ is at the wall and $n = N(r)$ is at the channel half-height for the mean velocity field; the other parameters are stored with a shift of a half grid step, see Appendix A). The expression in the denominator of (8) is necessary for normalizing the "columns" $\tilde{D}_{i,r} - D_{i,r}$, $i = 1, \ldots, 5$, $r = 1, \ldots, 4$, and bringing their contribution to the sum closer to equal. The coefficient w_i is defined to increase the contribution of the velocity to the norm: $w_1 = 10$, $w_2 = \ldots = w_5 = 1$.

In all simulations, the first off-wall grid point, at which the mean velocity value was stored, was placed at $y_1^+ \approx 0.5$, the grid step size increased in a geometric progression with a denominator $q \approx 1.04$. The overall number of grid points between the wall and the channel center, depending on the Reynolds number, is given in Table 1.

Table 1. The number of grid points in the turbulent channel flow test case.

r	Re_τ	$N(r)$
1	543.5	98
2	1001	114
3	1995	131
4	5186	155

After N_k was calculated, the developed channel flow was simulated using a set of coefficients corresponding to the centers of each face of a 16-dimensional box that surrounded the point $C^{(k)}$ and had a side length of $2\Delta C_i$, $i = 1, \ldots, 16$. Each point corresponds to a change in one of the constants C_i by $\pm\Delta C_i$, where $\Delta C_i = K \cdot C_i$ and the starting value of K is 0.01. The norm of the solution difference was calculated using Equation (8) for each point and the least one among these and N_k was chosen. If a simulation diverged for any of these points, then it was eliminated from consideration. If a norm of one of the face centers happened to be the least, a transition to this point was made and the k-th iteration ended. In a case where a norm of the initial point (N_k) was the least, there was no change in the coefficients, and the iteration was repeated with a 25% decrease in K, which is responsible for coefficient increments. The iterative process ended when the condition $K < 10^{-5}$ was satisfied, which corresponds to the 16-dimensional box being small enough. Thus, the process converged to one of the local minima of the norm of the difference between the solution with the model and the DNS data in the coefficient space.

It should be noted that the uncertainty in the DNS data reported in [35], namely the estimated standard deviation of the mean velocity, does not exceed 0.2%, and the estimated standard deviation of the Reynolds stresses does not exceed 3%. These errors are much smaller than the differences between the DNS data and the RANS simulations presented below, which implies no significant impact of the DNS data uncertainty on the calibration results.

2.4. Model Recalibration Process

The typical behavior of the iterative process is shown in Figure 2; the difference norm (8) dependence is divided by its initial value N_0 on the iteration, the number k is shown. Note that the initial value N_0 corresponds to the Jakirlić & Maduta model [24].

The process lasted for 2202 iterations. The norm decreased quickly during the first 1000 iterations, and after this point, it kept close to constant despite the ongoing coefficient change. Let us note that a relatively wide choice of coefficients with which the model accuracy is nearly the same could be used for future tuning for different types of flows. The final value of the norm N_k/N_0 was 0.42. Modified model coefficients that correspond to the post-calibration state are the following:

$$\begin{array}{llll} C_{11}=2.5, & C_{12}=0.305, & C_F=0.518, & Re_{t0}=266, \\ C_{11}^w=1.39, & C_{12}^w=0.185, & C_l=2.53, & C_{fw}=1.95, \\ C_{21}=0.745, & C_{21}^w=0.00141, & C_{\omega1}=0.266, & C_{\omega2}=1.03, \\ \sigma_R=1.63, & \sigma_\omega=2.77, & C_{\omega3}=0.934, & C_{cr2}=4.73. \end{array}$$

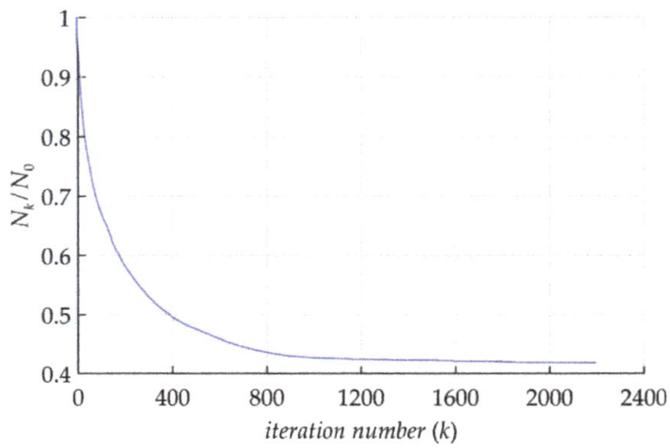

Figure 2. The behavior of the difference norm during the iteration process.

More than half of the coefficients experienced a substantial change. The coefficient C_{12}, which is used in the slow part of the redistribution term, was decreased by almost 70%; Re_{t0}, which affects the region where low-Reynolds number corrections are active, had an increase by 77%; C_{11}^w and C_{12}^w, which define the amplitude of the near-wall corrections to the slow part of the redistribution term, decreased by 20% and 38% correspondingly. C_{21}^w is close to zero after the calibration, which means that the near-wall correction to the rapid part of the redistribution term is essentially turned off. $C_{\omega1}$ and $C_{\omega2}$, which are responsible for the production and destruction of ω^h, were reduced by 40% and raised by 29% correspondingly. This imbalance is towards the destruction of ω^h was compensated by a heightened value of the cross-diffusion coefficient C_{cr2}, which became 17.2 times that of its initial value. Finally, the coefficients σ_R and σ_ω, which are inversely proportional to the intensity of the turbulent diffusion of R_{ij} and ω^h, noticeably grew by 48% and 2.5 times correspondingly. The change in coefficients during the difference norm N_k/N_0 decrease is presented in Figure 3.

The most important coefficients that are the first to change are C_l, Re_{t0} and $C_{\omega2}$. After that, C_{21}^w, σ_R, C_{11}^w, C_{21}, $C_{\omega1}$, C_{fw} and C_{12} join the trend. These ten coefficients are enough to lower the difference norm to 0.48. Only at the end of the calibration process, when the difference norm is lowered to 0.42 C_{cr2}, σ_ω, C_F, C_{12}^w and $C_{\omega3}$ start to change. As it turned out, C_{11} did not change during the calibration.

The physical implications of coefficient changes are as follows. First, the dissipation tensor anisotropy contribution to the slow part of the redistribution term seems to be much lower than was suggested in the baseline model. Second, the magnitudes of both parts of the near-wall corrections to the redistribution term appear to be overestimated in the baseline model, too. Most notably, the $\Phi_{ij,2}^w$ the term was nullified during the calibration, indicating its unimportance. Third, the diffusion intensities of R_{ij} and ω^h were reduced, possibly to compensate for the increase in velocity fluctuation magnitudes. Finally, the cross-diffusion term was found to be relevant at the late stages of the calibration. These observations may serve as guidelines for future developments of the near-wall DRSMs.

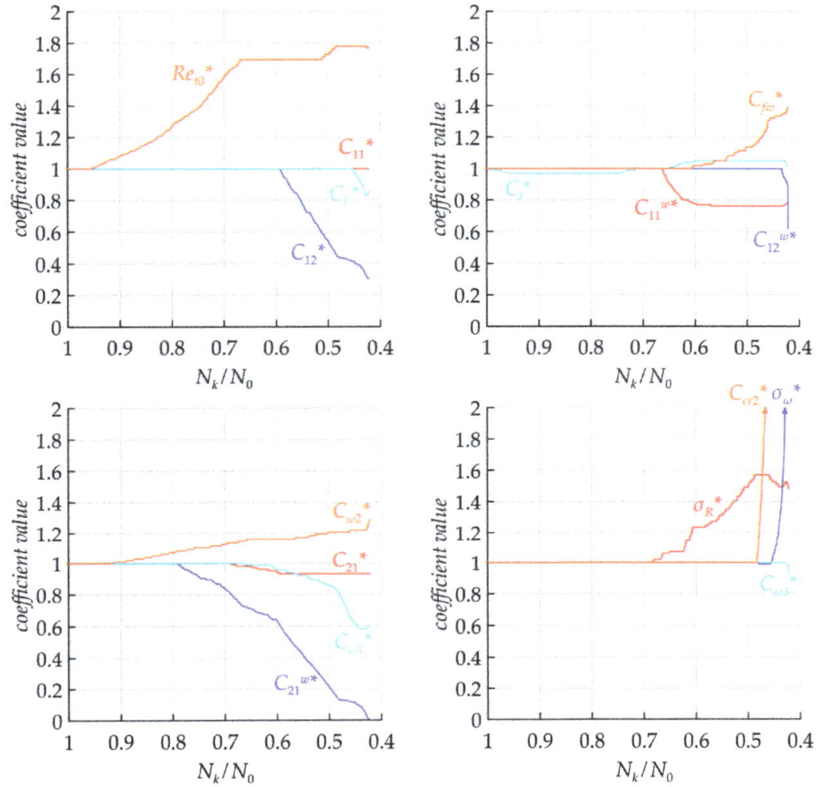

Figure 3. The behavior of the coefficients during the difference norm decrease. All coefficients are divided by their initial values, which is denoted by a star.

3. Testing within the RANS Framework

The Jakirlić & Maduta model [24] with baseline and modified sets of coefficients were implemented in an in-house CFD code *zFlare* [36]. In this program, using the finite-volume method, the complete system of Reynolds equations for a compressible gas is solved with the possibility of RANS/LES hybridization. Multiblock structured grids and an implicit time integration method DIRK22 (Diagonally Implicit Runge–Kutta Method, two stages, 2nd order of accuracy) [37] are used. In RANS simulations, the spatial approximation of convective terms was performed according to the WENO5 scheme (Weighted Essentially Non-Oscillatory, 5th order) [38] along the grid lines. For diffusion and source terms, 2-nd-order-accurate central difference formulas were used. To verify the software implementation of the models, the results of developed channel flow simulations at $Re_\tau \approx 2000$ using the *zFlare* program were compared with the results obtained from the auxiliary one-dimensional program calculation. This program was used in the calibration process. The computational grids corresponded to each other. Solutions for both the baseline and the modified models turned out to be practically indistinguishable between the two programs; the relative divergence in the mean velocity and Reynolds stresses profiles were within 10^{-3}. After this verification, the simulations described below were performed.

3.1. Turbulent Channel Flow at Friction Reynolds Number 2000

This subsection considers the developed turbulent channel flow at a Mach number 0.2 calculated by the bulk velocity (nearly incompressible flow). The basic computational grid contained $4 \times 65 \times 1$ cells, and the wall-adjacent cell height in inner variables was chosen as $y_1^+ \approx 0.5$. A half of the channel from the wall up to the symmetry plane was modeled. A general view of the grid and boundary conditions is shown in Figure 4. In the coordinat system of the figure, the flow is moving from left to right.

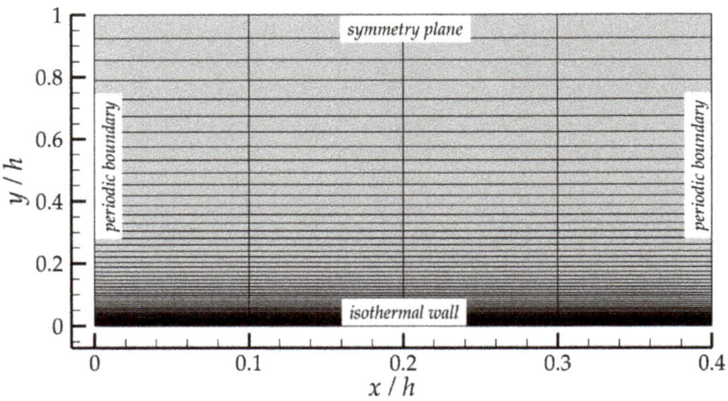

Figure 4. A computational grid used to simulate the turbulent channel flow and the boundary condition

To investigate the grid convergence, nested grids obtained through a consecutiv twofold refinement of the base grid in the wall-normal direction were used. In Figure 5, th mean velocity and Reynolds stress R_{xx} profiles obtained on these grids with the modifie model are shown. It can be seen that the solutions almost coincide.

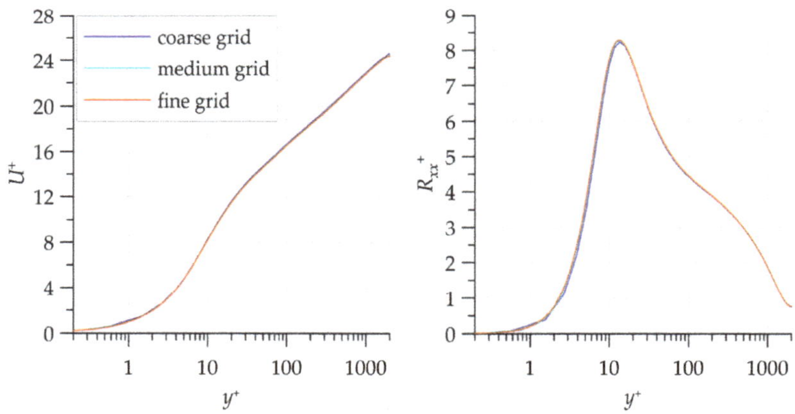

Figure 5. Grid convergence study in the turbulent channel flow test case.

Figure 6 compares the solutions obtained with the baseline and modified versions c the Jakirlić & Maduta model [24] with the DNS reference data [35]. The modified model i close to the DNS data for all the parameters. In particular, the inner peak of the Reynold stress R_{xx}, the behavior of the mean velocity in the buffer region and the distributio of the parameter ω^h across the channel are reproduced significantly better. Noticeabl differences from the reference data remain in the R_{zz} profile at $y^+ < 10$. The coefficien recalibration does not affect this distribution. This is caused not by the coefficient value but by the restrictions imposed through the structure of the closures (2), (3), (4). Improvin the description of R_{zz} may be an area for a separate study.

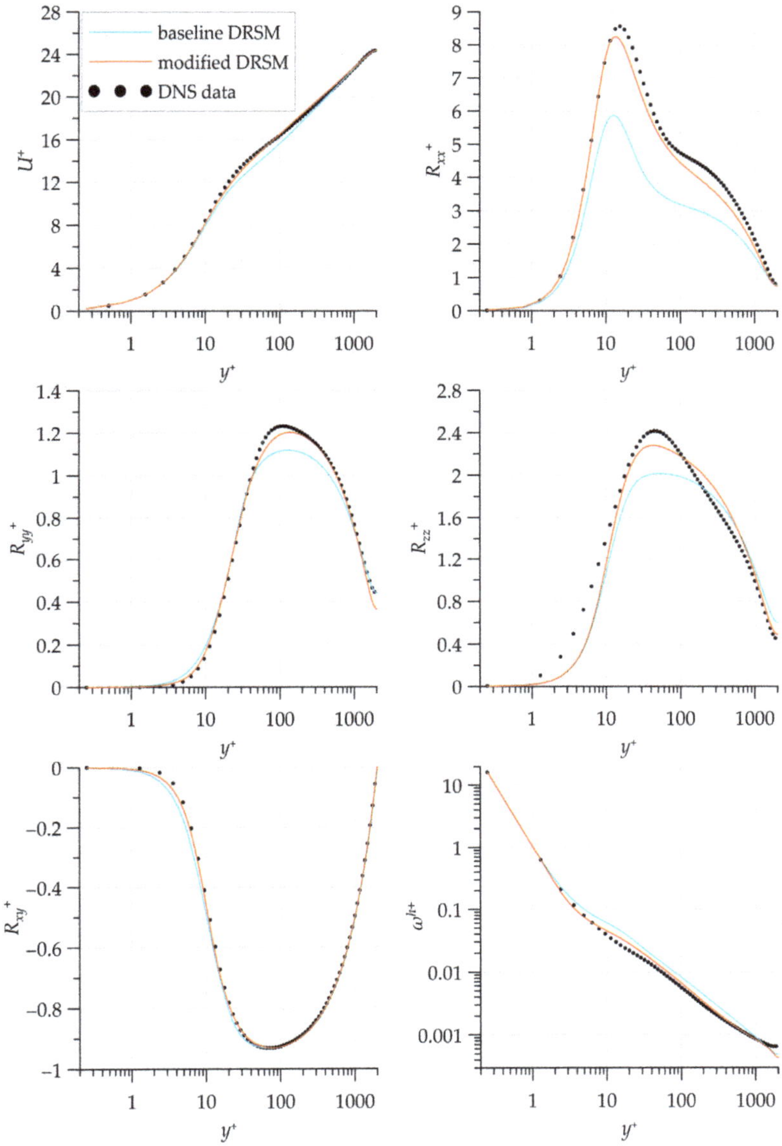

Figure 6. Turbulent channel flow solutions.

3.2. Flat Plate Turbulent Boundary Layer

Flat plate turbulent boundary layer simulations were performed in the set-up shown in Figure 7. The length scale L was chosen to be 1 m. The parameters of an inlet flow were prescribed so that the Mach number was equal to 0.2, and the Reynolds number based on L was 5×10^6. In a preliminary simulation, an extended computational domain was used: the plate leading edge was located at $x = 0$, and a buffer region was placed upstream at $-0.3 \leq x/L \leq 0$. However, it turned out that Jakirlić & Maduta-type models predict a delayed boundary layer transition from a laminar to a turbulent state. To avoid any associated difficulties, an auxiliary simulation using the SSG/LRR-ω DRSM [9] was carried out in the preliminary set-up. In this simulation, almost the entire boundary layer was turbulent. The cross-section $x/L = 0.02$ was extracted from this simulation (the Reynolds stress levels there already corresponded to a turbulent flow). The parameters from this section were used as inlet boundary conditions for simulations with the Jakirlić & Maduta models on a "cropped" grid, which had the left boundary located at $x/L = 0.02$. This exact computational domain is shown in Figure 7. A subsonic outflow condition with a fixed static pressure was set at the outlet boundary. On the upper boundary, a boundary condition based on the analysis of Riemann invariants was used.

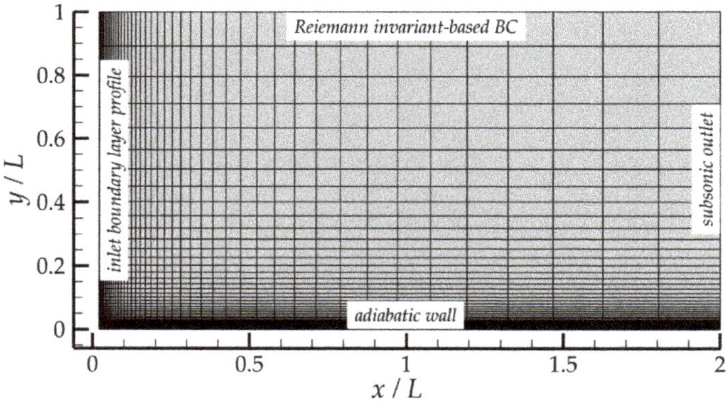

Figure 7. Computational grid used to simulate the flat plate turbulent boundary layer and the boundary conditions. Every 4-th grid line is shown.

The main grid contained $176 \times 384 \times 1$ cells. The wall-adjacent cell height in inner variables was within $0.092 \leq y_1^+ \leq 0.14$ (Figure 8, left, "base grid"). To perform the grid convergence study, simulations were also carried out using the modified model on a grid obtained by removing every second grid line from the original one. It was found that in the interval $0.02 \leq x/L \leq 1.6$, the streamwise distribution of the local momentum-thickness Reynolds number Re_θ is reproduced on the two grids within a 1.5% difference. The solutions deviate from each other more noticeably only at $x/L > 1.6$ due to the outlet boundary condition influence (Figure 8, right). Note that by definition, $Re_\theta = U_e \rho_e \theta / \mu_e$, where U_e, ρ_e, μ_e are the freestream velocity, density, and dynamic viscosity, respectively. Momentum thickness θ is expressed as

$$\theta = \int_0^\infty \frac{\rho U}{\rho_e U_e}\left(1 - \frac{U}{U_e}\right) dy.$$

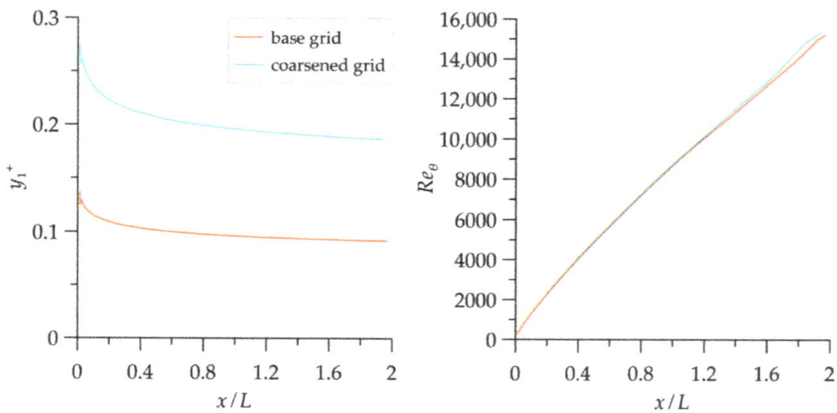

Figure 8. Dimensionless wall-adjacent cell height (**left**) and Re_θ distributions along the plate (**right**).

The integration is carried out across the boundary layer in the wall-normal direction.

Consider the wall-normal boundary layer parameter profiles extracted from the control section at $Re_\theta = 6500$ (Figure 9). The DNS results [39] were used as the reference data at the same value of Re_θ. It can be seen that the modified model improved the velocity distribution compared to the baseline model within the boundary layer. The Reynolds stresses and ω^h parameter distributions also approached the reference data, especially for the inner peak of R_{xx}.

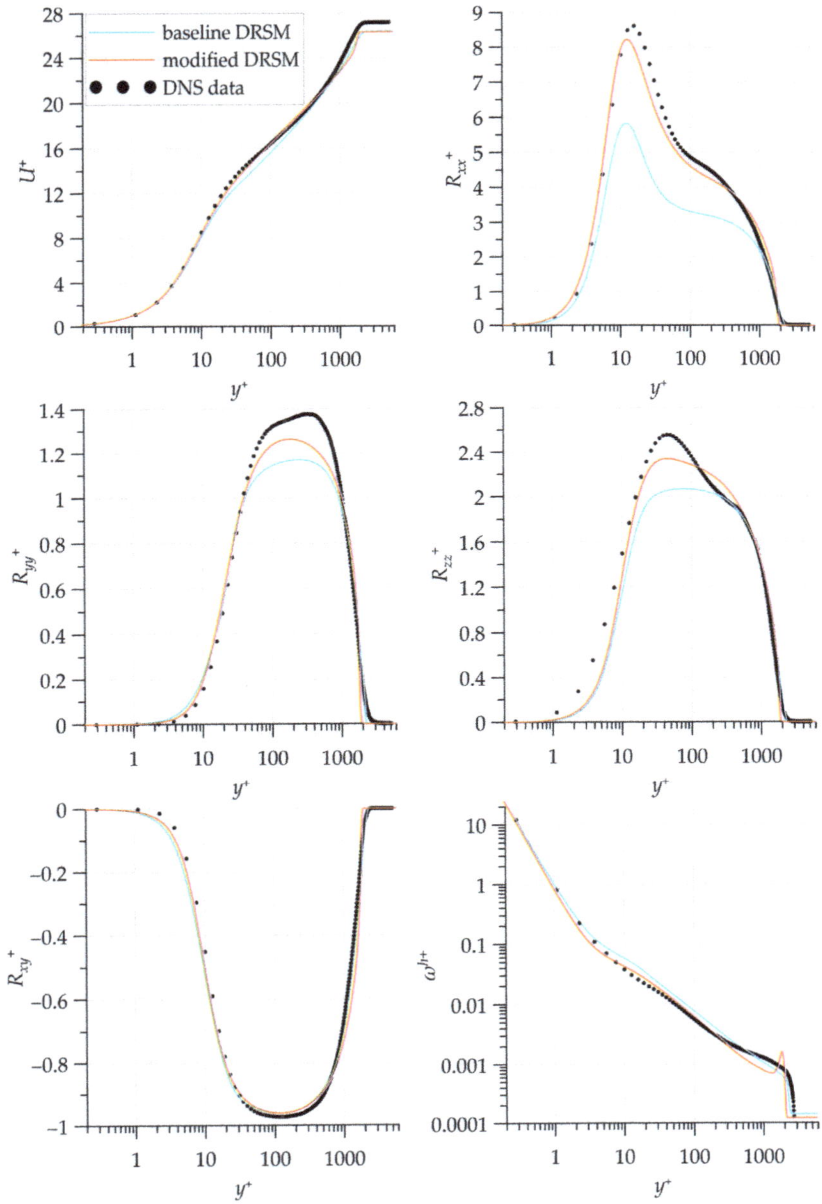

Figure 9. Flat plate turbulent boundary layer solutions.

It should be noted that at the outer boundary of the turbulent region, the tangent t the velocity profile became almost vertical, accompanied by a local peak of ω^h. This ma be due to a high value of the cross-diffusion coefficient C_{cr2}, which significantly contribute to the solution at boundaries of this type [40]. This occurred because there are no externa turbulent region boundaries in the channel flow set-up used for the model calibration. Thi does not cause an issue since the modified model is supposed to be used as part of hybri RANS/LES methods. In such simulations, the outer boundaries of turbulent regions ar within the LES region and are not affected by the RANS model.

To provide quantitative metrics, the difference norms were calculated usin Equation (8) without the summation of the Reynolds numbers. The cross-section co responding to $Re_\theta = 6500$ was used. The ratio of the modified to the baseline model' norms was 0.46, close to the channel flow value of 0.42.

Streamwise distributions of the friction coefficient along the plate $C_f = 2\tau_w/(\rho_e U_e^2$ obtained in the simulations and plotted as functions of Re_θ, are represented in Figure 1 They are compared with the Karman–Schoenherr (KS) empirical correlation [41]:

$$C_f = \left(17.08\left(\log_{10} Re_\theta\right)^2 + 25.11\left(\log_{10} Re_\theta\right) + 6.012\right)^{-1}. \quad (9$$

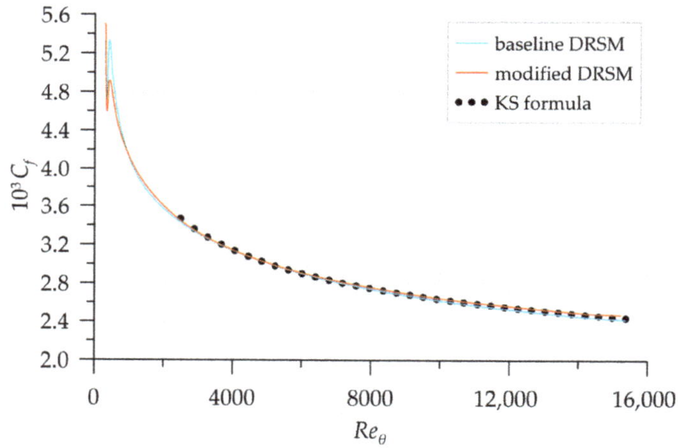

Figure 10. Friction coefficient distribution along the plate.

It can be seen that both the baseline and the modified models reproduce the distribu tion (9) satisfactorily, and their error level is approximately the same.

3.3. RANS Simulation of a Boundary Layer Separation from a Rounded Step

In this flow, the initially attached boundary layer undergoes separation in the channe expansion region. This test case allows us to evaluate the accuracy of the turbulenc model in a flow with a significant streamwise pressure gradient. The choice in favor of th Bentaleb, Lardeau & Leschziner set-up [42] was made due to the simplicity of its geometri formulation and the absence of side walls. The reference data for this flow are the freel available results of a wall-resolved LES simulation [43]. Their reliability is confirmed b considering two computational grids in [42], showing only minor differences in the mea flow fields.

The computational domain consisted of an inlet channel section, an expansion with rounded backstep and an outlet channel of a constant cross-section, see Figure 11. A flo with the inlet Mach number $M = 0.2$ (measured at the channel center) and the Reynold number $Re = \rho_\infty U_\infty H/\mu = 13{,}600$ was considered, where ρ_∞ and U_∞ are the densit and velocity at the inlet section center, correspondingly, H is a step height and μ is th constant dynamic viscosity. An adiabatic no-slip condition was set on the top and botton walls, and a constant static pressure was set at the outlet. The solution obtained in a preliminary simulation of the supplying channel was used at the inlet. In that simulation, the velocity and pressure data were taken from the reference LES and did not change during the simulation. The turbulence parameters R_{ij} and ω^h consistent with them were determined.

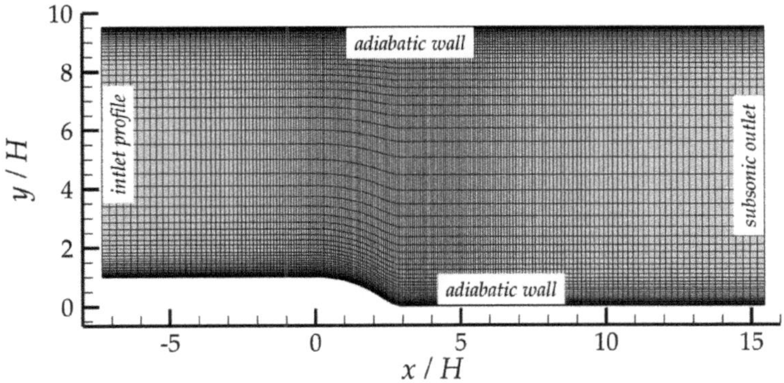

Figure 11. Computational grid used to simulate the rounded step flow and the boundary conditions. Every 4-th grid line is shown.

For the RANS simulations presented here, the grid obtained by modifying the one from the reference LES simulation was used. In the streamwise direction, the original node distribution was left unchanged. In the wall-normal direction, the node distribution of the lower half of the channel was mirrored in the upper half. It was done to use the no-slip condition on the flat (top) wall, whereas wall functions were used in the reference LES simulation. In the spanwise direction, the grid only had one cell due to the statistical spanwise homogeneity of the flow. As a result, the grid consisted of $768 \times 252 \times 1$ cells.

The grid convergence study showed that coarsening the grid twofold in the XY-plane leads to a change in the velocity profiles near the step region within 1% with the considered Jakirlić & Maduta models. This indicates a sufficient grid resolution.

In Figure 12, the profiles of the streamwise and transverse mean velocity components in the expanding channel are shown. It can be seen that in the chosen set-up, the recalibrated model distorts the solution more than the baseline model. The ratio of the modified to the baseline model's difference norms was calculated using Equation (8). The six cross-sections shown in Figure 12 were used. The ratio was found to be 1.51. This can be explained by the fact that the model was calibrated for flows without adverse pressure gradients in the current study. This seriously limits the applicability of the resulting model in a pure RANS setting. However, as mentioned above, the recalibrated model is intended for use as part of hybrid RANS/LES methods, where its role is limited by a description of the near-wall part of the boundary layer. In this region, the distribution of parameters, such as mean velocity, Reynolds stresses, dissipation rate of turbulence kinetic energy, etc., can be considered close to universal. This allows us to reasonably calibrate the model without using data of flows with a strong streamwise inhomogeneity of the parameters. The following section will demonstrate how the baseline and modified models perform in a hybrid RANS/LES mode in the same test case.

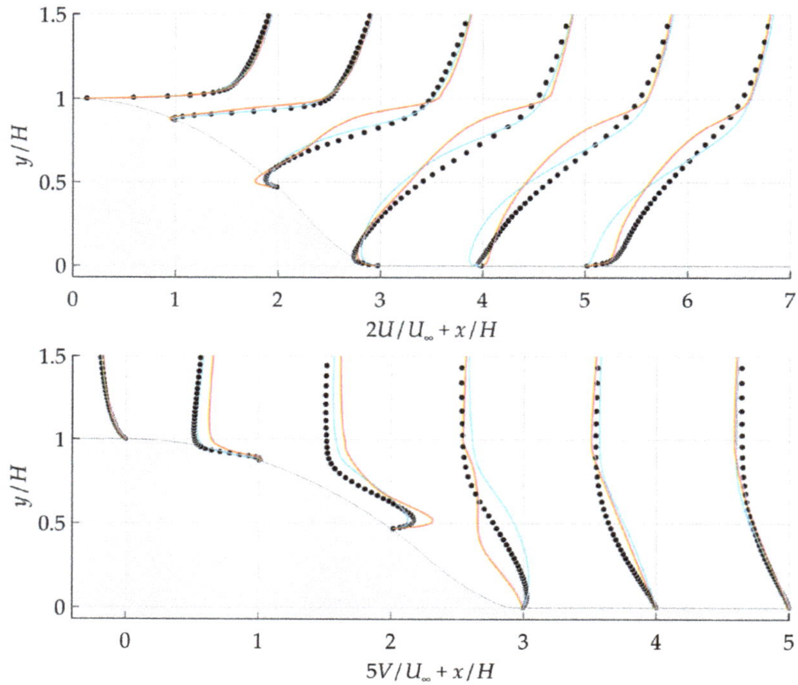

Figure 12. Mean streamwise (**upper**) and transverse (**lower**) velocity profiles. Baseline DRSM (blue line), modified DRSM (orange line), reference LES (black symbols). Step geometry is shown in grey.

4. Hybrid RANS/LES Simulations of the Curved Step Flow

To demonstrate the recalibrated model performance in the hybrid RANS/LES mode, i.e., in the framework for which it was originally designed, the problem of the boundary layer separation from a curved step was solved again using hybrid methods.

4.1. Formulation of the Hybrid RANS/LES Method Used

In this study, Jakirlić & Maduta type models were hybridized with a recently published differential subgrid stress model [44]. The hybridization was done according to the following formulas:

$$\frac{\partial R_{ij}}{\partial t}+\frac{\partial}{\partial x_k}\left[R_{ij}U_k+f_hT_{ijk}^{\mathrm{RANS}}+(1-f_h)T_{ijk}^{\mathrm{LES}}-\frac{\nu}{2}\frac{\partial R_{ij}}{\partial x_k}\right]=P_{ij}+f_h\left(\Phi_{ij}^{\mathrm{RANS}}-\varepsilon_{ij}^{h,\mathrm{RANS}}\right)+(1-f_h)\left(\Phi_{ij}^{\mathrm{LES}}-\varepsilon_{ij}^{\mathrm{LES}}\right),\quad(10)$$

$$\frac{\partial\omega^h}{\partial t}+\frac{\partial}{\partial x_k}\left[\omega^hU_k-\left(\frac{\nu}{2}+\frac{\nu_t}{\sigma_\omega}\right)\frac{\partial\omega^h}{\partial x_k}\right]=f_hC_{\omega1}\frac{\omega^h}{k}P-C_{\omega2}\left(\omega^h\right)^2+\frac{1}{k}P_{\omega3}+S_l+f_h\frac{2}{k}\left(0.55\frac{\nu}{2}+C_{cr2}\frac{\nu_t}{\sigma_\omega}\right)\frac{\partial k}{\partial x_k}\frac{\partial\omega^h}{\partial x_k}.\quad(11)$$

Here f_h is a blending function equal to 1 in the RANS sublayer and to 0 in the LES region. Its formulation is based on the f_b function of the IDDES method [28] and reads as follows:

$$f_h=\min\left\{2\exp\left(-9\left[0.25-C_\alpha\frac{d_w}{\Delta_{\max}}\right]^2\right),\ 1.0\right\},\quad(12)$$

where $\Delta_{\max}$ is the length of the longest cell edge. Compared to the function of the IDDES method, a coefficient $C_\alpha=0.5$ is introduced for increasing the RANS sublayer thickness to a value of about 0.1δ in the boundary layer upstream of the step.

In the Reynolds stress Equation (10), the tensors T_{ijk}^{RANS}, $\Phi_{ij}^{\mathrm{RANS}}$ and $\varepsilon_{ij}^{h,\mathrm{RANS}}$ are calculated according to the Formulas (2), (3) and (4), respectively. The closures for the tensors T_{ijk}^{LES}, Φ_{ij}^{LES} and $\varepsilon_{ij}^{\prime\mathrm{LES}}$ are given in Appendix B and correspond to the subgrid stress model [44] with one of the corrections proposed in [45]. In this model, the parameter ω^h is not used. To avoid numerical issues associated with solving the equation for ω^h in the LES region, the production term of ω^h, as well as the cross-diffusion term, are multiplied by f_h, i.e., are disabled outside the RANS sublayer.

4.2. Setup of the Hybrid Simulations and the Results

The grid for hybrid RANS/LES simulations was obtained by extruding the RANS grid described in Section 3.3 in z-axis direction by 96 cells, with the total width of the computational domain in this direction equal to $3.17H$, and by coarsening the grid twofold in the XY-plane. On the side planes, a periodic boundary condition was specified. At the inlet boundary, unsteady disturbances were superimposed on the profiles of the flow parameters according to the method [46], which quickly switched the simulation to the scale-resolving mode. In contrast to RANS simulations, spatial approximation of the convective fluxes was done using the hybrid scheme, which blends the upwind WENO5 approximation in the RANS sublayer and the 2-nd order central differences in the LES region using the formula proposed in [47]. The time step was equal to $0.01t_{char}$, where $t_{char} = H/U_\infty$ is the characteristic time of the flow. The period of flow settling was chosen to be equal to $52t_{char}$, after which the solution was averaged over a period of time $270t_{char}$. For illustration, an instantaneous flow field is shown in Figure 13, which was obtained in the hybrid simulation with the recalibrated Jakirlić & Maduta model in the RANS sublayer. It can be seen that an isolated separation is formed near the channel expansion region.

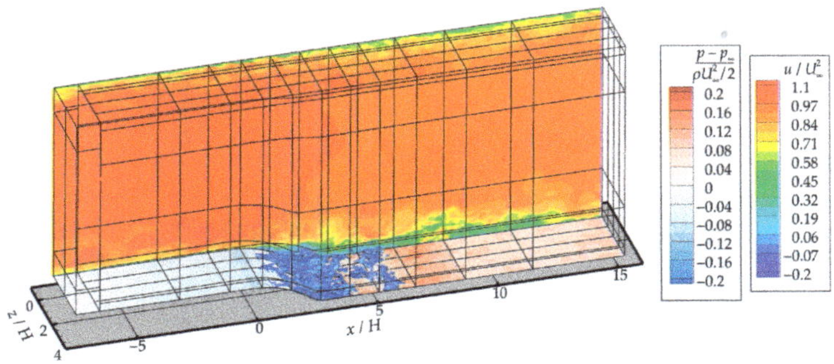

Figure 13. Instantaneous flow field in the curved step flow hybrid simulation. Streamwise velocity (vertical plane), pressure (lower surface) and zero streamwise velocity isosurface.

Figure 14 shows the profiles of streamwise and transverse mean velocity components obtained in the hybrid simulations. It is seen that, in contrast to the pure RANS setting, switching to the modified model reduced the size of the separation region to the reference value and caused almost complete agreement with the reference data on the streamwise velocity component. Only minor differences remained visible in the field of the transverse velocity component.

In Figure 15, the components of the Reynolds stress tensor obtained by summing up the resolved and subgrid stresses are compared. One can note an overall overestimation of the stress level by both methods, which is associated either with the features of the subgrid stress model requiring recalibration for shear flows or with an overestimated fluctuation amplitude introduced by the synthetic turbulence generator at the inlet. Note that the modified method captures the position of stress peaks and profile shapes more accurately despite this.

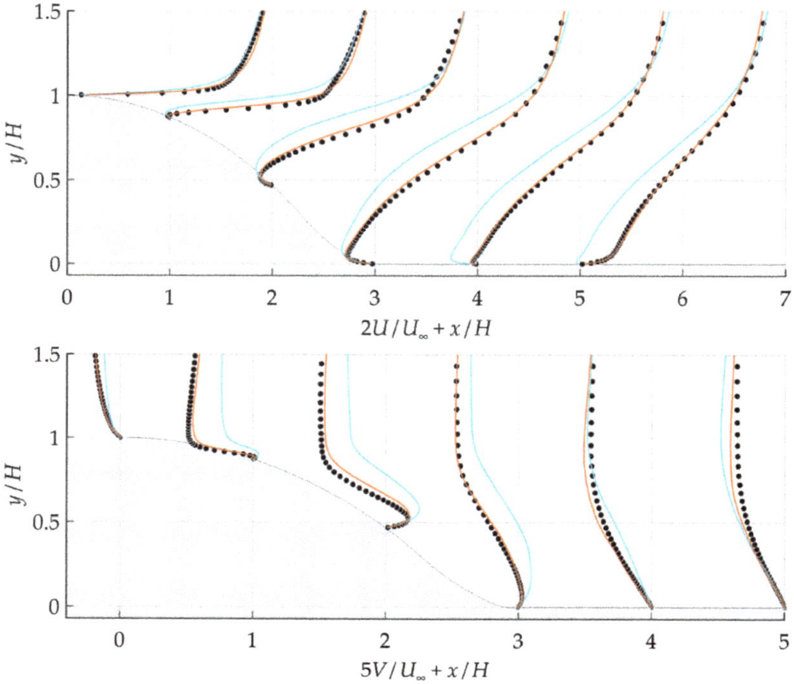

Figure 14. Mean streamwise (**upper**) and transverse (**lower**) velocity profiles. Hybrid simulation with the baseline model in the RANS subregion (blue line) and the modified model in the RANS subregion (orange line). Black symbols correspond to reference LES. Step geometry is shown in grey.

Interestingly, the ratio of the modified to the baseline model's difference norms calculated using Equation (8) over the six cross-sections shown in Figures 14 and 15 was 0.95. This indicates no significant superiority of the modified model over the baseline. However, this result is attributed to the large discrepancies in the Reynolds stress components, see Figure 15. Calculating the difference norms using only the mean velocity component U, a much better ratio is recovered, namely 0.21, consistent with Figure 14. Improving the prediction of the Reynolds stress components in this flow type is the subject of a future study.

Finally, in Figure 16, the streamwise distribution of the pressure coefficient $C_p = (\overline{p} - p_\infty)/(\rho U_\infty^2/2)$ is shown, where $\overline{p}$ is the mean pressure, p_∞ is the pressure at the center of the inlet section, as well as the streamwise distribution of the friction coefficient C_f. It can be seen from the C_p distribution that the modified hybrid method captures the drops in the vicinity of the step more accurately. The friction distribution reveals the problem of the synthetic turbulence generator, due to which it is significantly underestimated in the $x < 0$ region. However, in the separation region and downstream of it, the friction coefficient is recovered, and its level is reproduced satisfactorily, especially with the modified model.

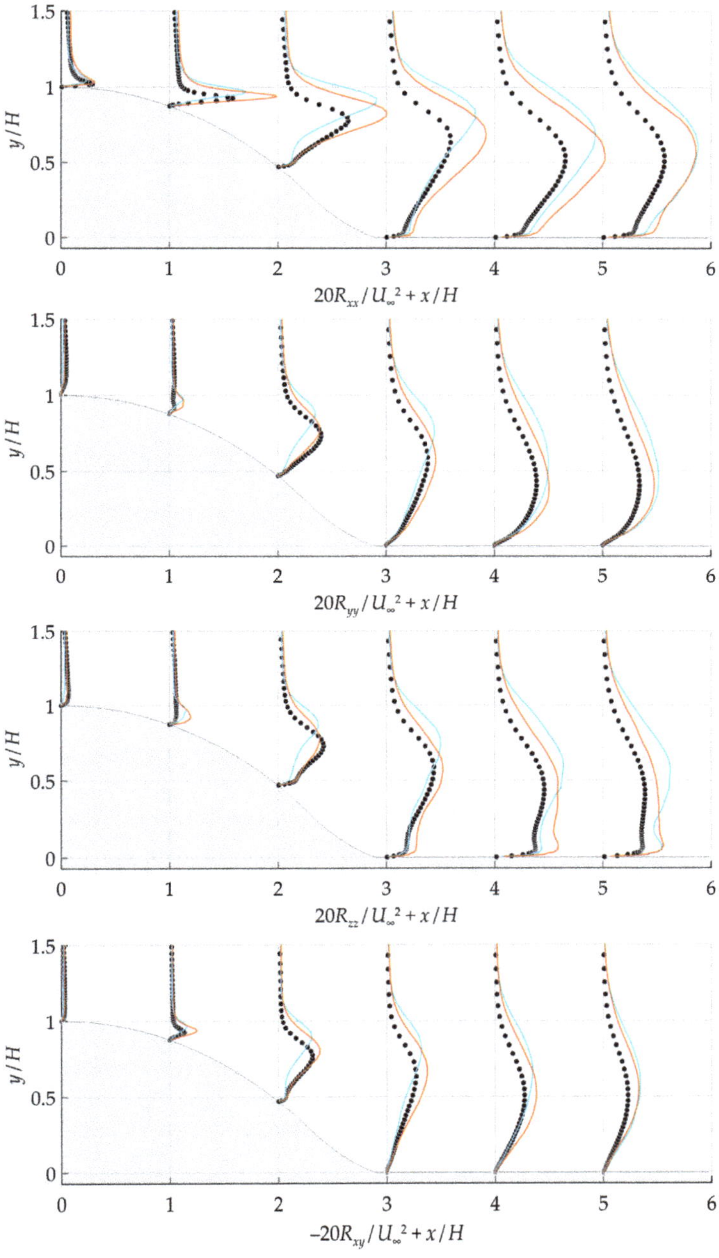

Figure 15. Reynolds stress profiles. Hybrid simulation with the baseline model in the RANS subregion (blue line) and the modified model in the RANS subregion (orange line). Black symbols correspond to reference LES. Step geometry is shown in grey.

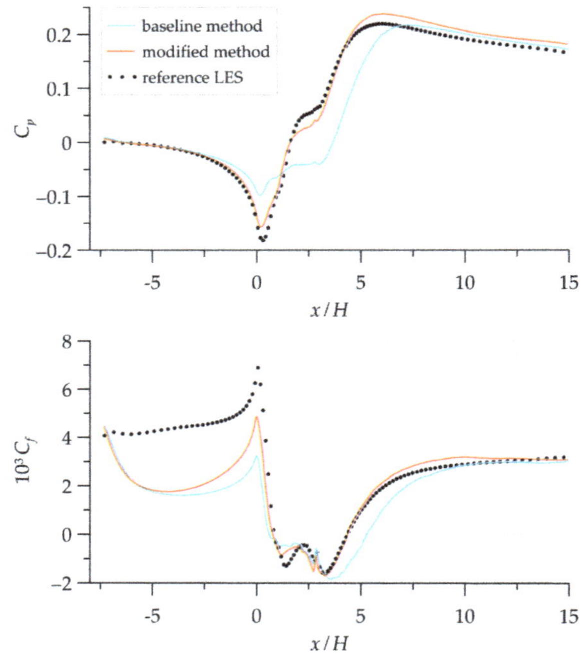

Figure 16. Pressure (**upper**) and friction (**lower**) coefficient distributions.

5. Conclusions

The low-Reynolds-number model of Jakirlić & Maduta [24] was recalibrated using an *a posteriori* procedure based on DNS data on developed channel flow at Reynolds numbers Re_τ in the range of 550–5200. The purpose of the calibration was to obtain a model focused on the description of the thin near-wall part of the boundary layer within the framework of hybrid RANS/LES methods. During the calibration process, the model error was reduced by 58%, which manifested in a better description of the mean velocity profile and the inner Reynolds stress peak and a more accurate distribution of the characteristic turbulence frequency ω^h in the developed channel flow compared to the base model. In the course of the study, an improved wall boundary condition for ω^h was proposed, which was used in all the presented simulations.

Validation of the model within the RANS framework showed that in a zero-pressure gradient flow—a turbulent boundary layer on a flat plate—the modified model gives the same improvement in the description of the flow field compared to the base model as in the developed channel flow. A minor inconsistency was found only at the outer boundary of the boundary layer. This is not a problem since the model is focused on hybrid simulations in which this boundary is located in the LES region. In the RANS simulation of the boundary layer separation from a smooth surface, the modified model showed deterioration in the region of the recirculation zone, which can be explained by the absence of data on flows under adverse pressure gradients in the calibration set. However, in the hybrid RANS/LES simulations, the transition from the base model to the modified one, on the contrary, significantly improved the solution, giving an almost complete coincidence of the mean streamwise velocity with the reference LES data. Thus, the applicability of the modified model to accurately describe the near-wall part of the boundary layer has been confirmed. At the same time, the presented results highlight the importance of the contribution of the RANS model to the overall accuracy of a hybrid simulation.

Author Contributions: Conceptualization, A.T.; methodology, A.T. and L.U.; programming, L.U.; data analysis and interpretation, L.U. and K.A.; writing—original draft preparation, L.U. and A.T.;

writing—review and editing, K.A. and V.S.; supervision, V.S. All authors have read and agreed to the published version of the manuscript.

Funding: The research was funded by a grant from the Russian Science Foundation No. 21-71-10105, https://rscf.ru/en/project/21-71-10105/ (accessed on 15 August 2023).

Data Availability Statement: The data presented in this study are available on request from the corresponding authors.

Conflicts of Interest: The authors declare no conflict of interest.

Nomenclature

Symbol	Description
Re	Reynolds number
Re_τ	friction Reynolds number
Re_t	turbulent Reynolds number
Re_θ	momentum-thickness Reynolds number
M	Mach number
u_τ	friction velocity (m/s)
U_i	i-th mean velocity component (m/s)
U_∞	velocity at the inlet section (m/s)
u_i	fluctuating part of the i-th velocity vector component (m/s)
τ_w	wall shear stress (N/m^2)
μ	dynamic viscosity (kg/(m·s))
ν	kinematic viscosity (m^2/s)
ν_t	eddy viscosity (m^2/s)
ρ	density (kg/m^3)
$\overline{p}$	mean pressure (N/m^2)
p_∞	pressure at the inlet section (N/m^2)
p	pressure fluctuation (N/m^2)
l_τ	near-wall length scale (m)
δ	boundary layer thickness (m)
h	channel half-height (m)
L	plate length (m)
L_t	turbulence length scale (m)
θ	momentum thickness (m)
d_w	distance to the nearest wall (m)
H	step height (m)
Δ	filter size (m)
$\Delta_{\max}$	length of the longest cell edge (m)
t_{char}	the characteristic time of the flow (s)
R_{ij}	Reynolds stress tensor (m^2/s^2)
k	turbulence kinetic energy (m^2/s^2)
ε	turbulence kinetic energy dissipation rate (m^2/s^3)
ω	turbulence frequency (1/s)
δ_{ij}	unity tensor
P_{ij}	Reynolds stress production tensor (m^2/s^3)
Φ_{ij}	redistribution tensor (m^2/s^3)
$\Phi_{ij,1}$	"slow" part of the redistribution tensor (m^2/s^3)
$\Phi^w_{ij,1}$	near-wall correction to the "slow" part of the redistribution tensor (m^2/s^3)
$\Phi_{ij,2}$	"rapid" part of the redistribution tensor (m^2/s^3)
$\Phi^w_{ij,2}$	near-wall correction to the "rapid" part of the redistribution tensor (m^2/s^3)
ε^h_{ij}	homogeneous part of the dissipation rate tensor (m^2/s^3)
$T^{(v)}_{ijk}$	turbulent transport by fluctuating velocity (m^3/s^3)
$T^{(p)}_{ijk}$	turbulent transport by fluctuating pressure (m^3/s^3)
T_{ijk}	sum of turbulent transport by fluctuating velocity and pressure (m^3/s^3)

S_{ij}	strain rate tensor (1/s)
Ω_{ij}	rotation rate tensor (1/s)
$P_{\omega 3}$	production of ω^h due to the second velocity derivatives, see Equation (5) (1/s^2)
S_l	length scale correction in the ω^h-equation, see Equation (5) (1/s^2)
P	turbulence kinetic energy production rate (m^2/s^3)
a_{ij}	stress-anisotropy tensor
A_2	second invariant of the stress-anisotropy tensor
A_3	third invariant of the stress-anisotropy tensor
A	Lumley's flatness parameter
e_{ij}	dissipation-anisotropy tensor
E_2	second invariant of the dissipation-anisotropy tensor
E_3	third invariant of the dissipation-anisotropy tensor
E	Lumley's flatness parameter analogue for the dissipation tensor
N_k	difference norm between the solution and the reference data, see Equation (8)
C_f	friction coefficient
C_p	pressure coefficient
f_h	RANS/LES blending function
n_k	k-th component of the normal vector
x, y, z	Cartesian coordinates
Subscripts and superscripts	
$+$	law-of-the-wall variables
h	"homogeneous" variables
e	freestream variables
RANS	formulas for tensors in the RANS sublayer
LES	formulas for tensors in the LES region
Acronyms	
RANS	Reynolds-averaged Navier–Stokes equations
LES	large eddy simulation
CFD	computational fluid dynamics
RSM	Reynolds stress model
DRSM	differential RSM
LRR	Launder–Reece–Rodi
SSG	Speziale–Sarkar–Gatski
SST	shear stress transport
JHh	Jakirlić and Hanjalić model with the homogeneous dissipation rate equation
GLVY	Gerolymos–Lo–Vallet–Younis
DNS	direct numerical simulation
IDDES	improved delayed detached eddy simulation
PANS	partially averaged Navier–Stokes
PITM	partially integrated transport modeling
SBES	stress-blended eddy simulation
DIRK22	diagonally implicit Runge–Kutta method, two stages, 2nd order of accuracy
WENO5	weighted essentially non-oscillatory 5th-order scheme
KS	Karman–Schoenherr

Appendix A. 1D Developed Channel Flow Equation System and Solution Method

To calibrate the Jakirlić & Maduta model [24], a C++ program that solved the one dimensional developed channel flow problem was written. This problem is described by the following equation system written in the inner variables:

$$\frac{\partial}{\partial y^+}\left[-\frac{\partial U^+}{\partial y^+} + R^+_{xy}\right] = \frac{1}{Re_\tau},$$

$$\frac{\partial}{\partial y^+}\left[-\frac{1}{2}\frac{\partial R^+_{ij}}{\partial y^+} + \left(\overline{u_i u_j u_y}\right)^+\right] = P^+_{ij} + \Phi^+_{ij} - \varepsilon^{h+}_{ij},\ ij \in \{xx,\ yy,\ zz,\ xy\},$$

$$\frac{\partial}{\partial y}\left[-\frac{1}{2}\frac{\partial\omega^{h+}}{\partial y^{+}}+T_y^{\omega+}\right]=S^{\omega+}.$$

The closures for Φ_{ij}^{+}, ε_{ij}^{h+}, $\left(\overline{u_i u_j u_y}\right)^{+}$, $T_y^{\omega+}$ and $S^{\omega+}$ correspond to the Jakirlić & Maduta model [24]. A superscript «+» denotes the values nondimensionalized using the inner scales of velocity $u_\tau=\sqrt{\tau_w/\rho}$ and length $l_\tau=\nu/u_\tau$.

In Figure A1, the arrangement of the staggered grid is shown. At the nodes with integer indices, the values of U^+ were stored. At the nodes with half-integer indices (located in the middle between the nodes with integer indices and denoted by the symbol P), the Reynolds stresses and ω^{h+} values were stored. The nodes indicated in the figure in gray were "external", black color denotes "internal" nodes. The nodes where the solution was constructed using modified formulas are marked in red and blue. Before the start of the calculation, all nodes, except for external ones, were filled with the initial condition values. At the beginning of the model calibration procedure, the initial conditions were taken from the DNS data. At subsequent iterations, calculations were carried out starting from the fields obtained at previous iterations. The equations were not solved at external nodes; instead, before the start of each time step, the following values were entered into them, obtained based on the data from within the computational domain:

$$(R_{xx}^{+})_{-1/2}=-(R_{xx}^{+})_{1/2};\ \left(R_{yy}^{+}\right)_{-1/2}=-\left(R_{yy}^{+}\right)_{1/2};\ (R_{zz}^{+})_{-1/2}=-(R_{zz}^{+})_{1/2};\ \left(R_{xy}^{+}\right)_{-1/2}=-\left(R_{xy}^{+}\right)_{1/2};$$

$$(R_{xx}^{+})_{N+1/2}=(R_{xx}^{+})_{N-1/2};\ \left(R_{yy}^{+}\right)_{N+1/2}=\left(R_{yy}^{+}\right)_{N-1/2};\ (R_{zz}^{+})_{N+1/2}=(R_{zz}^{+})_{N-1/2};\ \left(R_{xy}^{+}\right)_{N+1/2}=-\left(R_{xy}^{+}\right)_{N-1/2};$$

$$\omega_{-1/2}^{h+}=\omega_{1/2}^{h+};\ \ \omega_{N+1/2}^{h+}=\omega_{N-1/2}^{h+};\ \ U_0^{+}=0;\ \ U_{N+1}^{+}=U_{N-1}^{+}.$$

The values of ω^{h+} at the first three near-wall nodes (marked in blue) were specified using the Formula (7). At the internal nodes, the calculations were carried out according to a symmetric finite-difference spatial scheme of the 2nd order of accuracy.

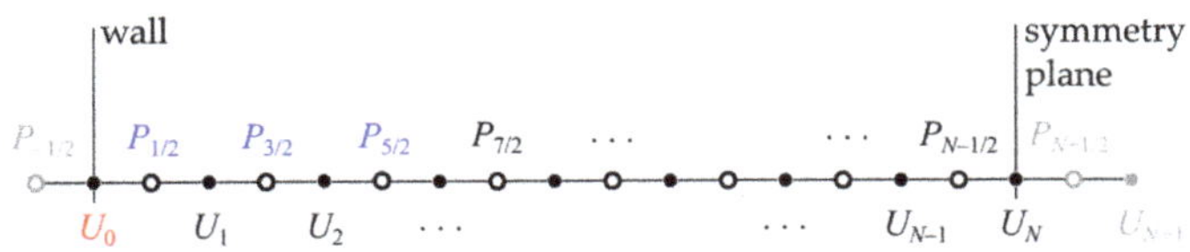

Figure A1. Staggered grid used in the one-dimensional channel flow simulations.

The problem was solved using the relaxation technique. Euler's forward method was used for time integration. The time step was taken to be equal to $\Delta t^{+}=C_{stab}/\max\omega^{h+}$, where the maximum was chosen over all nodes of the computational domain. C_{stab} coefficient was adjusted empirically for each Re_τ value and had an order of 10^{-3}. The calculation was run until five significant digits of the solution stabilized.

Appendix B. Differential Subgrid Stress Model

In the hybrid RANS/LES simulations presented in the current paper, a differential subgrid stress model proposed in [44] was used. LES terms of Equation (10) were closed according to the formulas below.

The model of the dissipation rate tensor:

$$\varepsilon_{ij}^{\mathrm{LES}}=\frac{2}{3}\varepsilon^{\mathrm{LES}}\delta_{ij},\qquad \varepsilon^{\mathrm{LES}}=\max\left\{C_E^{(0)}+C_E^{(1)}A_2+C_E^{(2)}A_3,\ C_E^{\mathrm{lim}}\right\}\frac{k^{3/2}}{\Delta}.$$

Due to the assumption of a large turbulent Reynolds number, the tensor $\varepsilon_{ij}^{\mathrm{LES}}$ is considered isotropic, and the total and "homogeneous" dissipation rates are considered to coincide. In the presented formula, the filter size Δ is equal to $0.4\Delta_{\max}$ and the coefficient values are as follows: $C_E^{(0)} = 0.8$, $C_E^{(1)} = -1.4$, $C_E^{(2)} = 2.1$, $C_E^{\text{lim}} = 0.21$.

Redistribution term model:

$$\mathbf{\Phi}^{\text{LES}} = \left(\Phi_{ij}^{\text{LES}}\right) = -C_{\Phi 1}\mathbf{a}\varepsilon^{\text{LES}} + C_{\Phi 2}^{(0)}k\mathbf{S} + C_{\Phi 2}^{(1)}\sqrt{k}\Delta(\mathbf{S\Omega} - \mathbf{\Omega S}) + C_{\Phi 2}^{(2)}\sqrt{k}\Delta\left(\mathbf{\Omega}^2 - \tfrac{1}{3}\text{tr}(\mathbf{\Omega}^2)\mathbf{I}\right)$$
$$+C_{\Phi 2}^{(3)}\Delta^2(\mathbf{S\Omega\Omega} + \mathbf{\Omega\Omega S} - \tfrac{2}{3}\text{tr}(\mathbf{S\Omega\Omega})\mathbf{I} - \text{tr}(\mathbf{\Omega}^2)\mathbf{S}) + C_{\Phi 3}k(\mathbf{aS} + \mathbf{Sa} - \tfrac{2}{3}\text{tr}(\mathbf{aS})\mathbf{I}) + C_{\Phi 4}k(\mathbf{\Omega a} - \mathbf{a\Omega}),$$
$$C_{\Phi 1} = 3.0,\quad C_{\Phi 2}^{(0)} = 1.04,\quad C_{\Phi 2}^{(1)} = -0.136,\quad C_{\Phi 2}^{(2)} = -0.058,\quad C_{\Phi 2}^{(3)} = -0.23,\quad C_{\Phi 3} = 0.34,\quad C_{\Phi 4} = 1.2.$$

Here, tensor multiplication obeys the matrix rule, $(\mathbf{AB})_{ij} = A_{ik}B_{kj}$. $\mathbf{I} = (\delta_{ij})$ is a unity tensor; $\mathbf{S}$ and $\mathbf{\Omega}$ are the resolved strain rate and rotation tensors, respectively:

$$S_{ij} = \frac{1}{2}\left(\frac{\partial U_i}{\partial x_j} + \frac{\partial U_j}{\partial x_i}\right),\qquad \Omega_{ij} = \frac{1}{2}\left(\frac{\partial U_i}{\partial x_j} - \frac{\partial U_j}{\partial x_i}\right).$$

According to the recommendation published in [45], the turbulence transport model is a weighted average of the model from [44] with the weight $\alpha_{\text{DSM}} = 0.8$ and of the simple gradient model with the weight $(1 - \alpha_{\text{DSM}})$:

$$T_{ijk}^{\text{LES}} = \alpha_{\text{DSM}}\left(\overline{u_i u_j u_k}^{\text{DSM}} + \frac{\overline{pu_i}^{\text{DSM}}\delta_{jk} + \overline{pu_j}^{\text{DSM}}\delta_{ik}}{\rho}\right) + (1 - \alpha_{\text{DSM}})\left(-\nu_{\text{Smag}}\frac{\partial R_{ij}}{\partial x_k}\right),$$

where

$$\overline{u_i u_j u_k}^{\text{DSM}} = \sum_{\{i\to j\to k\}}\left[-C_{T1}k^{\frac{1}{2}}\Delta\delta_{kl} + C_{T3}\Delta^2\frac{\partial U_k}{\partial x_l}\right]\frac{\partial\overline{u_i u_j}}{\partial x_l},$$

$$\frac{1}{\rho}\overline{pu_i}^{\text{DSM}} = -C_{PT}\overline{u_i u_m u_m}^{\text{DSM}},$$

$$\nu_{\text{Smag}} = (C_S\Delta_{\max})^2\cdot\sqrt{2S_{ij}S_{ij}}.$$

By $\sum_{\{i\to j\to k\}}$, a sum over the cyclic permutation of the indices is denoted: $\sum_{\{i\to j\to k\}} A_{ijk} = A_{ijk} + A_{jki} + A_{kij}$. The coefficient values in the formulas above are as follows: $C_{T1} = 0.019$, $C_{T3} = 0.064$, $C_{PT} = 0.42$, $C_S = 0.13$.

References

1. Chou, P.Y. On the velocity correlations and the solution of the equations of turbulent fluctuation. *Quart. Appl. Math.* **1945**, *3*, 38–54. [CrossRef]
2. Rotta, J. Statistische Theorie nichthomogener Turbulenz. *Z. Phys.* **1951**, *129*, 547–572. [CrossRef]
3. Eisfeld, B. (Ed.) *Differential Reynolds Stress Modeling for Separating Flows in Industrial Aerodynamics*; Springer: Cham, Switzerland, 2015; 101p. [CrossRef]
4. Hanjalić, K.; Launder, B. *Modelling Turbulence in Engineering and the Environment. Rational Alternative Routes to Closure*, 2nd ed.; Cambridge University Press: Cambridge, UK, 2022; 572p. [CrossRef]
5. Hanjalic, K.; Launder, B.E. A Reynolds stress model of turbulence and its application to thin shear flows. *J. Fluid Mech.* **1972**, *52*, 609–638. [CrossRef]
6. Launder, B.E.; Reece, G.; Rodi, W. Progress in the development of a Reynolds stress turbulence closure. *J. Fluid Mech.* **1975**, *68*, 537–566. [CrossRef]
7. Lumley, J.L. Computational modeling of turbulent flows. *Adv. Appl. Mech.* **1979**, *18*, 123–176. [CrossRef]
8. Speziale, C.G.; Sarkar, S.; Gatski, T.B. Modelling the pressure-strain correlation of turbulence: An invariant dynamical systems approach. *J. Fluid Mech.* **1991**, *227*, 245–272. [CrossRef]
9. Cecora, R.-D.; Radespiel, R.; Eisfeld, B.; Probst, A. Differential Reynolds-Stress Modeling for Aeronautics. *AIAA J.* **2015**, *53*, 739–755. [CrossRef]
10. Eisfeld, B.; Rumsey, C.L. Length-Scale Correction for Reynolds-Stress Modeling. *AIAA J.* **2020**, *58*, 1518–1528. [CrossRef]
11. Menter, F.R. Review of the shear-stress transport turbulence model experience from an industrial perspective. *Int. J. Comput. Fluid Dyn.* **2009**, *23*, 305–316. [CrossRef]
12. Eisfeld, B.; Rumsey, C.; Togiti, V. Verification and validation of a second-moment-closure model. *AIAA J.* **2016**, *54*, 1524–1541. [CrossRef]

3. Gibson, M.M.; Launder, B.E. Ground effects on pressure fluctuations in the atmospheric boundary layer. *J. Fluid Mech.* **1978**, *86*, 491–511. [CrossRef]
4. Craft, T.J.; Launder, B.E. New wall-reflection model applied to the turbulent impinging jet. *AIAA J.* **1992**, *30*, 2970–2972. [CrossRef]
5. Launder, B.E.; Li, S.-P. On the elimination of wall-topography parameters from second-moment closure. *Phys. Fluids* **1994**, *6*, 999–1006. [CrossRef]
6. Craft, T.J.; Launder, B.E. A Reynolds stress closure designed for complex geometries. *Int. J. Heat Fluid Flow* **1996**, *17*, 245–254. [CrossRef]
7. Craft, T.J. Developments in a low-Reynolds-number second-moment closure and its application to separating and reattaching flows. *Int. J. Heat Fluid Flow* **1998**, *19*, 541–548. [CrossRef]
8. Hanjalić, K.; Jakirlić, S.; Hadžić, I. Expanding the limits of "equilibrium" second-moment turbulence closures. *Fluid Dyn. Res.* **1997**, *20*, 25–41. [CrossRef]
9. Hanjalić, K.; Jakirlić, S. Contribution towards the second-moment closure modelling of separating turbulent flows. *Comp. Fluids* **1998**, *27*, 137–156. [CrossRef]
0. Jakirlic, S.; Hanjalic, K. A new approach to modelling near-wall turbulence energy and stress dissipation. *J. Fluid Mech.* **2002**, *539*, 139–166. [CrossRef]
1. Probst, A.; Radespiel, R. Implementation and Extension of a Near-Wall Reynolds-Stress Model for Application to Aerodynamic Flows on Unstructured Meshes. In Proceedings of the 46th AIAA Aerospace Sciences Meeting and Exhibit, Reno, NV, USA, 7–10 January 2008. [CrossRef]
2. Cécora, R.D.; Radespiel, R.; Jakirlić, S. Modeling of Reynolds-stress augmentation in shear layers with strongly curved velocity profiles. In *Differential Reynolds Stress Modeling for Separating Flows in Industrial Aerodynamics*; Eisfeld, B., Ed.; Springer: Cham, Switzerland, 2015; pp. 85–101. [CrossRef]
3. Morsbach, C. Reynolds Stress Modelling for Turbomachinery Flow Applications. Ph.D. Thesis, TU Darmstadt, Darmstadt, German, 2016.
4. Jakirlić, S.; Maduta, R. Extending the bounds of 'steady' RANS closures: Toward an instability-sensitive Reynolds stress model. *Int. J. Heat Fluid Flow* **2015**, *51*, 175–194. [CrossRef]
5. Gerolymos, G.A.; Vallet, I. Wall-normal-free Reynolds-stress closure for three-dimensional compressible separated flows. *AIAA J.* **2001**, *39*, 1833–1842. [CrossRef]
6. Gerolymos, G.A.; Sauret, E.; Vallet, I. Contribution to single-point closure Reynolds-stress modelling of inhomogeneous flow. *Theor. Comput. Fluid Dyn.* **2004**, *17*, 407–431. [CrossRef]
7. Gerolymos, G.A.; Lo, C.; Vallet, I.; Younis, B.A. Term-by-term analysis of near-wall second moment closures. *AIAA J.* **2012**, *50*, 2848–2864. [CrossRef]
8. Shur, M.L.; Spalart, P.R.; Strelets, M.K.; Travin, A.K. A hybrid RANS-LES approach with delayed-DES and wall-modelled LES capabilities. *Int. J. Heat Fluid Flow* **2008**, *29*, 1638–1649. [CrossRef]
9. Devenport, W.J.; Lowe, K.T. Equilibrium and non-equilibrium turbulent boundary layers. *Prog. Aerosp. Sci.* **2022**, *131*, 100807. [CrossRef]
0. Girimaji, S.S. Partially-averaged Navier–Stokes model for turbulence: A Reynolds-averaged Navier–Stokes to direct numerical simulation bridging method. *J. Appl. Mech.* **2006**, *73*, 413–421. [CrossRef]
1. Chaouat, B.; Schiestel, R. A new partially integrated transport model for subgrid-scale stresses and dissipation rate for turbulent developing flows. *Phys. Fluids* **2005**, *17*, 065106. [CrossRef]
2. Menter, F. Stress-blended eddy simulation (SBES)—A new paradigm in hybrid RANS-LES modeling. In Proceedings of the Sixth HRLM Symposium, Strasbourg University, Strasbourg, France, 26–28 September 2016; Available online: https://hrlm6.sciencesconf.org/118745/Menter_Abstract.pdf (accessed on 30 August 2023).
3. Jakirlić, S.; Jovanović, J. On unified boundary conditions for improved predictions of near-wall turbulence. *J. Fluid Mech.* **2010**, *656*, 530–539. [CrossRef]
4. Wilcox, D.C. *Turbulence Modeling for CFD*, 3rd ed.; DCW Industries: La Cañada, CA, USA, 2006; 522p.
5. Lee, M.; Moser, R.D. Direct numerical simulation of turbulent channel flow up to $Re_\tau \approx 5200$. *J. Fluid Mech.* **2015**, *774*, 395–415. [CrossRef]
6. Troshin, A.; Bakhne, S.; Sabelnikov, V. Numerical and physical aspects of large-eddy simulation of turbulent mixing in a helium-air supersonic co-flowing jet. In *Progress in Turbulence IX*; Örlü, R., Talamelli, A., Peinke, J., Oberlack, M., Eds.; Springer: Berlin/Heidelberg, Germany, 2021; pp. 297–302. [CrossRef]
7. Alexander, R. Diagonally Implicit Runge-Kutta Methods for Stiff O.D.E.'s. *SIAM J. Numer. Anal.* **1977**, *14*, 1006–1021. [CrossRef]
8. Jiang, G.-S.; Shu, C.-W. Efficient Implementation of Weighted ENO Schemes. *J. Comput. Phys.* **1996**, *126*, 202–228. [CrossRef]
9. Available online: https://torroja.dmt.upm.es/turbdata/blayers/high_re/profiles/ (accessed on 15 August 2023).
0. Kok, J.C. Resolving the Dependence on Freestream Values for the k-ω Turbulence Model. *AIAA J.* **2000**, *38*, 1292–1295. [CrossRef]
1. Schoenherr, K.E. Resistances of flat surfaces moving through a fluid. *Trans. SNAME* **1932**, *40*, 279–313.
2. Bentaleb, Y.; Lardeau, S.; Leschziner, M.A. Large-eddy simulation of turbulent boundary layer separation from a rounded step. *J. Turbul.* **2012**, *13*, N4. [CrossRef]
3. Available online: https://turbmodels.larc.nasa.gov/Other_LES_Data/curvedstep.html (accessed on 15 August 2023).

44. Balabanov, R.; Usov, L.; Troshin, A.; Vlasenko, V.; Sabelnikov, V. A Differential Subgrid Stress Model and Its Assessment in Larg Eddy Simulations of Non-Premixed Turbulent Combustion. *Appl. Sci.* **2022**, *12*, 8491. [CrossRef]
45. Balabanov, R.; Usov, L.; Nozdrachev, A.; Troshin, A.; Vlasenko, V.; Sabelnikov, V. Assessment of a Differential Subgrid Stres Model for Large-Eddy Simulations of Turbulent Unconfined Swirling Flames. *Fire* **2023**, *6*, 94. [CrossRef]
46. Shur, M.L.; Spalart, P.R.; Strelets, M.K.; Travin, A.K. Synthetic Turbulence Generators for RANS-LES Interfaces in Zona Simulations of Aerodynamic and Aeroacoustic Problems. *Flow Turbul. Combust.* **2014**, *93*, 63–92. [CrossRef]
47. Guseva, E.K.; Garbaruk, A.V.; Strelets, M.K. An automatic hybrid numerical scheme for global RANS-LES approaches. I Proceedings of the International Conference PhysicA.SPb/2016, Saint Petersburg, Russia, 1–3 November 2016. [CrossRef]

energies

MDPI

Article

Numerical Simulation of the Heat Transfer Inside a Shell and Tube Heat Exchanger Considering Different Variations in the Geometric Parameters of the Design

José Estupiñán-Campos [1], William Quitiaquez [1,2,*], César Nieto-Londoño [3] and Patricio Quitiaquez [2,4]

1 Department of Mechanical Engineering, Universidad Politécnica Salesiana, Quito EC170525, Ecuador; jestupinanc@est.ups.edu.ec
2 Productivity and Industrial Simulation Research Group (GIIPSI), Department of Master's Degree in Production and Industrial Operations Engineering, Universidad Politécnica Salesiana, Quito EC170525, Ecuador; rquitiaquez@ups.edu.ec
3 Escuela de Ingenierías, Universidad Pontificia Bolivariana, Medellín CO050031, Colombia; cesar.nieto@upb.edu.co
4 Department of Mechatronics Engineering, Universidad Politécnica Salesiana, Quito EC170525, Ecuador
* Correspondence: wquitiaquez@ups.edu.ec

Abstract: The present study aims to analyze the heat transfer variations in different models of shell and tube heat exchangers considering geometric variations in the baffle angles and in the tube's profiles. Each baffle configuration and geometric variation in the profiles were tested under different mass flow rates (0.25, 0.5, 0.75, and 1 $kg \cdot s^{-1}$) in the shell to study the heat transfer improvement. The models were simulated using a CFD simulation software ANSYS Fluent including an experimental geometry which was used to validate the simulation process. The experiment results are in good agreement with the CFD results. The analysis of the results shows that an angle of 60° in the baffles generated the highest heat flow (more than 40 kW) with an inclination to the cold flow inlet and a mixed distribution considering a mass flow rate of 1 $kg \cdot s^{-1}$ in the shell. In addition, the horizontal elliptic profile achieved a heat flow higher than 29 kW with a mass flow rate of 0.5 $kg \cdot s^{-1}$ in the shell.

Keywords: shell and tube heat exchangers; heat transfer; baffle; profiles; CFD

Citation: Estupiñán-Campos, J.; Quitiaquez, W.; Nieto-Londoño, C.; Quitiaquez, P. Numerical Simulation of the Heat Transfer Inside a Shell and Tube Heat Exchanger Considering Different Variations in the Geometric Parameters of the Design. *Energies* **2024**, *17*, 691. https://doi.org/10.3390/en17030691

Academic Editor: Artur Bartosik

Received: 31 December 2023
Revised: 18 January 2024
Accepted: 24 January 2024
Published: 31 January 2024

1. Introduction

Currently, heat exchange processes between fluids have gained great notoriety in industries such as electronics, aerospace, chemical, refrigeration, and in various modern engineering applications due to the high energy consumption they can generate. For this reason, the efficient use of energy is an alternative to reduce the scarcity of this resource worldwide, and it can be achieved using any type of energy-saving technique [1]. These techniques can involve algorithms to perform the calculation, and the comparison of thermal machine parameters, as presented by Mediaceja et al. [2].

In the case of shell and tube heat exchangers, efficient energy management can be carried out by considering more appropriate values for various design parameters, satisfying the energy needs and producing the best thermo-hydraulic behavior according to the objective function [3,4].

As a result of the numerical analysis, it is possible to obtain a large number of variables of interest to relate the behavior of the working fluids and thermal energy transfer. In this line of research, Ravikumar and Raj [5] present the effects of parameters such as temperature variation and pressure drop to establish a relationship with the heat flow and efficiency generated by changes in a shell and tube heat exchanger.

Due to the importance that the optimization of heat exchangers represents in this field, authors such as Pereira et al. [6] and Yang et al. [7] continue with the trend to redesign these systems with programming aimed at guiding, with certain guidelines, the development of models with greater efficiency and an adequate verification process. In this context, the work of Li et al. [8] stands out, where a change in the traditional design of the tube with circular profile for a tube with a twisted section was proposed, increasing the performance of the heat flow between 24 and 39%.

In this field, numerical analysis is used to determine the transport of thermal energy in a shell and tube heat exchanger (TC exchanger). Using this method, it is possible to establish the influence of the thermodynamic properties of various working fluids, as presented by Bahiraei et al. [9], where a three-dimensional model of the exchanger was generated and the passage of nanofluids was simulated to determine the temperature variation, generating a maximum increase of 13 °C when considering nanoparticles with different geometric shapes.

In this context, the study of improvements in the efficiency of heat exchange devices has been presented in research works such as the one developed by Abbasian and Uosofvand [10]. This work studies the variation in heat transfer between fluids by incorporating geometric modifications in the profile of the tube and baffle, using computational fluid dynamics (CFD) analysis, with an increase of 20.9% using an optimized tube with an elliptical angled arrangement.

Ren et al. [11] performed a numerical and experimental analysis of a thermal energy storage device using a shell and tube heat exchanger system, implementing changes in the tube layout to determine the variation in heat exchanger efficiency due to the influence of parameters such as tube diameter, tube length, flow rate, and tube layout, among others. In this way, they generated a sensitivity index of 0, 0.215, 0.2, and 0.35, respectively, quantifying the effects of the input variables (geometric changes) on the response (heat transfer) of the interaction of their effects.

CFD analysis as a method to study the transport of thermal energy in the working fluids in CT heat exchangers is widely used to determine pressure drops, changes in the heat transfer coefficient, and Nusselt and Reynolds numbers, as shown in the study by Quitiaquez et al. [12] and in the research developed by Li et al. [13]. In the latter, a numerical analysis was performed together with an experimental study varying the baffles of the device, to generate a maximum pressure drop inside the shell between 700 and 6500 Pa approximately, with a verification process showing a variation of 9.10 and 5.32%.

In this field of study, it is possible to relate the increase in heat flux to the improvement in heat transfer coefficient, device effectiveness, and the number of units transferred (NTU). In the study by Said et al. [14], the authors elaborated a numerical analysis of a TC exchanger with three geometric changes in the baffles, obtaining as maximum values and effectiveness of 0.37, an NTU of 0.625, and a maximum global heat transfer coefficient of 315 $W \cdot m^{-2} \cdot K^{-1}$, among others. The results showed an average variation between the global heat transfer coefficient and pressure drop of 8.44 and 8.77%, respectively, when comparing the experimental values with those obtained in the numerical analysis.

In order to improve the thermal effectiveness of the energy transfer devices, Biçer et al. [15] proposed a change in the design of the baffles to reduce the pressure drop on the shell walls while maintaining the thermal effectiveness of the heat exchanger. The generated model caused a pressure drop of less than 49%, with an increase in the tube surface temperature of 7%, using computational fluid dynamics analysis, compared with experimental results which showed a variation of 7.3%.

Nagib et al. [16] employed CFD simulations to evaluate a cross-flow heat exchanger incorporating partition plates to reduce the pressure drop by 20% and increase the Nusselt number, leading to an increase in the heat transfer coefficient of 7% considering flow regimes of 5500 and 14,500. To validate the outlet temperature of heat exchangers with various mass flow rates, Vivekanandan et al. [17] performed a CFD analysis, with experimental results showing a similar trend to those obtained in the numerical analysis, with an error of 15% and an overall heat transfer coefficient of 449.2 $W \cdot m^{-2} \cdot K^{-1}$. A similar development can be seen in works such as the one developed by Krishna et al. [18], when comparing the results obtained by numerical analysis with experimental results, with a variation of 5.32%, studying the temperature difference along the heat exchanger device with a mass flow rate between 0.05 and 0.25 kg·s^{-1}.

A similar validation was developed in the research of Aliaga et al. [19], showing a difference of less than 6.2% between the experimental results and those obtained by CFD in an optimized heat exchanger prototype to achieve an efficiency of 95%. A study developed on changes in geometrical design parameters in shell and tube heat exchangers with baffle plates connected to a central tube, showed results with a variation of 15% between the experimental model and the numerically analyzed one, with a maximum efficiency of 0.775 with flow rates between 2 and 6 lpm in the tube, together with values between 2 and 8 lpm in the shell of the device [20].

The present research work aims to perform a CFD analysis in ANSYS 2023 R1 of the behavior of the fluids in a heat exchanger that presents geometric design variations to determine the arrangement that generates a greater temperature variation between the inlet and outlet of the working fluids in the device, resulting in an increase in the heat flow. In the introduction, a review of the research that has been carried out in this field is presented, and the CFD analysis process is subsequently addressed in the materials and methods section. Finally, the results obtained for each model analyzed are shown together with the most relevant conclusions that they generated.

2. Materials and Methods

In the present study, a comparative analysis was performed with the temperature variation in the working fluids at the ends of the shell and tube heat exchanger. This study considered a base case in which geometrical changes in the ducts and baffles were incorporated. The temperature variations were determined by means of a CFD analysis process that started by creating the designs within a modeling software, then the meshing was generated, which in turn involved a mesh independence analysis and quality verification. Subsequently, the simulation parameters were entered, the mathematical calculation models were selected, and the results were obtained and compared with the proposed base case, as shown in Figure 1.

2.1. Parameters for Analysis

In order to develop the analysis of the thermal energy transfer between the working fluids in the various models of the proposed heat exchanger, several operability factors were considered, as presented in Table 1, to establish values for comparison with a base case.

Table 1. Exchanger operating conditions.

	Input Parameters Mass Flow Rate [kg·s^{-1}]	Temperature Inlet [°C]
Cold fluid	0.25, 0.5, 0.75, 1	20
Hot fluid	0.3	70

Similarly, to perform the CFD analysis, the thermodynamic properties of the working fluid (water) and the material of the heat exchange devices were considered, as shown in Table 2.

Table 2. Exchanger operating conditions 2.

		Water		
Temperature [°C]	Density [kg·m^{-3}]	Thermal Conductivity [W·m^{-1}·K^{-1}]	Specific Heat Capacity [J·kg^{-1}·K^{-1}]	Dynamic Viscosity [kg·m^{-1}·s^{-1}]
20	998.2	0.5861	4183	0.001002
30	995.7	0.603	4183	0.0007977
40	992.2	0.6178	4182	0.0006533

Table 2. *Cont.*

Water				
Temperature [°C]	**Density [kg·m^{-3}]**	**Thermal Conductivity [W·m^{-1}·K^{-1}]**	**Specific Heat Capacity [J·kg^{-1}·K^{-1}]**	**Dynamic Viscosity [kg·m^{-1}·s^{-1}]**
50	988	0.6305	4181	0.0005471
60	983.2	0.641	4183	0.0004666
70	977.8	0.6495	4187	0.0004041

Copper		
Density [kg·m^{-3}]	**Thermal Conductivity [W·m^{-1}·K^{-1}]**	**Specific Heat Capacity [J·kg^{-1}·K^{-1}]**
8978	387.6	381

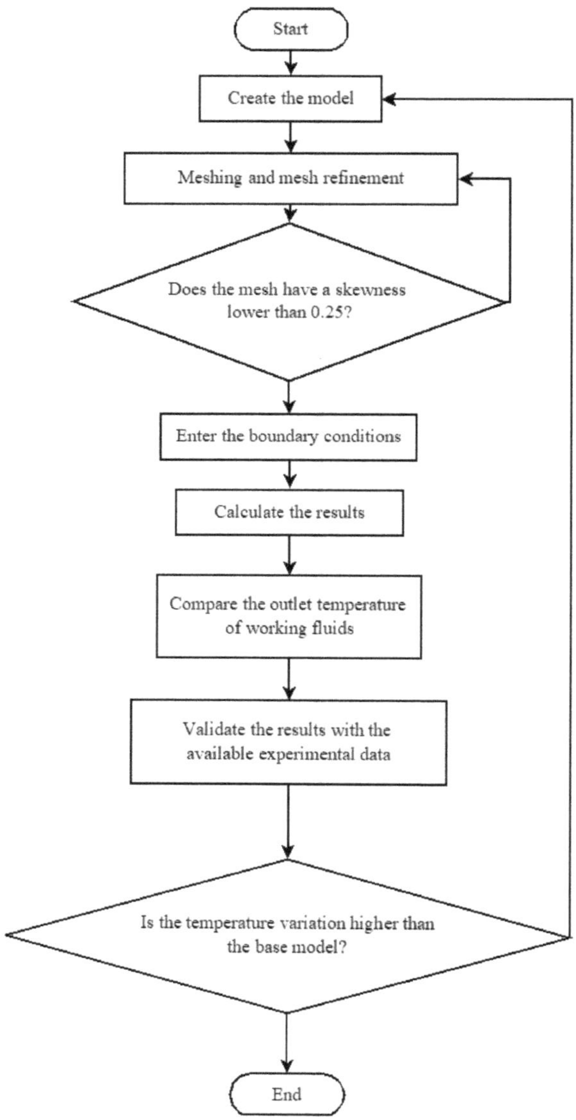

Figure 1. Flow chart of the simulation process.

2.2. CFD Analysis

Computational fluid dynamics analysis was used to determine the heat transfer in the various models of shell and tube heat exchangers proposed, which presented results such as the temperature at the outlets of the device and the surface temperature in the ducts.

2.3. Meshing

This research performed a numerical analysis of several models of shell and tube heat exchangers, starting from a base model that was selected through a comparative study with initial models that presented baffle walls with different heights. The mesh quality in each model was measured by skewness and is presented in Table 3.

Table 3. Exchanger operating conditions 3.

Model	Number of Elements Average	Skewness Average
Exchanger for validation	3,509,554	0.2899
Initial model 1 (baffles with 25% of the height)	467,003	0.238
Initial model 2 (baffles with 50% of the height)	471,875	0.2405
Starting model 3 (baffles with 75% of the height)	448,733	0.2417
Models with inclined baffle (30°)	471,552	0.2404
Models with inclined baffle (45°)	479,986	0.2406
Models with inclined baffle (60°)	500,137	0.2422
Inner tube models with arches and inclined baffles (60°)	483,386	0.2413
Elliptical (vertical) inner tube models with inclined baffles (60°)	497,216	0.2735
Elliptical (horizontal) inner tube models with inclined baffles (60°)	492,203	0.2767
Octagonal inner tube models with inclined baffles (60°)	495,627	0.253

In order to determine the convergence (average output temperature), the analysis of the base model was performed with different meshes that increased the number of elements, varying the mesh quality. Thus, it was shown that when exceeding 350,000 elements, the output temperature converges to 31.9 °C (Figure 2), validating the mesh quality used.

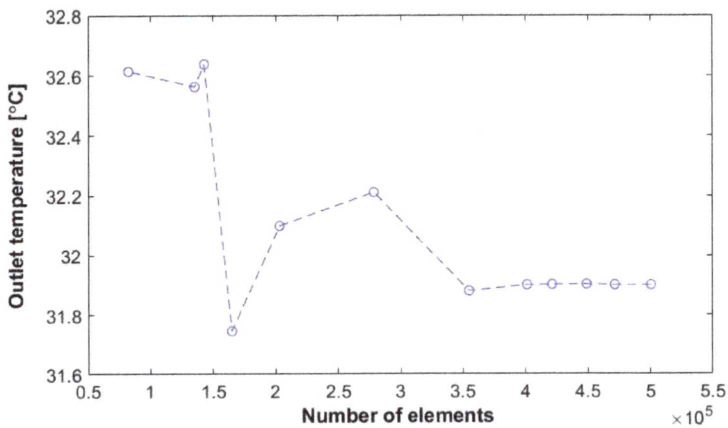

Figure 2. Convergence of the mesh in the exchanger (initial model 3).

The meshing of the analyzed models presented tetrahedral elements that adapted to the curvatures of the ducts, as in the case of the initial model 3 shown in Figure 3.

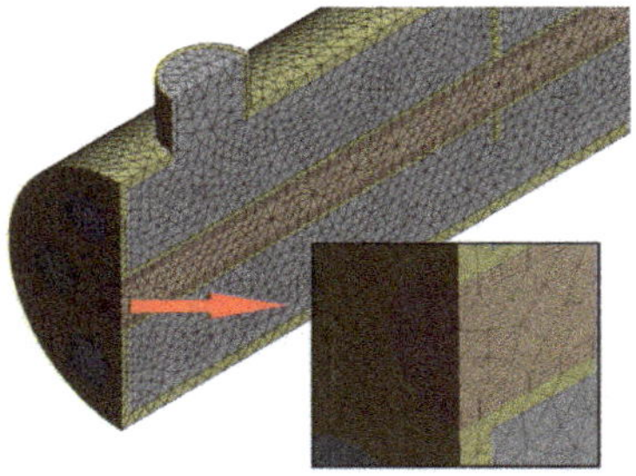

Figure 3. Initial model 3 meshing.

2.4. Governing Equations

Navier–Stokes equations, which involve the conservation of continuity (mass), the quantity of motion, and energy, as seen in Equations (1), (2), and (3), respectively, were used to determine the behavior of the working fluids during the heat exchange [21,22]. Mathematical models can be selected according to the characteristics of the study to calculate the influence of the conditions in the heat transfer process through walls, as it was presented in the work of Chihab et al. [23].

$$\frac{\partial \rho}{\partial t} + \nabla \cdot \left(\rho \vec{V}\right) = S_m \tag{1}$$

$$\frac{\partial}{\partial t}\left(\rho \vec{V}\right) + \nabla \cdot \left(\rho \vec{V}\vec{V}\right) = -\nabla P + \nabla \cdot \left(\overline{\overline{\tau}}\right) + \rho \vec{g} + \vec{F} \tag{2}$$

$$\frac{\partial}{\partial t}(\rho E) + \nabla \cdot \left[\vec{V} \cdot (\rho E + P)\right] = -\nabla \cdot \left(\sum_j h_j J_j\right) + S_h \tag{3}$$

In order to study the behavior of the fluids in a turbulent state inside curved ducts, studies like the investigation of Wang and Suen [24] refer to turbulence models (the standard k-ε model, the RNG k-ε model with/without swirl option, and the standard k-ω model, among others) that produce an adequate similarity between simulated and experimental results. Considering the flow regime of the fluids in the present analysis, the turbulence model chosen was the standard k-ε turbulence model (Equations (4) and (5)), which allows us to obtain results with high accuracy according to the experimental data [25].

$$\frac{\partial}{\partial t}(\rho k) + \frac{\partial}{\partial x_i}(\rho k u_i) = \frac{\partial}{\partial x_j}\left[\left(\mu + \frac{\mu_t}{\sigma_k}\right)\frac{\partial k}{\partial x_j}\right] + G_k + G_b - \rho\varepsilon - Y_M + S_k \tag{4}$$

$$\frac{\partial}{\partial t}(\rho \varepsilon) + \frac{\partial}{\partial x_i}(\rho \varepsilon u_i) = \frac{\partial}{\partial x_j}\left[\left(\mu + \frac{\mu_t}{\sigma_\varepsilon}\right)\frac{\partial \varepsilon}{\partial x_j}\right] + C_{1\varepsilon}\frac{\varepsilon}{k}(G_k + C_{3\varepsilon}G_b) - C_{2\varepsilon}\rho\frac{\varepsilon^2}{k} + S_\varepsilon \tag{5}$$

As part of the numerical analysis for the heat exchanger model, the pressure–velocity coupling scheme and the spatial discretization are presented in Table 4.

Table 4. Pressure–velocity coupling and spatial discretization.

Pressure–Velocity Coupling	
Schematic	**Coupling**
Spatial Discretization	
Pressure	Second order
Momentum	Second order
Energy	Second order
Kinetic energy turbulence	Second order
Turbulence dissipation rate	Second order

3. Results

The results obtained in the present study involve similar operating conditions in shell and tube heat exchanger devices that present geometrical variations from a base model to establish the changes in heat flow between the working fluids used.

3.1. Validation

To perform the validation of the computational fluid dynamics analysis in the heat exchange device, it is possible to compare the values of certain parameters as results obtained in the simulation with the experimental values in factors such as outlet temperature, effectiveness, or heat flow rate, with variations of the results between 5.32 and 15%, as presented in the research of Vivekanandan et al. [20] and Elgendy et al. [26], respectively.

For this reason, when comparing the experimental results obtained in the study of a shell and tube heat exchanger with different baffles developed by Biçer et al. [15] with a similar model simulated by the authors, it can be appreciated that an average error of 3.15% was obtained, as observed in the trend of Figure 4, validating the numerical analysis process used.

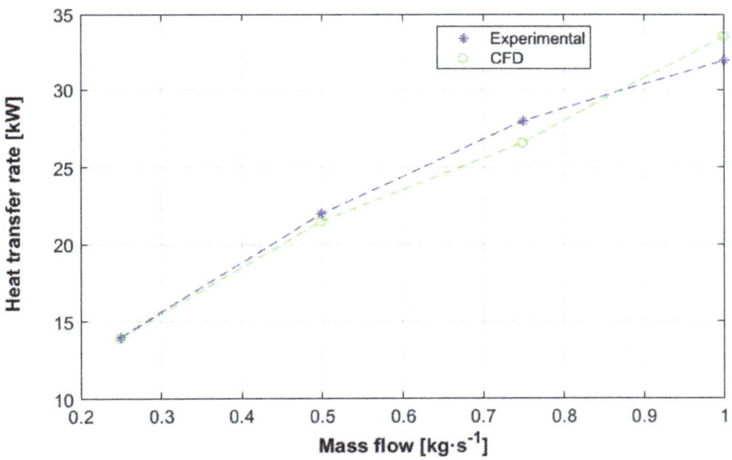

Figure 4. Heat transfer rates at different mass flow rates.

3.2. Base Model

A base model to establish an initial proposal was made for a shell and inner tubes, similar to the concept of the geometry used for the validation, as shown in the research work of Biçer et al. [15], with a reduced number of tubes and dimensions. The model's dimensions are presented in Table 5.

Table 5. Heat exchanger dimensions.

Length [m]	Internal Diameter of the Pipes [m]	External Diameter of the Pipes [m]	Internal Diameter of the Shell [m]	External Diameter of the Shell [m]	Distance between Tubes [m]	Internal Diameter of External Pipe [m]	Distance between Baffles [m]
0.35	0.008	0.010	0.038	0.042	0.012	0.012	0.07

In the proposed device, three initial models were considered where the baffle height was varied (25, 50, and 75%) concerning the inner diameter of the shell, in order to compare the variation in heat flux in each case in relation to the change in the average outlet temperature, as shown in Figure 5.

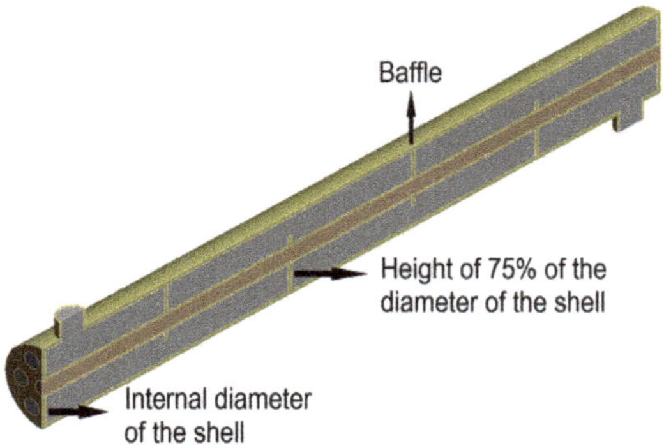

Figure 5. Initial model of the heat exchanger with 75% baffle height.

During the behavior analysis of the fluids in the initial models of the heat exchange with baffles with 25% of the inner diameter of the shell (Model 1), 50% (Model 2), an 75% (Model 3), it was determined that the highest heat flow (related to the outlet tempera ture) occurred in Model 3, as can be appreciated in Figure 5 (part A), with a temperatur change along the heat transfer device seen in Part B of Figure 6. For this reason, this mode was selected as the base model.

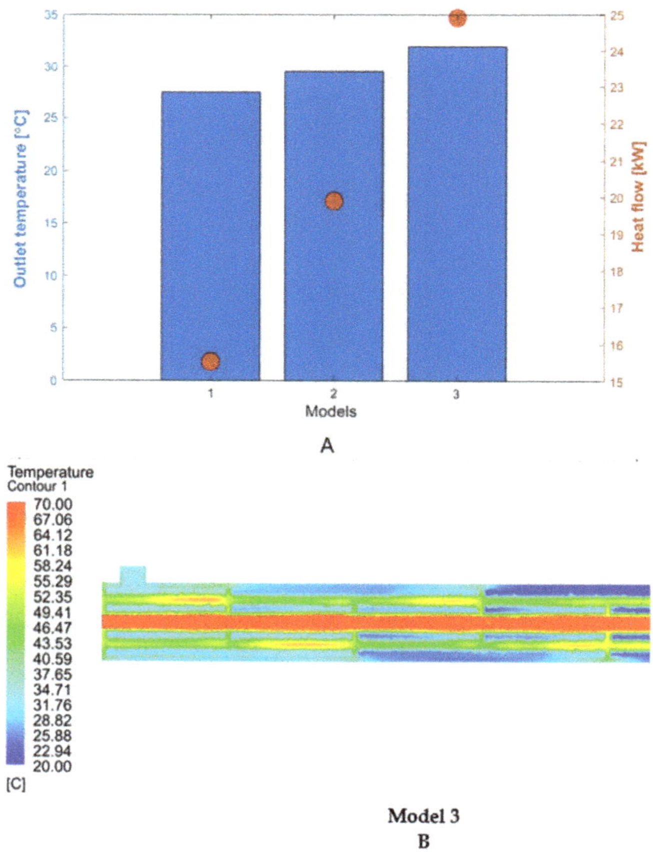

Figure 6. Change of temperature versus heat flow in the three models (**A**), and temperature contou in the heat exchanger section of Model 3 (**B**).

3.3. Changes in Inclination

From the base model, as a proposal to improve heat transfer, the considered geometries presented inclinations in the baffle walls varying their orientation for the inlet and outlet of the cold fluid, as was observed in several investigations [9,27].

3.3.1. Inclination of 30°

The behavior of the fluids in the heat exchanger was simulated by incorporating an inclination in the baffles of 30° concerning the vertical axis, where one end was directed toward the cold fluid outlet (A), the cold fluid inlet (B), and alternately (C), as shown in Figure 7.

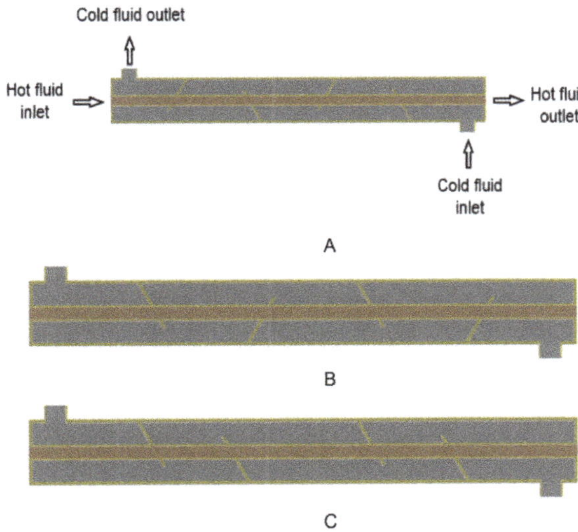

Figure 7. Heat exchangers with a 30° inclination in the baffles.

The 30° inclination of the baffle walls generated changes in the heat flux in the heat transfer device; when the results between the inclination cases considering different mass flow rates of the cold fluid were compared, it was determined that Case B results in a higher heat flux (39.3 kW) for a mass flow of 1 kg·s^{-1}, followed by the other cases and the base model, as shown in Figure 8. In addition, it can be seen that Case A generates the lowest heat flux, with a value of 15.44 kW for a mass flow of 0.25 kg·s^{-1}.

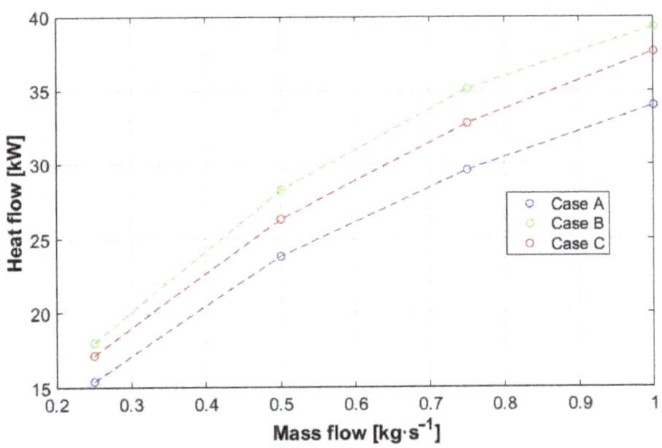

Figure 8. Heat flow for baffles with a 30° inclination in cases A, B, and C.

3.3.2. Inclination of 45°

Similarly, the behavior of the working fluid in the device was analyzed considering a 45° inclination of the baffles concerning the vertical axis, with one end oriented towards the cold fluid outlet (A), the cold fluid inlet (B), and alternately (C), as presented in Figure 9.

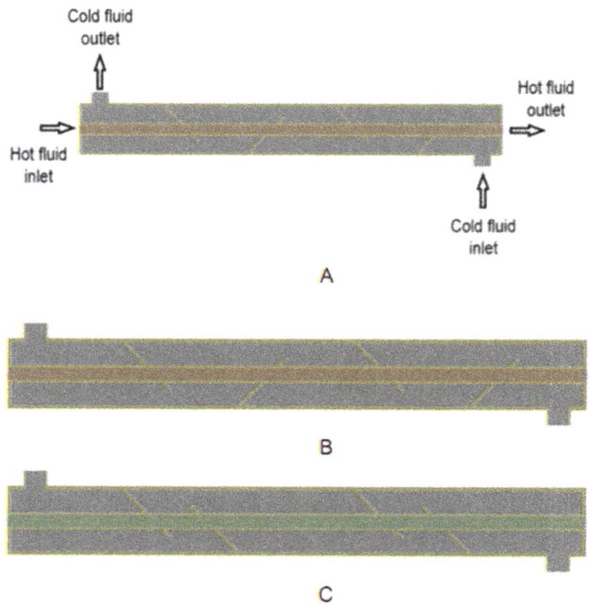

Figure 9. Heat exchangers with a 45° inclination in the baffles.

The study of the change in heat flow between the working fluids of the exchanger for an inclination of 45° for the different cases analyzed with various mass flows in the cold fluid, showed that Case B, when it was simulated with 1 kg·s^{-1}, presented the highest heat transfer (40.03 kW) compared to the other cases and the base model, as shown in Figure 10. In this context, when a similar case was studied with the same mass flow, Case A generated the lowest heat flux (32.78 kW).

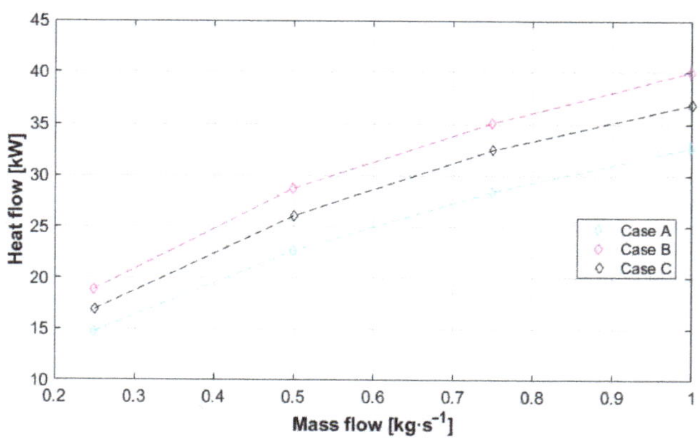

Figure 10. Heat flow for baffles with a 45° inclination in cases A, B, and C.

3.3.3. Inclination of 60°

Additionally, the analysis of the change in heat flow in the heat exchanger was performed by incorporating 60° inclinations in the baffles for the vertical axis along with end orientations towards the cold fluid outlet (A), cold fluid inlet (B), and alternately (C), as shown in Figure 11.

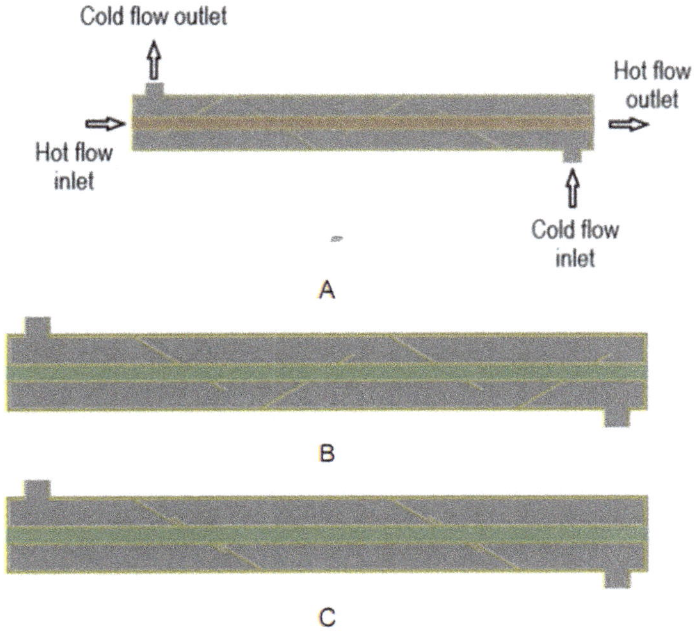

Figure 11. Heat exchangers with a 60° inclination in the baffles.

As a result of the simulation of the proposed cases, it was determined that both Cases B and C presented a notable increase in the heat flow between the working fluids, which was higher than that obtained with Case A and with the base model, as shown in Figure 12. This marked difference between the cases can be confirmed by comparing the heat flow of Case B (40.52 kW) and Case C (40.55 kW), with increases of 28.68 and 28.77%, respectively, considering the higher value of the mass flow analyzed.

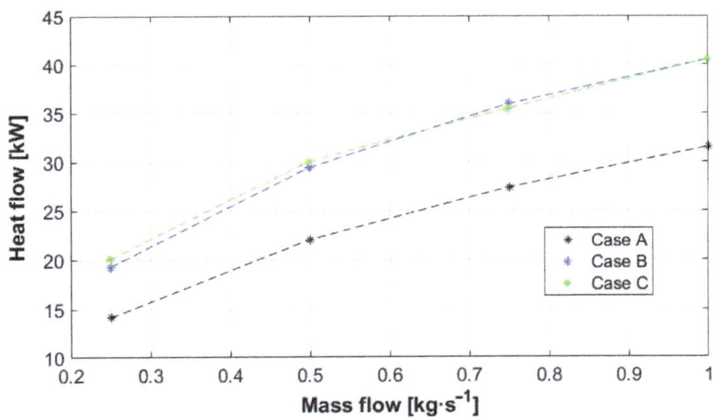

Figure 12. Heat flow for baffles with a 60° inclination in cases A, B, and C.

3.4. Changes in the Profile

Since the exchangers that incorporated baffles with an inclination of 60° with th ends oriented towards the cold fluid inlet and in an alternating manner presented a highe heat transfer, the models that were originally proposed with central tubes with circula profiles were used to analyze the heat flow with different profiles, considering a mass flo in the shell of 0.5 kg·s^{-1}. For this study, a concept similar to the research developed b Olabi et al. [28] and Hadibafekr et al. [29] was considered, where the temperature chang in the working fluid was studied with proposed cross sections incorporating variou curvatures in the heat exchanger ducts.

3.4.1. Ellipse Profiles

An analysis of thermal energy flow in heat exchanger devices with profiles incorpo rating elliptical curves has been previously performed, as presented in the research c Andrade et al. [30]. However, the influence of this type of profile can vary according to th design of the exchanger analyzed, which is why, to determine the influence of the chang in the heat flux present in the proposed device, a vertically and horizontally oriente ellipse-shaped profile was considered, as shown in Figure 13.

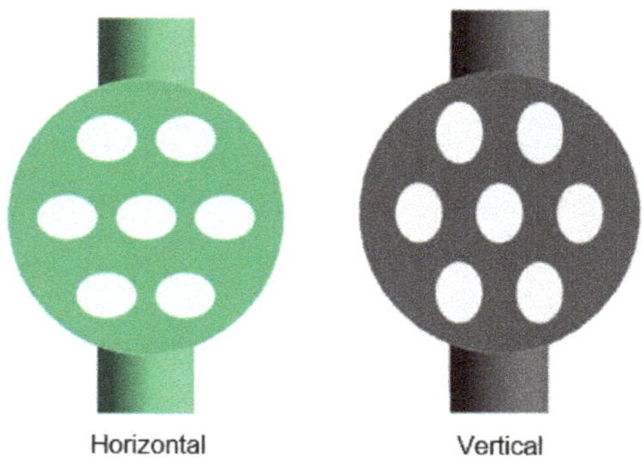

Figure 13. Internal tubes with elliptical profiles.

The analysis developed in the exchanger with horizontal elliptical profiles resulte in a higher heat flow in the device with inclined baffles and ends oriented towards th cold fluid outlet (Case B), compared to the baffles with alternating oriented ends (Case C with a heat flow of 29.02 and 27.14 kW, respectively, as shown in Figure 14A. Similarl when obtaining the results of the simulations of the heat exchangers with vertical ellipticа profiles, it was determined that Case B (25.79 kW) generated a greater increase in heat flov when compared to Case C (24.15 kW), as shown in Figure 14B.

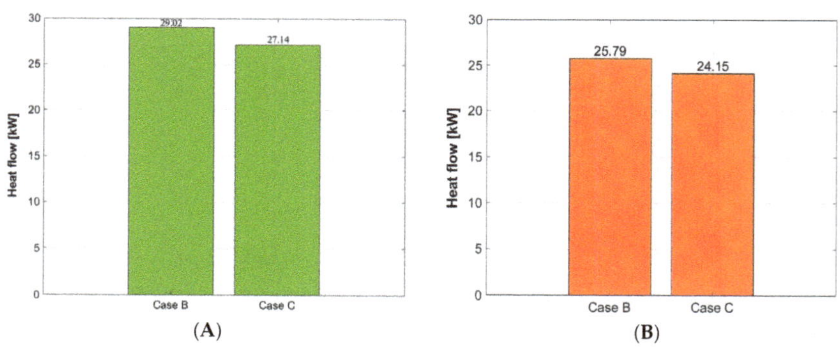

Figure 14. Heat flow considering horizontal (**A**) and vertical (**B**) elliptical profiles.

3.4.2. Profiles with Arches

As can be seen in studies such as those developed by Li et al. [8] and Abbasian et al. [10], the use of curvatures in the heat exchanger duct profiles generates an increase in heat flow. For this reason, as another alternative proposal to the circular profile in the outer zone, the behavior of the heat exchanger device was analyzed by incorporating profiles with four arches in the internal tubes in the outer part and circular inner profiles, as shown in Figure 15.

Figure 15. Internal tubes with arched profiles.

Similar to the study with elliptical profiles, simulations of the heat exchangers with arched profiles were carried out for the cases of baffles inclined at 60° with the ends oriented toward the cold fluid outlet (Case B) and alternately (Case C). When comparing the results of these cases, the presence of a higher heat flow in Case B (28.29 kW) was evident, as shown in Figure 16.

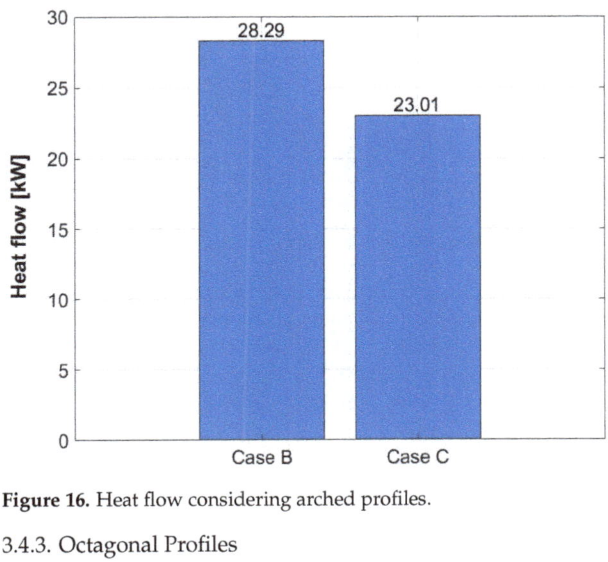

Figure 16. Heat flow considering arched profiles.

3.4.3. Octagonal Profiles

As a final change proposal for the heat exchanger internal tubes profiles and to show a higher heat flow in tubes with curved profiles, an octagonal profile was considered in the external area and a circular profile in the internal region of the tubes, as shown in Figure 17.

Figure 17. Inner tubes with octagonal profiles.

These profiles were analyzed under similar conditions as the previous cases, maintaining the inclination of the baffles and the orientation of their ends (Case B and C), generating, as a result, a higher heat flux for Case B (27.96 kW) than Case C (27.57 kW), with a reduced difference when compared to the previously simulated profiles, as presented in Figure 18.

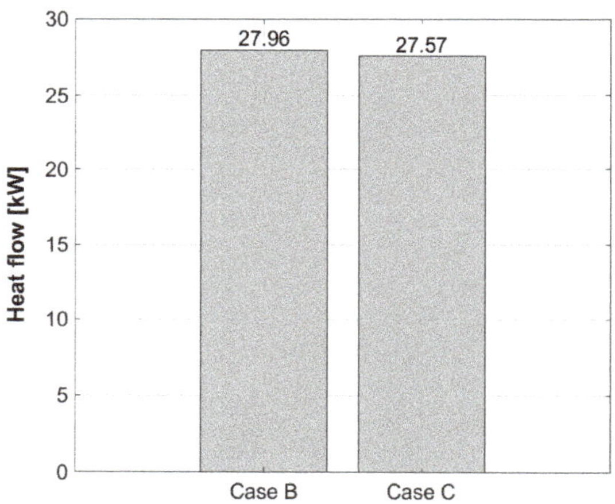

Figure 18. Heat flow considering octagonal profiles.

There was an increase in heat transfer in the fluids when varying the size of the baffle walls along with their inclination. This phenomenon may be associated with the directionality of the cold fluid inside the device by presenting a more extensive trajectory with greater contact with the surface of the hot fluid ducts [31], and with the possible presence of a greater amount of turbulence vortices in the areas adjacent to the baffle walls.

The change in the surface sections of the hot fluid ducts showed the influence on the orientation of the geometric shapes for the path of the fluid to be heated along with a change in the heat transfer that occurred by incorporating flat surfaces and a greater surface area of contact according to the geometric profile of the ducts.

4. Discussion

There was an increase in heat transfer in the fluids as the size of the baffle walls varied along with their inclination. This phenomenon may be associated with the directionality of the cold fluid inside the device as it presents a longer path with greater contact with the surface of the hot fluid ducts and the presence of turbulence vortices in the areas adjacent to the deflector walls.

The change in the surface sections of the hot fluid ducts showed the influence on the orientation of the geometric shapes with respect to the fluid path, along with a change in heat transfer that occurred when incorporating flat surfaces and a greater surface area of contact according to the geometric profile of the ducts.

The changes in the ducts' geometric shape demonstrated that this fact could influence the heat flux of the device which is directly proportional to the energy efficiency. A deeper analysis can determine a geometric shape able to generate a significant increase in the heat flux.

5. Conclusions

By developing the computational fluid dynamics analysis of various heat exchanger models and modifying geometrical parameters in their design, the following conclusions were obtained:

1. The influence of the increase in the height of the baffle walls concerning the internal diameter of the shell was shown with the heat exchange in the device where average cold fluid outlet temperatures of 27.42, 29.51, and 31.91 °C were obtained for baffles with 25, 50, and 75% of the height, respectively.
2. The inclination of the deflector walls that generated the greatest temperature difference in the device was 60° for the horizontal axis, with heat fluxes of 29.44 and 30.07 kW for the ends oriented towards the cold fluid outlet and alternating, respectively. This demonstrates the positive influence of the baffles' inclination in a higher heat transfer process.
3. The heat flux in a heat exchanger is directly proportional to the energy efficiency, considering that the lowest heat flux (24.15 kW) was generated by the combination of profiles with an elliptical shape vertically oriented and baffle walls with an inclination of 60°. This means that the combination cannot be recommended.
4. The duct's geometric shape that generated the greatest temperature difference between the inlets and outlets of the device, expressed by an increase in heat flow (29.02 kW), was the one that incorporated baffle walls with an inclination of 60° with the ends oriented toward the cold fluid outlet with a horizontal elliptical profile on the inner tubes.
5. Comparing the highest heat fluxes obtained by the changes in the baffles' inclination (29.44–30.07 kW) with the best heat flux presented in the combination of baffle inclination and different duct shape (29.02 kW), the baffle inclination emerges as the strongest geometric factor in the energy efficiency of the heat exchanger.

Author Contributions: J.E.-C. and W.Q.: conceptualization, methodology, validation, and writing—review and editing. W.Q. and C.N.-L.: conceptualization, methodology, software, and writing—original draft. P.Q.: data curation and formal analysis. W.Q. and C.N.-L.: supervision, and validation. W.Q. and J.E.-C.: writing—review and editing. All authors have read and agreed to the published version of the manuscript.

Funding: This work was financially funded by the Engineering, Productivity, and Industrial Simulation Research Group (GIIPSI) of the Universidad Politécnica Salesiana and Salesian Institutions of Higher Education (IUS).

Data Availability Statement: The data presented in this study are available on request from the corresponding author. The data are not publicly available due to future research.

Acknowledgments: The Universidad Politécnica Salesiana and Universidad Pontificia Bolivariana supported this work via the Engineering, Productivity, and Industrial Simulation Research Group (GIIPSI).

Conflicts of Interest: The authors declare no conflicts of interest.

Nomenclature

Symbol	Description
ρ	Density; [kg·m^{-3}]
t	Time; [s]
V	Velocity; [m·s^{-1}]
S_m	Mass source; [kg·m^{-3}·s^{-1}]
P	Pressure; [Pa]
μ	Dynamic viscosity; [Pa·s]
F	Force; [N]
E	Total energy; [J]
∇	Grad operator
S_h	Defined energy source; [W·m^{-3}]
J	Mass flow; diffusion flow; [kg·m^{-2}·s^{-1}]
k	Kynetic energy turbulence; [m^2·s^{-2}]
u	Velocity magnitude; [m·s^{-1}]
x	Axial coordinate
μ_t	Dynamic turbulence viscosity; [kg·m^{-1}·s^{-1}]
σ_k	Prandtl turbulence number
G_k	Kinetic turbulence energy generation
G_b	Flotability kinetic turbulence energy generation
ϵ	Disipation rate; [m^2·s^{-3}]
S_k	Kinetic turbulence source; [kg·m^{-1}·s^{-3}]

References

1. Gupta, S.K.; Verma, H.; Yadav, N. A review on recent development of nanofluid utilization in shell & tube heat exchanger fo saving of energy. *Mater. Today Proc.* **2021**, *54*, 579–589. [CrossRef]
2. Mediaceja, Y.R.; Alfonso, H.L.L.; Sánchez-Escalona, A.A.; Camaraza-Medina, Y.; Corrales, M.F.S.; Urgelles, M.L.; Leyva, E.C Análisis termoenergético del sistema de generación de vapor de una central térmica de 49 MW. *Enfoque UTE* **2020**, *11*, 87–10 [CrossRef]
3. Lara-Montaño, O.D.; Gómez-Castro, F.I.; Gutiérrez-Antonio, C. Comparison of the performance of different metaheuristi methods for the optimization of shell-and-tube heat exchangers. *Comput. Chem. Eng.* **2021**, *152*, 107403. [CrossRef]
4. Ünverdi, M. Prediction of heat transfer coefficient and friction factor of mini channel shell and tube heat exchanger usin numerical analysis and experimental validation. *Int. J. Therm. Sci.* **2022**, *171*, 107182. [CrossRef]
5. Ravikumar, M.; Raj, Y.A. Investigation of fin profile on the performance of the shell and tube heat exchanger. *Mater. Today Pro* **2021**, *45*, 7910–7916. [CrossRef]
6. Pereira, I.P.; Bagajewicz, M.J.; Costa, A.L. Global optimization of the design of horizontal shell and tube condensers. *Chem. Eng Sci.* **2021**, *236*, 116474. [CrossRef]
7. Yang, Z.; Ma, Y.; Zhang, N.; Smith, R. Optimization of Shell and Tube Heat Exchangers Sizing with Heat Transfer Enhancemen *Comput. Aided Chem. Eng.* **2020**, *48*, 937–942. [CrossRef]
8. Li, X.; Wang, L.; Feng, R.; Wang, Z.; Liu, S.; Zhu, D. Study on shell side heat transport enhancement of double tube heat exchanger by twisted oval tubes. *Int. Commun. Heat Mass Transf.* **2021**, *124*, 105273. [CrossRef]
9. Bahiraei, M.; Naseri, M.; Monavari, A. A CFD study on thermohydraulic characteristics of a nanofluid in a shell-and-tube hea exchanger fitted with new unilateral ladder type helical baffles. *Int. Commun. Heat Mass Transf.* **2021**, *124*, 105248. [CrossRef]
10. Arani, A.A.A.; Uosofvand, H. Double-pass shell-and-tube heat exchanger performance enhancement with new combined baffl and elliptical tube bundle arrangement. *Int. J. Therm. Sci.* **2021**, *167*, 106999. [CrossRef]
11. Ren, H.; He, M.; Lin, W.; Yang, L.; Li, W.; Ma, Z. Performance investigation and sensitivity analysis of shell-and-tube phas change material thermal energy storage. *J. Energy Storage* **2020**, *33*, 102040. [CrossRef]
12. Quitiaquez, W.; Estupinan-Campos, J.; Isaza-Roldan, C.A.; Nieto-Londono, C.; Quitiaquez, P.; Toapanta-Ramos, F. Numerica simulation of a collector/evaporator for direct-expansion solar-assisted heat pump. In Proceedings of the 2020 IEEE ANDESCON Quito, Ecuador, 13–16 October 2020; pp. 1–6.
13. Li, N.; Chen, J.; Cheng, T.; Klemeš, J.J.; Varbanov, P.S.; Wang, Q.; Yang, W.; Liu, X.; Zeng, M. Analysing thermal-hydrauli performance and energy efficiency of shell-and-tube heat exchangers with longitudinal flow based on experiment and numerica simulation. *Energy* **2020**, *202*, 117757. [CrossRef]
14. El-Said, E.M.; Elsheikh, A.H.; El-Tahan, H.R. Effect of curved segmental baffle on a shell and tube heat exchanger thermohydrauli performance: Numerical investigation. *Int. J. Therm. Sci.* **2021**, *165*, 106922. [CrossRef]
15. Biçer, N.; Engin, T.; Yaşar, H.; Büyükkaya, E.; Aydın, A.; Topuz, A. Design optimization of a shell-and-tube heat exchanger wit novel three-zonal baffle by using CFD and taguchi method. *Int. J. Therm. Sci.* **2020**, *155*, 106417. [CrossRef]

16. Elmekawy, A.M.N.; Ibrahim, A.A.; Shahin, A.M.; Al-Ali, S.; Hassan, G.E. Performance enhancement for tube bank staggered configuration heat exchanger—CFD Study. *Chem. Eng. Process. Process Intensif.* **2021**, *164*, 108392. [CrossRef]
17. Vivekanandan, M.; Saravanan, G.; Vijayan, V.; Gopalakrishnan, K.; Krishna, J.P. Experimental and CFD investigation of spiral tube heat exchanger. *Mater. Today Proc.* **2020**, *37*, 3689–3696. [CrossRef]
18. Kola, P.V.K.V.; Pisipaty, S.K.; Mendu, S.S.; Ghosh, R. Optimization of performance parameters of a double pipe heat exchanger with cut twisted tapes using CFD and RSM. *Chem. Eng. Process. Process Intensif.* **2021**, *163*, 108362. [CrossRef]
19. Aliaga, D.; Feick, R.; Brooks, W.; Mery, M.; Gers, R.; Levi, J.; Romero, C. Modified solar chimney configuration with a heat exchanger: Experiment and CFD simulation. *Therm. Sci. Eng. Prog.* **2021**, *22*, 100850. [CrossRef]
20. Vivekanandan, M.; Venkatesh, R.; Periyasamy, R.; Mohankumar, S.; Devakumar, L. Experimental and CFD investigation of helical coil heat exchanger with flower baffle. *Mater. Today Proc.* **2020**, *37*, 2174–2182. [CrossRef]
21. Wen, T.; Lu, L.; He, W.; Min, Y. Fundamentals and applications of CFD technology on analyzing falling film heat and mass exchangers: A comprehensive review. *Appl. Energy* **2020**, *261*, 114473. [CrossRef]
22. Quitiaquez, W.; Estupiñán-Campos, J.; Nieto-Londoño, C.; Quitiaquez, P. CFD Analysis of Heat Transfer Enhancement in a Flat-Plate Solar Collector/Evaporator with Different Geometric Variations in the Cross Section. *Energies* **2023**, *16*, 5755. [CrossRef]
23. Chihab, Y.; Garoum, M.; Laaroussi, N. Dynamic thermal performance of multilayer hollow clay walls filled with insulation materials: Toward energy saving in hot climates. *Energy Built Environ.* **2024**, *5*, 70–80. [CrossRef]
24. Wang, Z.; Suen, K. Numerical comparisons of the thermal behaviour of air and refrigerants in the vortex tube. *Appl. Therm. Eng.* **2020**, *164*, 114515. [CrossRef]
25. Jiang, H.; Niu, F.; Ma, Q.; Su, W.; Wang, E.; He, J. Numerical analysis of heat transfer between air inside and outside the tunnel caused by piston action. *Int. J. Therm. Sci.* **2021**, *170*, 107164. [CrossRef]
26. Elgendy, H.; Czerski, K. Numerical Study of Flow and Heat Transfer Characteristics in a Simplified Dual Fluid Reactor. *Energies* **2023**, *16*, 4989. [CrossRef]
27. Arsenyeva, O.; Tovazhnyanskyy, L.; Kapustenko, P.; Klemeš, J.J.; Varbanov, P.S. Review of Developments in Plate Heat Exchanger Heat Transfer Enhancement for Single-Phase Applications in Process Industries. *Energies* **2023**, *16*, 4976. [CrossRef]
28. Olabi, A.; Wilberforce, T.; Sayed, E.T.; Elsaid, K.; Rahman, S.A.; Abdelkareem, M.A. Geometrical effect coupled with nanofluid on heat transfer enhancement in heat exchangers. *Int. J. Thermofluids* **2021**, *10*, 100072. [CrossRef]
29. Hadibafekr, S.; Mirzaee, I.; Khalilian, M.; Shirvani, H. Thermohydraulic performance intensification of lobed heat exchangers using tube axis corrugation: Numerical assessment on geometrical configurations. *Chem. Eng. Process. Process Intensif.* **2022**, *171*, 108763. [CrossRef]
30. Cando, A.X.A.; Sarzosa, W.Q.; Toapanta, L.F. CFD Analysis of a solar flat plate collector with different cross sections. *Enfoque UTE* **2020**, *11*, 95–108. [CrossRef]
31. Quitiaquez, P.; Cocha, J.; Quitiaquez, W.; Vaca, X. Investigation of geometric parameters with HSS tools in machining polyamide 6 using Taguchi method. *Mater. Today Proc.* **2022**, *49*, 181–187. [CrossRef]

energies

MDPI

Article

Thermo-Hydraulic Performance of Pillow-Plate Heat Exchangers with Secondary Structuring: A Numerical Analysis

Reza Afsahnoudeh [1], Andreas Wortmeier [1], Maik Holzmüller [2], Yi Gong [2], Werner Homberg [2] and Eugeny Y. Kenig [1,*]

[1] Chair of Fluid Process Engineering, Paderborn University, 33098 Paderborn, Germany; reza.afsahnoudeh@upb.de (R.A.); awortmei@upb.de (A.W.)
[2] Chair of Forming and Machining Technology, Paderborn University, 33098 Paderborn, Germany; mh@luf.uni-paderborn.de (M.H.)
* Correspondence: eugeny.kenig@upb.de

Abstract: Pillow-plate heat exchangers (PPHEs) represent a suitable alternative to conventional shell-and-tube and plate heat exchangers. The inherent waviness of their channels promotes fluid mixing in the boundary layers and facilitates heat transfer. The overall thermo-hydraulic performance of PPHEs can further be enhanced by applying secondary surface structuring, thus increasing their competitiveness against conventional heat exchangers. In this work, various secondary structures applied on the PPHE surface were studied numerically to explore their potential to enhance near-wall mixing. Computational fluid dynamics (CFD) simulations of single-phase turbulent flow in the outer PPHE channel were performed and pressure drop, heat transfer coefficients, and overall thermo-hydraulic efficiency were determined. The simulation results clearly demonstrate a positive impact of secondary structuring on heat transfer in PPHEs.

Keywords: pillow-plates; heat exchanger; heat transfer; surface structuring; secondary structures

Citation: Afsahnoudeh, R.; Wortmeier, A.; Holzmüller, M.; Gong, Y.; Homberg, W.; Kenig, E.Y. Thermo-Hydraulic Performance of Pillow-Plate Heat Exchangers with Secondary Structuring: A Numerical Analysis. *Energies* **2023**, *16*, 7284. https://doi.org/10.3390/en16217284

Academic Editors: Gianpiero Colangelo and Artur Bartosik

Received: 29 July 2023
Revised: 18 October 2023
Accepted: 25 October 2023
Published: 26 October 2023

Correction Statement: This article has been republished with a minor change. The change does not affect the scientific content of the article and further details are available within the backmatter of the website version of this article.

1. Introduction

Progressing climate change and a growing shortage of resources require a transformation of industrial processes towards increasing energy efficiency and reduced raw material consumption. In this regard, optimization of heat exchangers as central elements of numerous industrial plants can make a decisive contribution. For decades, conventional shell-and-tube heat exchangers (STHEs) have been the primary choice for the process industry due to their robustness, reliability and design flexibility. However, their drawback is a lack of compactness. Plate heat exchangers (PHEs) with cross-corrugated channels, on the other hand, are more compact, lighter, and offer higher heat transfer coefficients. Yet, their temperature and pressure range are limited, and they may face leakage problems when constructed with gaskets. Additionally, fully welded PHEs are challenging, if not impossible, to clean (cf. [1]).

Pillow-plate heat exchangers (PPHEs) offer a promising alternative to conventional STHEs and PHEs. PPHEs consist of multiple pillow plates (PPs) assembled together into a single unit. These PPs are created through spot welding two steel sheets, seam welding along the edges, and then undergoing hydroforming, which results in their characteristic pillow-like surface (Figure 1). The seam welding along the edges ensures leak tightness. Due to the flat structure of the PP, it can be treated as a panel. As a result, constructing a PPHE involves assembling several PP panels together. In this way, two channels are built: one inside the welded PP panel (inner channel) and one between two adjacent welded PPs (outer channel).

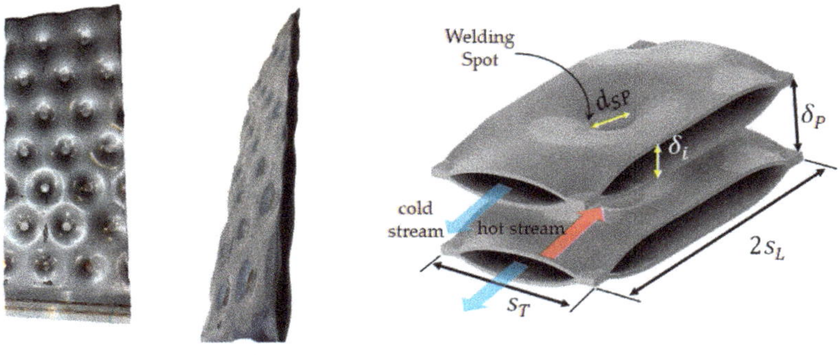

(a) (b) (c)

Figure 1. Pillow plate representation: front view (**a**), side view (**b**) and a digital image with the characteristic geometry parameters (**c**). Cold fluid flows through inner channels, while hot fluid flows through outer channels.

PPHEs are lighter, more compact, and efficient than STHEs. However, in order to match the efficiency and compactness of PHEs, PPHEs must undergo further optimization.

Piper et al. [2] conducted a CFD-based study of turbulent flow in the outer channel of PPs. They studied fluid dynamics and heat transfer and revealed that the periodical alteration of the cross-section of the outer channel leads to the corresponding acceleration and deceleration of flow, causing adverse pressure gradient. This leads to a boundary layer separation, which occurs upstream of the welding spots, resulting in large, flat-shaped recirculation zones occupying roughly 30% of the PP surface area. These zones are the primary cause of form drag, and this is responsible for about 50% of the Darcy friction factor.

Moreover, the surface area covered by such recirculation zones is less effective for heat transfer. Therefore, higher heat transfer efficiencies and reduced form drag contribution to the overall pressure loss are expected when the size of these zones is decreased. This can be achieved through intensified turbulent mixing near the surface of the PP, which would retard boundary layer separation. Thus, a further optimization of the PPHEs is possible.

Bergles et al. [3] reviewed different methods to intensify heat exchanger performance. These methods can be subdivided into two groups, namely, "active" and "passive" techniques [3,4]. Active techniques use external power enhancing heat transfer, for instance, different frequency vibrations of a heat exchanger or of a fluid in it. In contrast, passive techniques are based on improved heat transfer through unit surface modifications or the use of fluid additives (nanofluids). The number of publications on passive techniques is considerably higher than those on active ones [5,6].

PPHE performance enhancement was mostly studied in terms of inner channels [7–9]. Piper et al. [10] investigated fluid dynamics and heat transfer in the outer channels of a PPHE with secondary dimple structures. The application of secondary structures represents a passive technique extending the heat transfer area. It is capable of reducing the negative impact of the recirculation zones and, hence, shows a significant potential for performance enhancement of the studied PPHE. Such secondary structures can be realized using the method of Electrohydraulic Incremental Forming (EHIF) [10]. This idea has to be explored with respect to different geometrical forms and fluid-dynamic conditions. This paper presents such an investigation based on comprehensive CFD simulations.

2. Geometry Generation and Surface Structuring of Studied PPHEs

To carry out CFD simulations of PPHEs, a digital image of the PP geometry that defines the computational domain is necessary. A realistic reconstruction of the PP channel is crucial for a more accurate description of fluid dynamics in PPs. For this reason, we utilized a forming simulation method, as detailed in [11], to generate the basic PP geometry. This geometry does not contain secondary structures and is used as a *benchmark* for comparison with more complex geometries. This benchmark is already optimized with respect to th thermo-hydraulic performance.

In this study, we focused on configurations in which the PPs are arranged parallel t each other and distanced by δ_P = 13 mm. The other specific PP properties are an inflatio height of δ_i = 8 mm and a typical triangular welding spot pattern with a diameter o d_{SP} = 10 mm. The longitudinal and transverse welding spot pitches are s_L = 36 mm an s_T = 42 mm, respectively.

In principle, the choice of a secondary form is nearly unlimited. We have preliminar explored a wide range of different possibilities and selected two following building type dimples and ellipsoidal forms. Dimples are chosen due to their simplicity. On the othe hand, ellipsoids are more streamlined than dimples, thus promising lower form dra Figure 2 shows PPs with dimpled secondary structures, while Figure 3 depicts PPs wit ellipsoidal surface structuring. The digital images of PPs with secondary structures wer generated using CAD software SOLIDWORKS.

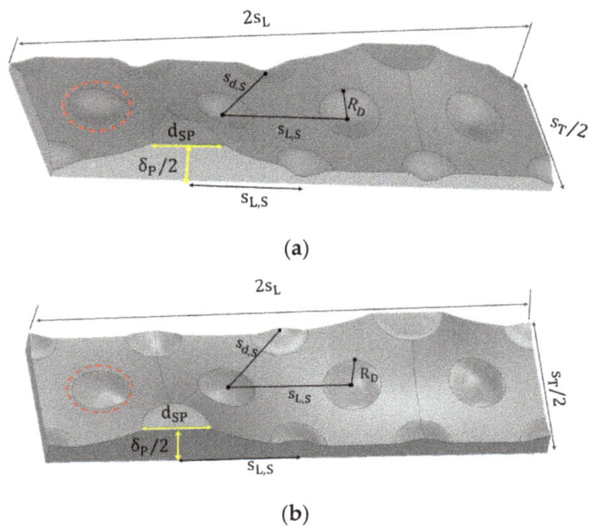

Figure 2. Periodic section of a PP outer channel with dimpled surface: negative (**a**) and positive (**b** Red dash line denotes one of the dimples.

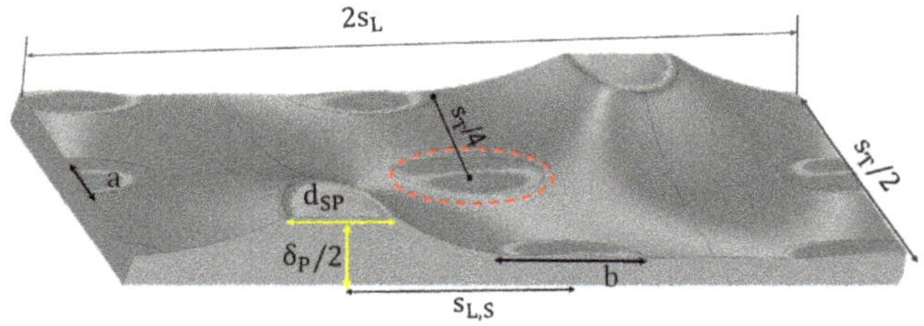

Figure 3. Periodic section of PP outer channel with ellipsoidal surface structuring. Red dash lin denotes one of the ellipsoids.

The secondary structured PPs analyzed in this study are categorized according t the projected geometry and arrangement of their secondary structures. The structure that result in a larger volume of the outer channel are referred to as "positive", while th structures resulting in a smaller volume of the outer channel are called "negative". Th dimples differ not only by their direction but also by their depth δ_S and radius R_D. Th projected surface area of a single ellipsoid is equal to that of a single dimple with a radiu of 4.1 mm. The dimensions of the secondary-structured surfaces used in this study ar summarized in Table 1. The abbreviations in this table denote the important geometry characteristics of the secondary structures. Letter D stands for dimple, while the following number gives the radius of the dimple. Letter E denotes ellipsoidal secondary structures with the following numbers a × b according to Figure 3. The number after the slash shows the depth of the secondary structures; here, the plus sign denotes the positive structures, while the minus sign means the negative structures.

Table 1. Geometrical parameters of investigated pillow plates with secondary structures.

$s_{L,S}$ (mm)	$s_{d,S}$ (mm)	R_D (mm)	a (mm)	b (mm)	δ_S (mm)	Abbreviation
-	-	-	-	-	-	Benchmark
18	$\sqrt{\left(\frac{s_T}{4}\right)^2+\left(\frac{s_{L,S}}{2}\right)^2}=13.83$	4.1	-	-	+0.5	D4.1/+0.5
18	13.83	4.1	-	-	+1.0	D4.1/+1.0
18	13.83	4.1	-	-	+1.5	D4.1/+1.5
18	13.83	4.1	-	-	−0.5	D4.1/−0.5
18	13.83	4.1	-	-	−1.0	D4.1/−1.0
18	13.83	4.1	-	-	−1.5	D4.1/−1.5
18	13.83	4.4	-	-	+0.5	D4.4/+0.5
18	13.83	4.4	-	-	+1.0	D4.4/+1.0
18	13.83	4.4	-	-	+1.5	D4.4/+1.5
-	-	-	4.72	14.2	+0.5	E4.72 × 14.2/+0.5
-	-	-	4.72	14.2	+1.0	E4.72 × 14.2/+1.0
-	-	-	4.72	14.2	+1.5	E4.72 × 14.2/+1.5

3. CFD Simulation

3.1. Model and Solution Method

The flow was modeled as single phase, incompressible, and turbulent, with constant fluid properties and neglected external forces. This was accomplished using the Reynolds-Averaged Navier–Stokes (RANS) equations [12]. Using Einstein's notation, these equations are written as follows:

$$\frac{\partial \overline{u}_k}{\partial t}+\frac{\partial}{\partial x_j}\left(\overline{u}_j\overline{u}_k\right)=-\frac{1}{\rho}\frac{\partial \overline{p}}{\partial x_k}+2\nu\frac{\partial}{\partial x_j}\left(\frac{\partial \overline{u}_k}{\partial x_j}+\frac{\partial \overline{u}_j}{\partial x_k}\right)-\frac{\partial}{\partial x_j}\left(\overline{u'_k u'_j}\right) \quad (1)$$

where x_k (k = 1, 2, 3) are Cartesian coordinates and $\overline{u}_k$ are the corresponding velocity components.

Equation (1) is supplemented by the continuity equation:

$$\frac{\partial \overline{u}_j}{\partial x_j}=0 \quad (2)$$

The term $\overline{u'_i u'_j}$ in Equation (1) is called the specific Reynolds stress tensor and it has to be modeled to close the RANS equation (cf. [12]). One such possibility is given by the Boussinesq approximation, resulting in so-called turbulent viscosity [13]. To determine the turbulent viscosity, specially developed turbulence models are required.

Due to the formation of the recirculation zones at the upstream of the welding spot in the outer channel and possible additional recirculation generated by secondary structures [10], it is crucial to choose an appropriate turbulence model that can accurately resolve the flow separation (cf. [12]). In contrast to Piper et al. [2] and Zibart et al. [14], who computed the Reynolds stresses with the elliptic blending k-ε model, we applied the k-ω-SST model. This is a two-equation-based model. Due to its ability to account for both near-wall and far-field behavior of the flow [15–17], it has an advantage of a better prediction of boundary-layer and separated flows as compared to the other turbulence models. In our work, the k-ω-SST model suggested in [16] is applied.

The computational effort is reduced by utilizing flow periodicity. It was accomplished by applying periodic boundary conditions for the velocity vector at the inlet (x = 0) and outlet (x = 2 s_L) boundaries of the channel. This means that the outlet velocity profile was mapped to the inlet after twice the longitudinal pitch (2 s_L). Furthermore, symmetric boundary conditions at y = 0, y = 0.5 s_T and z = 0 were applied. At the wall, a no-slip boundary condition (u = 0) was set.

To determine the temperature field in the studied PPs, the energy equation for the turbulent flow must be solved. By using Reynolds averaging, a corresponding equation for the description of the average temperature field $\overline{T}(x, y, z)$ can be obtained [13]:

$$\rho C_p\left(\frac{\partial \overline{T}}{\partial t} + \frac{\partial}{\partial x_j}(\overline{u}_j\overline{T})\right) = q_v + \frac{\partial}{\partial x_j}\left(\lambda\frac{\partial \overline{T}}{\partial x_j}\right) - \rho C_p\frac{\partial}{\partial x_j}(\overline{u'_jT}) \quad (3)$$

The last term on the right-hand side of Equation (3) governs turbulent heat transfer; similar to the specific Reynolds stress tensor, it has to be modeled [13].

For the energy equation, a constant surface-averaged temperature was set at the inlet (x = 0) of a periodic element. In the case of a thermally developed flow, the shape of temperature profiles at the successive streamwise location distanced by the period length (2 s_L) is similar [18]. In our work, the shape of the temperature profile at the outlet (x = 2 s_L) is used to generate a new inlet temperature profile. To ensure that the original boundary condition at x = 0 is satisfied, this outlet profile is rescaled to the predefined average temperature at the inlet. At the wall, a constant temperature is set.

Figure 4 depicts the simulation domain, while Table 2 summarizes the boundary conditions. In the simulations, second-order approximation schemes for the discretization of convective and diffusive terms were used. The convergence is evaluated based on two criteria. First, the normalized residual for all equations must be below 10^{-5}. Second, the calculated characteristic quantities, such as mean wall shear stress, heat transfer coefficient and mass flow at the periodic inlet and outlet must reach a constant value.

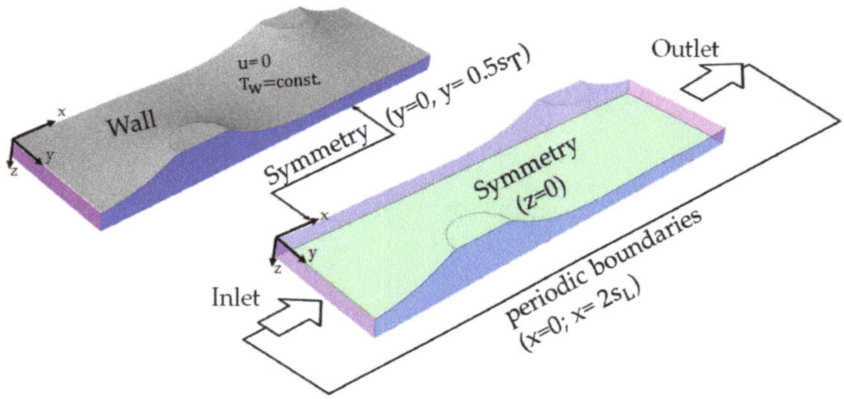

Figure 4. Periodic computational domain and boundary conditions.

Table 2. Applied boundary and initial conditions.

	x = 0	x=2 s_L	y = 0	y=0.5 s_T	z = 0	Wall	Initial
u	periodic	periodic	symmetry	symmetry	symmetry	no-slip (u = 0)	0 m/s
T	$T_{x=0,m} = 323$ K	zero gradient	symmetry	symmetry	symmetry	$T_W = 303$ K	323 K
p	-	1 bar	symmetry	symmetry	symmetry	-	1 bar

The above equations were solved numerically using the open-source CFD simulation software OpenFOAM 8, which is based on the finite volume method (FVM).

3.2. *Grid Generation*

The grid was generated with Ansys ICEM CFD software, as it provides more options for block-structured grids than the meshing tool by OpenFOAM 8. A mesh refinement was applied to capture the steep gradients in the vicinity of the walls and along the edge of the welding spot by increasing the mesh density perpendicular to the wall and in the direction of the radius of the welding spot (Figure 5).

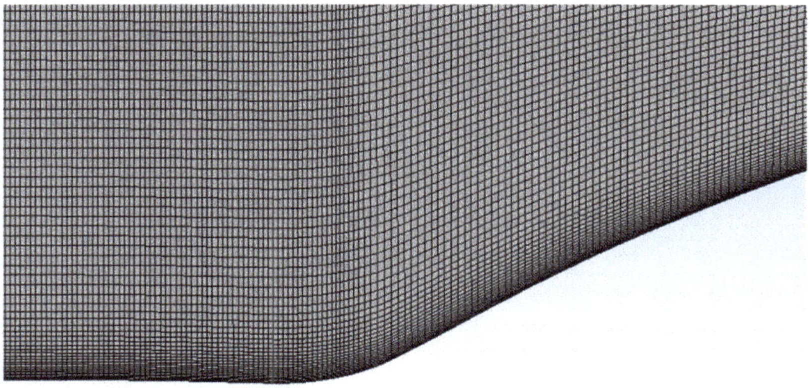

Figure 5. Representation of the mesh near the welding spot.

A grid independence study was carried out for all of the investigated geometries. Depending on the geometry, different grids were obtained. The grid was accepted as final when doubling the number of cells resulted in just insignificant differences in the pressure drop and heat transfer coefficient (less than 0.1%). Consequently, the number of grid cells ranged from about 17.5 million for the benchmark to 30 million for D4.4/+1.5.

To resolve the boundary layers accurately, the grid construction included the following steps. First, the distance between the wall (heat transfer surface) and the grid node closest to it was estimated using dimensionless wall distance (y^+):

$$y^+ = \frac{u_\tau \tilde{y}}{\nu} \text{ with } u_\tau = \sqrt{\frac{\tau_w}{\rho}} \quad (4)$$

where u_τ and $\tilde{y}$ represent shear velocity and the coordinate normal to the wall, respectively. In all cases, grid independence was achieved for y^+ values below 0.5. To ensure a smooth transition from viscous sublayer to the buffer layer, at least 4–5 further cells are included in the viscous sublayer. Finally, the cell thickness ($\Delta\tilde{y}_n$) was determined by linearly stretching in the direction normal to wall, with stretching factor ($\frac{\Delta\tilde{y}_{n+1}}{\Delta\tilde{y}_n}$) of 1.1.

4. Process Parameters

To evaluate the thermo-hydraulic characteristics of the flow in PPs, several process parameters are used. The average Reynolds number of the flow in the PP channel is defined with the following expression:

$$Re = \frac{u_m d_h}{\nu} \quad (5)$$

where u_m is the mean stream velocity in the channel of a PP and d_h is its hydraulic diameter, which can be calculated by the method proposed in [11].

The overall pressure drop Δp is calculated from the difference between the surface-averaged pressure values at the inlet and outlet boundaries. Afterwards, the pressure drop coefficient $\xi_{\Delta p}$ (also known as the Darcy friction factor) can be evaluated by th Darcy-Weisbach equation:

$$\xi_{\Delta p} = \frac{2d_h \Delta p}{\rho u_m^2 2s_L} \quad (6$$

The overall pressure drop can be expressed as a sum of the pressure drops due to dra and friction (cf. [19]):

$$\Delta p = \Delta p_{Drag} + \Delta p_{Friction} \quad (7$$

In accordance with Buckingham's π theorem (dimensional analysis), the drag an friction pressure drop can be expressed as [19]:

$$\Delta p_{Drag} = \frac{\rho u_m^2 2s_L \xi_D}{2d_h} \quad (8$$

$$\Delta p_{Friction} = \frac{\rho u_m^2 2s_L \xi_R}{2d_h} \quad (9$$

where ξ_D is the form drag coefficient and ξ_R is the Fanning friction factor, which can b determined as follows:

$$\xi_R = \frac{8\tau_w}{\rho u_m^2} \quad (10$$

Inserting Equations (7)–(10) in Equation (6), the following equation for the evaluatio of the form drag coefficient is obtained:

$$\xi_D = \xi_{\Delta p} - \xi_R \quad (11$$

The Nusselt number is defined by:

$$Nu = \frac{h_m d_h}{\lambda} \quad (12$$

where h_m represents the surface-averaged heat transfer coefficient, which can be calculate as follows:

$$h_m = \frac{\dot{Q}_w}{A_w \Delta T_{LMTD}} \quad (13$$

Here $\dot{Q}_w$ is the heat transferred through the wall and ΔT_{LMTD} is the logarithmic mea temperature in the channel:

$$\Delta T_{LMTD} = \frac{\Delta T_{x=0,w} - \Delta T_{x=2s_L,w}}{\ln\left(\frac{\Delta T_{x=0,w}}{\Delta T_{x=2s_L,w}}\right)} \quad (14$$

In Equation (5), the differences $\Delta T_{x=0,wall}$ and $\Delta T_{x=2s_L,wall}$ are calculated as follows:

$$\Delta T_{x=0,w} = T_{x=0,\,m} - T_w = \left(\frac{\int_{A_{cs,\,x=0}} uTdA}{\int_{A_{cs,\,x=0}} udA}\right) - T_w \quad (15$$

$$\Delta T_{x=2s_L,w} = T_{x=2s_L,\,m} - T_w = \left(\frac{\int_{A_{cs,\,2s_L}} uTdA}{\int_{A_{cs,\,2s_L}} udA}\right) - T_w \quad (16$$

The thermo-hydraulic ε efficiency is defined as a ratio of the transferred heat to th pumping power:

$$\varepsilon = \frac{\left|\dot{Q}_w\right|}{\dot{V}\Delta p} \quad (17$$

The evaluation of individual secondary structures is based on relative deviations of different quantities from the relevant benchmark quantities defined as follows:

$$Z^* = \frac{Z\ (\text{secondary structures}) - Z\ (\text{benchmark})}{Z\ (\text{benchmark})} \times 100 \quad (18)$$

5. Model Validation

For the validation of the CFD model, we used the PP studied by Piper et al. [2], for which experimental data are available. This PP slightly differs from the selected benchmark: it has an inflation height of δ_i = 7 mm and a welding spot diameter of d_{SP} = 12 mm. In Figure 6, the specific pressure drop determined by the CFD simulations is shown together with the data obtained from experiments carried out by Piper et al. [2]. The agreement is good over the entire range of investigated Reynolds numbers (7000–14,000), with a relative error below 5%.

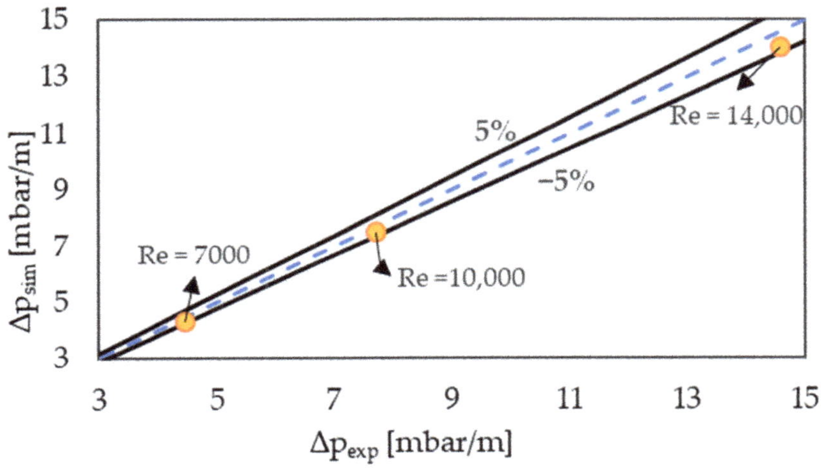

Figure 6. Comparison of simulated pressure drop in the outer PP channel with the data measured by Piper et al. [2], for 7000 ≤ Re ≤ 14,000.

6. Results and Discussion

With the validated CFD model, simulations were performed with the aim of investigating the effect of secondary structures on the flow characteristics and heat transfer. The results were obtained for a constant Reynolds number, Re = 5000, which was chosen according to the efficiency analysis carried out in [2].

Figure 7 illustrates the flow streamlines and heat transfer phenomena for the benchmark case. In Figure 7a, the formation of recirculation zones upstream of the welding spot is visible; its size can be evaluated by the region of positive wall shear stress in flow direction ($\tau_{w,x}$). When relating Figure 7b, which shows the normalized wall heat flux ($\dot{q}_w/\dot{q}_{w,max}$), to Figure 7a, it becomes obvious that the heat transfer is very slow within this area. Similar results were obtained and evaluated for all investigated PPs with secondary structures. The results are summarized in Table 3.

From the analysis of the data given in Table 3, it can be concluded that deeper secondary structures enhance heat transfer due to the intensified near-wall mixing. However, deeper secondary structures also generate larger and more energetic recirculation zones that cause a higher pressure drop. The near-wall mixing is stronger for dimples with negative structures, thus enhancing heat transfer. Figure 8 shows the near-wall streamlines and recirculation zones for the geometry D4.1/−1.5, which yields the fastest heat transfer and the highest pressure drop value.

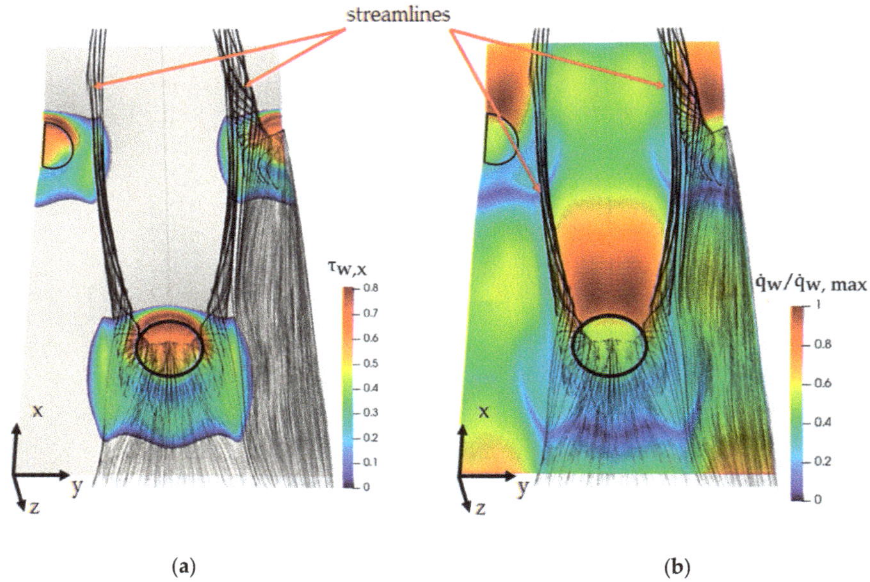

Figure 7. Simulation results for the benchmark: characteristic vortex and boundary layer separation (**a**) and normalized wall heat flux (**b**). The welding spots are indicated by black circles.

Table 3. Thermo-hydraulic characteristics of the flow in the outer channel of the studied PPs at Re = 5000.

PPHE	Δp $\left(\frac{\text{mbar}}{\text{m}}\right)$	Δp *	$\lvert\dot{Q}_W\rvert$ (W)	$\dot{Q}_W$ *	h $\left(\frac{\text{W}}{\text{m}^2\text{K}}\right)$	h *	A_W (mm^2)	A_W *	d_h (mm)	ε *
Benchmark	5.9	-	126	-	3209	-	1547	-	13.8	-
D4.1/−0.5	6.7	15	130	3.8	3322	3.5	1551	0.27	13.6	−9.8
D4.1/−1.0	8.9	51	147	16.9	3727	16.1	1562	0.94	13.4	−23.3
D4.1/−1.5	12.5	113	169	34.2	4264	32.9	1577	1.93	13.1	−38.3
D4.1/+0.5	6.7	14	133	5.5	3346	4.3	1550	0.16	13.9	−7.8
D4.1/+1.0	7.7	30	141	12.2	3546	10.5	1558	0.71	13.9	−14.7
D4.1/+1.5	8.3	42	143	14.0	3580	11.5	1578	1.96	14.0	−21.1
D4.4/+0.5	6.8	16	135	7.5	3396	5.8	1550	0.15	13.9	−7.4
D4.4/+1.0	7.6	29	138	10.0	3475	8.3	1558	0.67	14.0	−15.3
D4.4/+1.5	8.2	40	151	20.1	3713	15.7	1572	1.62	14.0	−15.9
E4.72 × 14.2/+0.5	5.4	−8.5	121	−3.6	3053	−4.9	1550	0.16	13.8	5.1
E4.72 × 14.2/+1.0	5.4	−7.4	123	−2.4	3099	−3.4	1558	0.71	13.8	4.7
E4.72 × 14.2/+1.5	5.7	−3.5	126	0.0	3134	−2.4	1571	1.54	13.8	2.0

* Relative deviation to the benchmark in percent.

Another factor that contributes to the higher pressure drop in the case of negative structures is the enlargement of the wall surface area, resulting in growing friction. This effect is accompanied by a cross-section decrease further increasing the pressure drop. In contrast, the heat transfer is significantly more intensive due to enhanced near-wall mixing. For a better judgment of the heat exchanger performance, the thermo-hydraulic efficiency ε can be used. In the last column of Table 3, the relative value of this parameter is reported, which is calculated as a difference between the benchmark and secondary structure efficiencies divided by the corresponding value for the benchmark. With increasing depth of the secondary structures, heat transfer improves, yet the ε value drops and is worst for D4.1/−1.5. Therefore, we excluded all negative structures from the following investigations and focused on positive structures.

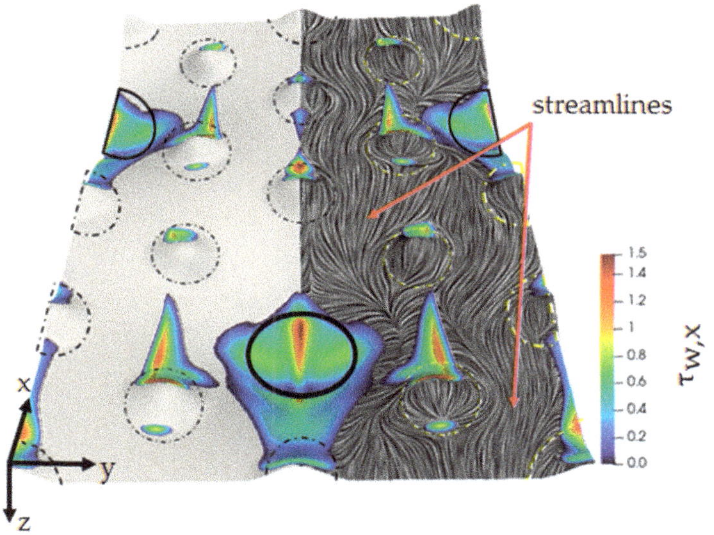

Figure 8. Boundary separation and near-wall streamlines for geometry D4.1/−1.5. Recirculation zones are represented by $\tau_{w,x} \geq 0$. The welding spots are indicated by black circles. The dotted circles show the location of secondary structures.

In Figure 9, the pressure drop values for different positive secondary structures are shown. For all positive-structured dimpled surfaces, the pressure drop is higher than for the benchmark. With increasing dimple depth, the corresponding pressure drop also increases. On the other hand, when the radius of the dimple elements (R_D) is increased while keeping its depth constant, heat transfer is enhanced and the pressure drop decreases slightly. PPHEs with ellipsoidal secondary structures, with their more streamlined form and arrangement, demonstrate a lower pressure drop than the benchmark.

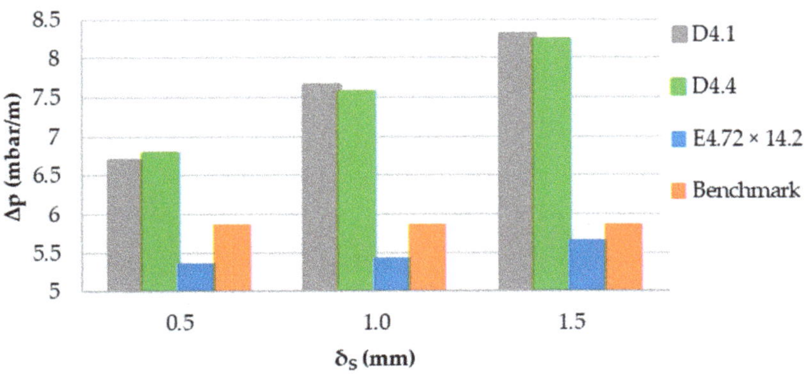

Figure 9. Pressure drop in PPHEs with different secondary structures and positiv δ_S.

In order to gain a deeper understanding of the impact of secondary structures on the pressure drop, it is helpful to compare the friction coefficients. Figure 10 shows the friction factors for different PPHE configurations. For dimpled secondary structures, the Fanning friction factor is larger than that of the benchmark (Figure 10b). This is as expected, taking into account that the secondary structures increase the surface area of the PPs. The form drag coefficient depends on the shape of the secondary structures and has the smallest value for the E4.72 × 14.2/+0.5 configuration (Figure 10c). In fact, all friction coefficients, and thus the pressure drop values, for all of the ellipsoidal secondary structures are smaller than those of the benchmark and the other studied secondary structures.

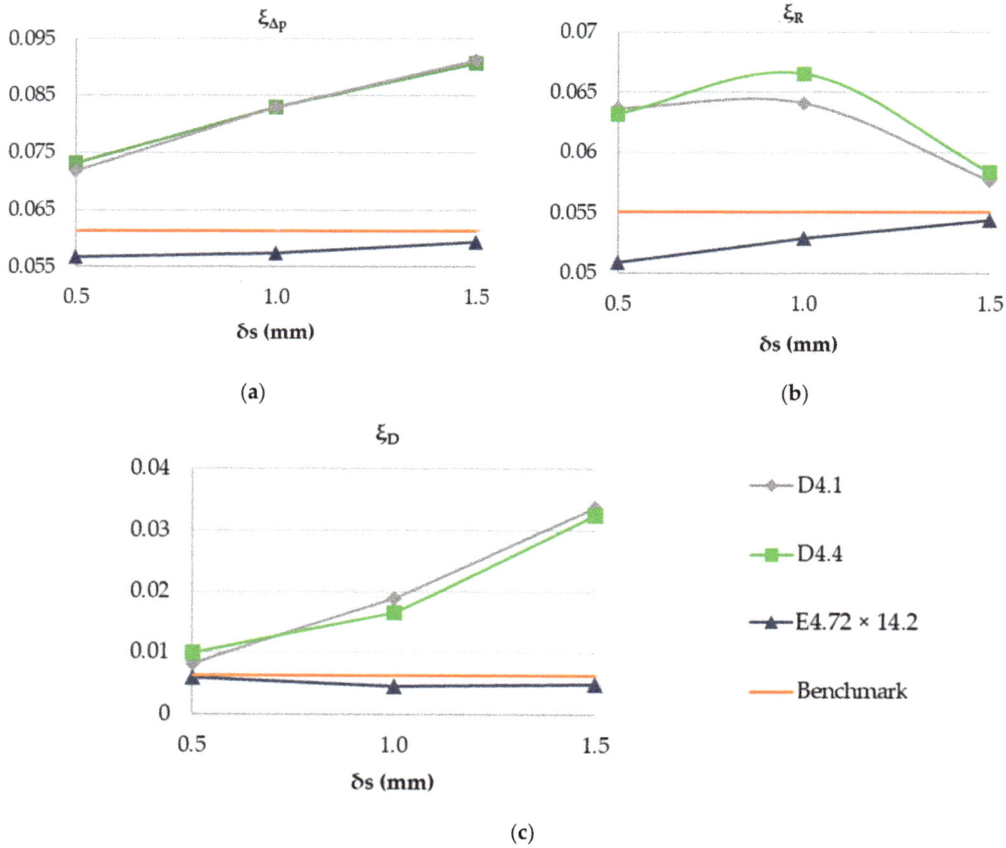

Figure 10. Dimensionless friction coefficients for different secondary structures with positive δ_S: Darcy friction factor (**a**), Fanning friction factor (**b**) and form drag coefficient (**c**).

The lower friction factor values for ellipsoidal secondary structures can be attributed to their more streamlined shape and their impact on the shape of the vortex. Such vortices can be detected when negative values of the velocity component in the mainstream direction ($u_x \leq 0$) arise. Figure 11 shows the vortices upstream the welding spot in the benchmark. Due to the location and shape of the vortices, the fluid has to bypass these areas, which results in higher form drag and an increased flow path length. This brings about a higher friction. Figure 12 shows the vortices for the case E4.72 × 14.2/+1.5, which has the same $\dot{Q}_W$ value as the benchmark. Here, the vortices are smaller and cover the welding spot leaving a larger portion of the channel volume free for the flowing fluid. As a consequence all friction coefficients are smaller. Also, smaller vortices are less energetic, which results in a lower pressure drop. On the other hand, because of the streamlined shape of the ellipsoid structures, the near-wall mixing reduces. As a result, the heat transfer is slower than for dimpled structuring. Yet, due to the lower pressure drop, the thermo-hydraulic efficiency increases. In fact, it is only ellipsoidal secondary structures that permit the value of ε to increase.

In Figure 10b, a maximum Fanning friction factor is visible for the dimpled surface at δ_S = 1 mm. One possible explanation for this maximum is that adding the dimples increases both the surface area of the PP and the turbulence. For smaller δ_S and R_D, the influence of increasing surface area on Fanning friction is dominant, leading to an increase in friction. For larger dimples, however, the influence of the local turbulence becomes dominant, resulting in a decreasing ξ_R. The combination of these two opposite effects leads to a maximum Fanning friction factor value. On the other hand, higher turbulence leads to higher form drag. The more streamlined secondary structures can compensate the increase in form drag. Consequently, larger and more streamlined secondary structures can reduce both the Fanning friction factor and the form drag.

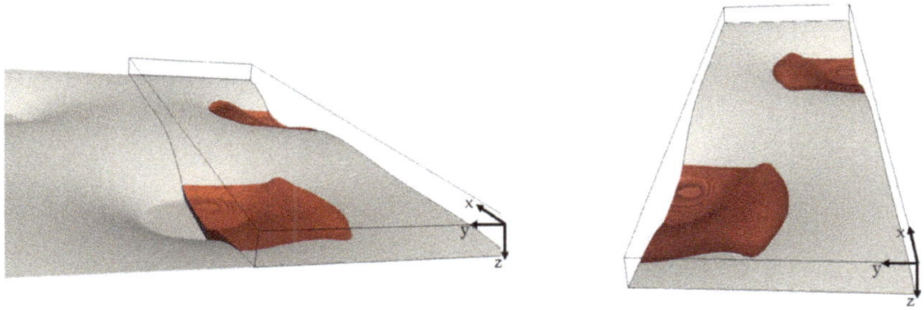

(a) (b)

Figure 11. Recirculation zones in the benchmark: side view on the periodic element together with its mirror image (**a**) and top view (**b**). Tornado-like vortices upstream the welding spots are shown in dark color. The black lines represent the edges of the computational domain.

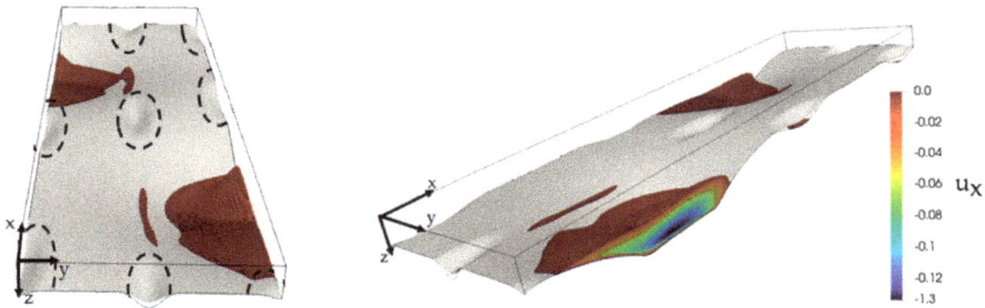

(a) (b)

Figure 12. Illustration of recirculation zones of E4.72 × 14.2/+1.5. Top view of periodic element (**a**) and side view of simulation domain (**b**). The black lines represent the edges of simulation domain and the dash lines show the location of the secondary structures.

7. Conclusions

In this work, a CFD-based study was carried out to investigate the fluid flow and heat transfer in the outer channel of pillow plates (PPs) with various secondary surface structuring. Two basic forms, dimples and ellipsoids, with different geometrical parameters, were selected as elements of secondary structures. Dimples were chosen due to their simple form, while ellipsoids were selected for their more streamlined shape. The structures resulting in a larger volume of the outer channel were denoted "positive", whereas structures resulting in a smaller volume of the outer channel were called "negative". For dimples, both negative and positive structures were investigated, while for ellipsoids, only positive structures were studied.

To evaluate the performance of the PPs with secondary structures, it was compared with the performance of a benchmark representing a previously optimized conventional PP geometry.

Deeper secondary structures were found to enhance near-wall mixing and improve heat transfer, yet at the cost of a higher pressure drop. Secondary structures increasing the volume of the outer channel were judged as more promising. Secondary structures with elements having a larger projected surface area and more streamlined arrangement offered enhanced heat transfer with just minimal pressure drop increase, compared to the benchmark.

The evaluation was carried out based on the thermo-hydraulic efficiency. Generally, it decreases with the increasing depth of secondary structures, while it increases with larger projected surface area of secondary structures. This is true for all studied PPs. Due to the preliminary optimization of the benchmark geometry, it was hardly possible to improve both heat transfer and pressure drop characteristics, i.e., there was a common trade-off between them. Only for PPs with ellipsoidal secondary structures, the thermo-hydraulic efficiency was higher than that of the benchmark. This means that further improvement can potentially be achieved with ellipsoidal or more streamlined secondary structures having elements with a larger projected surface area.

It appears reasonable that secondary structuring has an impact on the thermo-hydraulic characteristics in both outer and inner channels. Therefore, the following study will focus on the investigation of the flow and heat transfer in the inner channels of the PPs with secondary structuring.

Author Contributions: Conceptualization, R.A. and E.Y.K.; methodology, R.A.; software, R.A. and Y.G.; validation, R.A. and A.W.; investigation, R.A. and M.H.; resources, E.Y.K.; writing—original draft preparation, R.A.; writing—review and editing, E.Y.K.; supervision, E.Y.K. and W.H. All authors have read and agreed to the published version of the manuscript.

Funding: The authors are grateful for the financial support from the German Research Foundation (DFG, Project number KE 837/46-1). Furthermore, the authors gratefully acknowledge the funding of this project by computing time provided by the Paderborn Center for Parallel Computing (PC 2).

Data Availability Statement: Data sharing not applicable.

Conflicts of Interest: The authors declare no conflict of interest.

Nomenclature

A_w	surface area of the channel wall m^2)	λ	thermal conductivity ($Wm^{-1}K^{-1}$)
a	minor axis of an ellipsoid element (m)	μ	dynamic viscosity (kg $m^{-1}s^{-1}$)
b	major of an ellipsoid element axis (m)	ν	kinematic viscosity (m^2s^{-1})
C_p	specific heat capacity ($Jkg^{-1}K^{-1}$)	ξ	friction factor (-)
d	diameter (m)	ω	specific turbulence dissipation (s^{-1})
h	heat transfer coefficient ($Wm^{-2}K^{-1}$)	τ	shear stress (Pa)
k	kinetic turbulent energy (m^2s^{-2})	**Subscripts**	
Nu	Nusselt number (-)	D	dimple
p	pressure (Pa)	d	diagonal
Pr	Prandtl number (-)	h	hydraulic
$\dot{Q}$	heat flow rate (W)	i	inner
$\dot{q}_w$	wall heat flux (Wm^{-2})	j	cartesian system index
q_v	energy source/sink (Wm^{-3})	k	cartesian system index
R	radius (m)	L	longitudinal
Re	Reynolds number (-)	m	mean
$s_{d,S}$	diagonal pitch of neighboring dimples (m)	max	maximum
s_L	longitudinal welding spot pitch (m)	min	minimum
$s_{L,S}$	longitudinal pitch of secondary structures (m)	n	grid node number
s_T	transversal welding spot pitch (m)	o	outer
$\dot{V}$	volumetric flow rate (m^3s^{-1})	P	plate
T	temperature (K)	S	secondary structures
u	velocity (ms^{-1})	SP	welding spot
x, y, z	Cartesian coordinates	**Superscripts**	
Greek symbols		*	relative deviation to the benchmark
δ	height (m)	**Other Symbols**	
Δp	pressure drop (Pa)	$\overline{}$	Reynolds-averaged value
ε	thermo-hydraulic efficiency (-)	$'$	fluctuation quantity

eferences

Shah, R.K.; Sekulic, D.P. *Fundamentals of Heat Exchanger Design*; John Wiley & Sons: Hoboken, NJ, USA, 2003.
Piper, M.; Tran, J.M.; Kenig, E.Y. A CFD Study of the Thermo-hydraulic Characteristics of Pillow-Plate Heat Exchangers. In Proceedings of the ASME Summer Heat Transfer Conference, Washington, DC, USA, 10–14 July 2016.
Bergles, A.E.; Nirmalan, V.; Junkhan, G.H.; Webb, R.L. *Bibliography on Augmentation of Convective Heat and Mass Transfer*; Heat Transfer Laboratory, Iowa State University: Ames, IA, USA, 1983.
Webb, R.L. *Principles of Enhanced Heat Transfer*, 1st ed.; John Wiley & Sons: New York, NY, USA, 1994.
Arsenyava, O.; Tovazhnyanskyy, L.; Kapustenko, P.; Klemeš, J.J.; Varbanov, P.S. Review of developments in plate heat exchanger heat transfer enhancement for single-phase applications in process industries. *Energies* **2023**, *16*, 4976. [CrossRef]
Mastani Joybari, M.; Selvnes, H.; Sevulat, A.; Hafner, A. Potentials and challenges for pillow-plate heat exchangers: State-of-the-art review. *Appl. Therm. Eng.* **2022**, *214*, 118739. [CrossRef]
Shirzad, M.; Mousavi Ajarostaghi, S.S.; Aghajani Delavar, M.; Sedighi, K. Improve the thermal performance of the pillow plate heat exchanger by using nanofluid: Numerical simulation. *Adv. Powder Technol.* **2019**, *30*, 1356–1365. [CrossRef]
Piper, M.; Zibart, A.; Tran, J.M.; Kenig, E.Y. Numerical investigation of turbulent forced convection heat transfer in pillow plates. *Int. J. Heat Mass Transf.* **2016**, *94*, 516–527. [CrossRef]
Ghasemi, K.; Tasnim, S.; Mahmud, S. Second law analysis of pillow plate heat exchanger to enhance thermal performance and its simulation studies. *Heat Mass Transf.* **2023**, *59*, 55–66. [CrossRef]
). Piper, M.; Zibart, A.; Djakow, E.; Springer, R.; Homberg, W.; Kenig, E.Y. Heat transfer enhancement in pillow-plate heat exchangers with dimpled surfaces: A numerical study. *Appl. Therm. Eng.* **2019**, *153*, 142–146. [CrossRef]
l. Piper, M.; Olenberg, A.; Tran, J.M.; Kenig, E.Y. Determination of the geometric design parameters of pillow-plate heat exchangers. *Appl. Therm. Eng.* **2015**, *91*, 1168–1175. [CrossRef]
2. Wilcox, D.C. *Turbulence Modelling for CFD*, 3rd ed.; DCW Industries: La Cañada, CA, USA, 2006.
3. Schlichting, H.; Gersten, K. *Boundary-Layer Theory*, 9th ed.; Springer: Berlin, Germany, 2017.
4. Zibart, A.; Kenig, E.Y. Numerical investigation of conjugate heat transfer in a pillow-plate heat exchanger. *Int. J. Heat Mass Transf.* **2021**, *165*, 120567. [CrossRef]
5. Menter, F.R. Two-equation eddy-viscosity turbulence models for engineering applications. *AIAA J.* **1994**, *32*, 1598–1605. [CrossRef]
5. Menter, F.R.; Kuntz, M.; Langtry, R. Ten Years of Industrial Experience with the SST Turbulence Model. *Turbul. Heat Mass Transf.* **2003**, *4*, 625–632.
7. Menter, F.R. Review of the shear-stress transport turbulence model experience from an industrial perspective. *Int. J. Comput. Fluid Dyn.* **2009**, *23*, 305–316. [CrossRef]
3. Patankar, S.V.; Liu, C.H.; Sparrow, E.M. Fully developed flow and heat transfer in ducts having streamwise-periodic variations of cross-sectional area. *ASME J. Heat Transf.* **1977**, *99*, 180–186. [CrossRef]
). Zierep, J.; Bühler, K. *Grundzüge der Strömungslehre, Grundlage, Statik und Dynamik der Fluide*, 10th ed.; Springer Vieweg: Wiesbaden, Germany, 2015. [CrossRef]

energies

MDPI

Article

Effect of the Solid Particle Diameter on Frictional Loss and Hea Exchange in a Turbulent Slurry Flow: Experiments and Predictions in a Vertical Pipe

Artur S. Bartosik

Department of Production Engineering, Kielce University of Technology, Al. Tysiaclecia P.P. 7, 25-314 Kielce, Poland; artur.bartosik@tu.kielce.pl

Abstract: The study deals with experiments and predictions on turbulent flow and heat exchang in a fully developed slurry flow in a vertical upward pipe. Four slurries were considered: two wit glass spheres particles with diameters of 0.125 mm and 0.240 mm, respectively, and two with san spheres particles with diameters of 0.470 mm and 0.780 mm, respectively. The volume concentratio of the particles was changed in the range of 10% to 40%. This study has indirectly demonstrated th existence of turbulence suppression to a degree dependent on the diameter of the solid particles. mathematical model for heat transfer between slurry and pipe was developed using the two-equatio turbulence model and a specially designed wall function, including particle diameter and soli concentration. The model assumed a constant wall temperature and heat flux. The study's objectiv was to determine the influence of the diameter of the solid particles on the heat exchange. The Nusse number was found to change sinusoidal, reaching a maximum for a slurry with d = 0.125 mm, and minimum for d = 0.470 mm. The higher the solid concentration, the lower the Nusselt number. Th novelty and value of this study lies in the deeper characterisation and understanding of the influenc of the diameter of solid particles on heat exchange.

Keywords: vertical slurry flow; damping of turbulence; influence of solid particle on heat exchang

Citation: Bartosik, A.S. Effect of the Solid Particle Diameter on Frictional Loss and Heat Exchange in a Turbulent Slurry Flow: Experiments and Predictions in a Vertical Pipe. *Energies* **2023**, *16*, 6451. https://doi.org/10.3390/en16186451

Academic Editor: Dmitry Eskin

Received: 29 July 2023
Revised: 3 September 2023
Accepted: 4 September 2023
Published: 6 September 2023

1. Introduction

Slurry flow has interested scientists and engineers for centuries [1]. Slurries occu widely in chemical and mining industries. The influence of temperature on slurry frictio is of relevance for the mining industry. In some cases, mineral extraction is carried out at temperature of 200 °C and the tailings from this process are pumped at a temperature o 70 °C [2]. Many engineering companies design, construct, and operate long-distance slurr pipelines [3,4].

Slurry flow depends on Reynolds number, solid concentration, bulk and settlin velocities, pipe diameter, physical properties of the solid and liquid phases, particle siz distribution, and flow geometry [5,6]. As a result of this complexity, engineers are constantl looking for new methods to better understand solid–liquid and solid–solid interaction They are interested in a simple useful mathematical model for parametric studies [7–9].

Solid particles in the liquid affect the shear stresses on the wall and the erosion an abrasion of the pump and pipeline components [10–12]. Solid particles also affect the carrie liquid's turbulence and the heat exchange intensity [13,14]. It is well known that slurrie with fine solid particles are responsible for increased viscosity and yield shear stress. I such a case, rheological measurements are necessary to determine the viscosity, yield stres and proper rheological model. However, when considering a slurry with medium or coars solid particles, it is accepted that the mathematical model includes the viscosity of th carrier phase, as the rheological properties cannot be measured because the sedimentatio process is too intensive [5,15]. This phenomenon is pronounced if the density of the soli particle increases. If the density of medium or coarse solid particles is higher than tha of carrier liquid, the particles settle under gravity, but they can be fully suspended for sufficiently high velocity.

The importance of the diameter of the solid particles in the carrier liquid remains a challenge for many scientists [16,17]. Experiments on solid–liquid flow are complex, time consuming, and rarely available in the literature. Such experiments require careful preparation of solid particles, including screening or sieving, determining the median particle diameter, and slurry concentration. Running the flow requires degassing the slurry, determining the settling velocity to avoid flow blockages, keeping a constant temperature, determining the mass of the solid particles delivered, and finally, proper measurement instruments and control and monitoring systems. Measurement instruments can be intrusive or non-intrusive [18,19]. Intrusive methods have limitations because measurements at high concentrations of the solid phase result in the risk of damage or contamination of the probe. Moreover, such measures are almost impossible near a pipe wall. Non-intrusive methods, like optics, ultrasounds, or magnetic resonance do not affect the flow structure but they have limitations too. When optical methods are considered, the particles induce attenuation of the beams. However, if the refractive index matches the suspended particles and the carrier fluid, it is possible to perform measurements of particle velocities but for limited solid concentrations. For these reasons, most measurements refer to gas–liquid flows. The newest techniques, like ultrasounds or magnetic resonances, also have limitations. Although these methods propose a relatively good spatial resolution, this is still insufficient to measure the fluctuating parts of the solid and liquid phases [20–23], which is essential for developing more reliable turbulence models. Based on the above, one can say that there is still an alternative to using standard turbulence models together with a specially designed wall damping function. The wall damping function can be developed based on comparing predictions and measurements of global parameters, such as frictional loss and velocity and temperature distributions. This approach was used in the current studies.

The main objective of the study is to determine the influence of the diameter of the solid particles on the heat exchange between a pipe and a turbulent slurry flow in the upward vertical pipeline. Four slurries were studied—two slurries with glass spheres and two slurries with sand particles. The median particle diameters of the respective slurries were 0.125 mm, 0.240 mm, 0.470 mm, and 0.780 mm. Solid volume concentrations changed in the range of 10% to 40%.

The novelty and value of this study lies in the deeper characterisation and understanding of the influence of the diameter of solid particles on heat exchange. In addition, the novelty of this article lies in proposing a simple physical and mathematical model, which includes the median particle diameter and solids concentration. This model is suitable for parametric studies, which allow us to simulate the effect of solid particle diameter, solid concentration, heat flux, wall temperature, and physical and thermal properties of solid and liquid phases on heat exchange efficiency between pipe and turbulent slurry flow. The study presents the results and analysis of measurements and numerical predictions, which include $dp/dx = f(U_b)$, $T = f(y)$, $Nu = f(Re)$, and $Nu = f(d)$.

1.1. Literature Review: Isothermal Flow

Solid particles in a carrier liquid result in a slip velocity between the solid and liquid phases. Yianneskis and Whitelaw [24] pioneered LDA measurements for fully developed turbulent pipe flow for a slurry with solid particles of a median diameter of 0.27 mm. The authors noted that there are no significant differences in the velocity of the liquid and solid phases. Furthermore, the importance of slip velocity decreased with increasing solid concentration. However, slip velocity is generally essential in dilute slurries, that is, in slurries with low solid concentration or in slurries with sufficiently high particle diameter and density [25,26].

Some researchers conducted studies on the behaviour of solid particles near a pipe wall [27–30]. They commonly concluded that the ejection-sweep cycle near a pipe wall is strongly affected by solid particles. Furthermore, the authors emphasised that solid particles can enhance or suppress turbulence. Researchers such as Gore and Crowe [31], Jianren et al. [32], Eaton et al. [33], Fessler and Eaton [34], and Li et al. [35] emphasised the influence of solid particles on the modulation of turbulence. In general, we have obtained sufficient research results that confirm that the behaviour of solid particles in a liquid can led to the increase or suppression of turbulence. However, no function has been determined allowing us to clearly state whether we deal with the suppression or enhancement of turbulence.

Li et al. [35] analysed the effect of adding large solid particles to a flat plate boundary layer that was spatially developed on the transport characteristics of slurries. The authors compared the slurry transport characteristics at different flow rates and different solid concentrations of glass spheres with a median particle size of 0.125 mm before and after adding the same volume fraction of glass spheres of 0.44 mm. They analysed the particle velocity distribution, the turbulent kinetic energy of the small particles, and the wall shear forces. The authors found that adding 0.44 mm particles decreased the turbulent kinetic energy of the small particles and significantly reduced the total resistance of the pipe by 10% at the bulk velocity of 4 m/s and the solid volume concentration by 20%. Matousek et al. [36] formulated similar conclusions with a slurry containing solid particles with sizes of 0.22 mm and 1.56 mm.

Javed et al. [37,38] performed experiments on the frictional behaviour of the slurry in a horizontal flow. The authors considered two grades of sand and glass spheres with median particle diameters of 0.103, 0.447, and 0.5 mm, respectively, and straw with a nominal particle length of 19.2 mm. The friction loss of the slurry measured in the vertical pipe of bulk velocities in the range of 0.5–4.2 m/s and mass concentrations in the range of 5–25% showed a reduction of 19% in drag at the solid concentration of 25% and at the bulk velocity of 3.5 m/s [37,38].

Experiments on medium- and coarse-solid particles suspended in a carrier liquid demonstrate the significant influence of particle diameter on the solid concentration distribution across a pipe. Nasr-El-Din et al. [39] and Sumner et al. [40] proved that in a vertical flow, the distribution of solids depends on the diameter of the particle and decreases when approaching the pipe wall. When the average diameter of the solid particles increases, the concentration profile becomes more inhomogeneous, showing a clear tendency to reach a minimum value near the pipe wall. In this case, the frictional loss was lower than expected. The authors also demonstrated that solid phase concentration profiles are uniform if fine solid particles are used [39,40].

The results mentioned above coincide with those obtained by other researchers, such as Eaton and Fessler [41] and Eskin and Miller [42]. For example, Eskin and Miller [42] proposed an original approach to describe the dynamics of a solid particle at the microscopic level for the case where a slurry flows through a gap. Their model showed that the dynamics of the solid particles in the carrier liquid is due to the fluctuation of the solid particles in the region of high shear rate. They found that solid particles migrate from the area of high to the area of low shear rate, resulting in the flow being characterised by a nonuniform distribution of the solid particles in the cross-section. The authors also found that the reduced concentration of the solid phase at the crack wall, due to particle migration, reduces friction loss comparing to a flow in which the distribution of the solid particles was homogeneous. In conclusion, the flow of the slurry through the gap was characterised by a lower pressure drop than expected [42].

Silva et al. [43] and Cotas et al. [44] used commercial CFD software to simulate eucalyptus and pine pulps fully developed turbulent flow; as such, slurries exhibit drag reduction. They applied a pseudo-homogeneous approach. The authors adopted the turbulence model of Chang et al. [45], in which they used the turbulence-damping function proposed by Bartosik [46]. They validated the mathematical model by comparing the measured and predicted friction pressure loss with decent accuracy.

Based on the gathered knowledge, concerning an isothermal slurry flow with medium or coarse solid particles, one can say that the median particle diameter has a significant influence on turbulence. However, this phenomenon is complex because many parameters, such as particle diameter, solid concentration, flow conditions, flow geometry, and properties of the solid and liquid phases, affect turbulence. This complexity leads to the need for research on medium and coarse slurry flows and mathematical models that are simple and relevant for engineering applications. The majority of mathematical models, which use a single- or two-phase approach, apply a standard two-equation turbulence model together with a commercial code [47,48]. Too little attention is paid to the phenomenon of slurry flow and the importance of physical features, which usually results in solving 3D governing equations, even though it is not justified. It is useless to apply a mathematical model to predict heat exchange in a slurry flow if the model was not positively validated for the isothermal flow first.

1.2. Literature Review: Non-Isothermal Flow

Most available data on heat exchange in slurry flows refer to low and moderate solid concentrations, and laminar flow. In contrast, data for high solid concentrations, which are of particular interest in the long-distance transportation of minerals, are difficult to access [49,50]. Reliable prediction first requires reliable measurement. Therefore, experiments on medium and coarse slurry flows are still expected, and the ability to simulate a slurry flow to predict frictional losses, velocity, and temperature distribution remains one of the main challenges of computational fluid dynamics.

The influence of the diameter of solid particles on heat exchange in a turbulent slurry flow has been experimentally investigated by several researchers [51–55]. For example, Mandal et al. [51] theoretically analysed the effect of particle size on fluid flow and heat transfer using the commercial code Fluent. The authors considered a fly ash slurry flow with median particle diameters of 0.012, 0.020, 0.028, and 0.034 mm in a horizontal pipe. They performed computations at a constant bulk density equal to 4 m/s and a volume concentration of 40%. The authors found that granular pressure and heat transfer increase with particle size. Other experiments on the heat transfer coefficient in a slurry transported in a horizontal pipe were performed by Rozenblit et al. [55]. The authors focused on the flow of the acetyl–water mixture on a moving bed with solid volume concentrations ranging from 5% to 15%. The authors used an electro-resistance sensor and infrared imaging to measure the temperature profile on the heated wall. The authors observed that the local heat transfer coefficient changed from its lowest value at the bottom of the pipe to the highest value at the upper heterogeneous layer [55].

Based on the literature, it can be noted that most of the research on heat transfer in solid–liquid flow refers to ice slurry and slurries with fine solid particles and colloidal suspensions, called nanofluids. Nanofluids have been known for several decades and were designed to improve heat exchange in some engines and heat exchangers [56–59]. Wang et al. [60] carried out experiments on the thermal conductivity of Al_2O_3 and CuO nanoparticles mixed with water, vacuum pump liquid, engine oil, and ethylene glycol [60]. The authors used Al_2O_3 with a median particle diameter of 28 nm and CuO with a median particle diameter of 23 nm. All particles were loosely agglomerated in the chosen liquid. The authors proved that the thermal conductivity of the mixture depends on the properties of the liquid and solid phases and increases with increasing solid volume concentration [60].

Recently, Cieslinski [47] reviewed the achievements in numerical modelling of forced convection in nanofluids in round and smooth tubes. The author noted that most mathematical models assume constant heat flux acting on a chosen pipe length and constant wall temperature. In the single-phase approach, the mixture of nanoparticles and carrier liquid is treated as homogeneous; therefore, the thermophysical properties are those of a nanofluid. In the two-phase approach, the thermophysical properties of the nanoparticles and carrier liquid are applied independently. In the review of mathematical models, the author selected only those that have been validated. The mathematical models mainly described suspension of nanoparticles of Al_2O_3, CuO, Cu, SiO_2, TiO_2, ZrO_2 and water as a carrier liquid. The author recognised 30 mathematical models for single-phase flows and 23 for two-phase flows. Cieslinski [47] stated that some researchers, such as, for example, Lotfi et al. [61] and Mokmeli and Saffar-Avval [62] observed that the precision of a single-phase approach is similar to that of a two-phase approach. The problem of closing the governing equations was solved using the standard k-ε turbulence model. The author stated that some available simulations indicated an improvement in heat transfer caused by carefully selected solid particles, while some did not [47].

Summarising the literature review on isothermal and non-isothermal slurry flow, one can say that the research is focused on fine slurry or laminar flows [48], while data for slurries with medium and coarse solid particles are rarely available. Furthermore, the mathematical models used for the prediction of slurry flow mainly rely on standard turbulence models and do not take into account the influence of the particle diameter and solid concentration on turbulence. Due to the limited access to experimental data on the behaviour of solid particles at the pipe wall in a broad range of solid particles and solid concentrations, we still face limitations in predicting velocity and temperature distributions. In the absence of reliable data for turbulence, especially near a pipe wall over a wide range of solid concentrations and particle diameters, it is difficult to suggest a new turbulence model. However, it is possible to modify existing turbulence models to predict the pressure drop, velocity, and temperature distributions in a slurry flow.

2. Method of Analysis

For the purpose of this study, the measurements of the isothermal slurry flow with glass spheres, carried out by Shook and Bartosik [63], and sand particles, carried out by Sumner [40], were chosen. The experiments were carried out in a closed vertical loop made of clear PVC in the upward direction. The vertical pipe was straight and smooth, and its inner diameter was 26 mm. The experimental test rig of these experiments is presented in Figure 1. All solid particles used in experiments were closely sized; that is, the particles had a narrow diameter distribution. The particles suspended in water were round, smooth and rigid. Four slurries were studied—two of them contained medium, while two other coarse solid particles. The following solid particles were used in this study:

a. Medium solid particles:
- glass spheres with median diameter d = 0.125 mm and particle density ρ_P = 2440 kg/m^3;
- glass spheres with median diameter d = 0.240 mm and particle density ρ_P = 2440 kg/m^3.

b. Coarse solid particles:
- sand with median diameter d = 0.470 mm and particle density rP = 2650 kg/m^3
- sand with median diameter d = 0.780 mm and particle density rP = 2650 kg/m^3

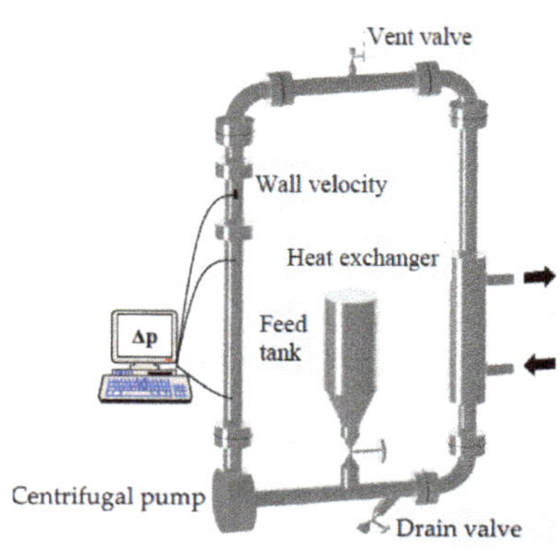

Figure 1. Experimental test rig; upward vertical slurry flow [63]. Source: own elaboration.

2.1. Physical Model

To develop a mathematical model for heat exchange in a slurry containing the aforementioned solid particles, a physical model must be formulated first. To build a simple mathematical model, it is better to consider a vertical pipe flow, because such a flow is axially symmetric, which simplifies the model. Therefore, in this study, a straight vertical pipeline was used, and the slurry flow was in the upward direction. The cylindrical coordinate system has been used in the mathematical model (x, r, θ), where '0x' is the symmetry axis of the pipe and represents the main flow direction, while the radial coordinate '0r' represents the distance from the symmetry axis. In this study, the coordinate '0θ', which represents the angle around the symmetry axis, will not be considered, which is the result of assumptions.

The physical model of slurry flow in the upward vertical pipeline was developed based on the following assumptions:

- The slurry flow is carried out in a straight vertical pipeline in the upward direction; the diameter of the inner pipe is equal to 0.026 m.
- The flow is steady, turbulent, axially symmetric ($V = 0$ and $\partial V/\partial r = 0$) and without circumferential eddies ($W = 0$ and $\partial W/\partial \varphi = 0$).
- The flow of the slurry is fully developed hydrodynamically, which means that the time-averaged velocity component U does not change in the flow direction ($\partial U/\partial x = 0$).
- The wall temperature is constant and equal to $T_W = 293.15$ K.
- The heat flux is constant, homogeneous, and equal to q = 1500 W/m; the heat flux acts from the pipe wall towards the symmetry axis.
- The slurry flow is fully thermally developed, which means that the temperature changes linearly in the flow direction 'ox', $\partial T/\partial x = \text{const} \neq 0$.
- Heat production in a slurry is negligible.
- Heat conduction through the pipe wall and radiation were neglected.
- The viscosity of the carrier liquid (water) was determined for the temperature of the pipe wall.
- The thermal properties of the solid particles are constant.
- The distribution of solid concentration in the cross-section of the pipe is constant.
- The single-phase approach is applied.
- The density of the slurry, the conduction coefficient, and the specific heat are constant and were determined from the temperature of the wall using Equation (1). Therefore, it was arbitrarily assumed that $(T_W - T_b) < 5$ K.

$$\Phi_m = C\Phi_S + (1 - C)\Phi_L \tag{1}$$

where T_b is the bulk temperature (averaged over the pipe cross-section); Φ is a general dependent variable that represents the density, heat conduction, or specific heat of the slurry; C is the volume concentration of solids, while the subscripts m, S, L refer to slurry (mixture), solid, and liquid, respectively.

The physical properties of the solid and liquid phases are presented in Table 1, while the physical properties of the slurries are shown in Table 2.

Table 1. The physical properties of solid and liquid phases.

Name	d, mm	ρ, kg/m³	μ, Pa s	c_p, J/(kg K)	λ, W/(m K)
Carrier liquid	0.000	998.32	0.0010046	4180.90	0.598
Glass spheres particles	0.125 0.240	2440.00		729.00	1.047
Sand particles	0.470 0.780	2650.00		712.00	1.088

where d is the median diameter of solid particles; ρ is the density of carrier liquid or slurry; μ is the dynamic viscosity of the carrier liquid; c_p is specific heat at constant pressure; λ is the thermal conductivity of the carrier liquid or solid particles.

Table 2. The physical properties of slurries.

Name	C, %	ρ_m, kg/m^3	$(c_p)_m$, J/(kg K)	λ_m, W/(m K)
Slurry with glass spheres particles	10	1142.49	3835.71	0.643
	30	1430.82	3145.33	0.733
	40	1574.99	2800.14	0.778
Slurry with sand particles	10	1163.49	3834.01	0.647
	30	1493.82	3140.23	0.745
	40	1658.99	2793.34	0.794

where ρ_m is the slurry density; $(c_p)_m$ is the specific heat of the slurry at constant pressure, and λ_m is the thermal conductivity of the slurry.

2.2. Mathematical Model

Taking into account the rules of time-averaging and neglecting terms with fluctuating components of density and viscosity, one can formulate the general governing equations of mass, momentum, and internal energy for steady vertical flow in the upward direction.

$$\frac{\partial}{\partial x_i}(\overline{\rho}\overline{U}_i) = 0 \quad (2)$$

$$\frac{\partial}{\partial x_j}(\overline{\rho}\overline{U}_i\overline{U}_j) = -\frac{\partial \overline{p}}{\partial x_i} + \frac{\partial}{\partial x_j}\left(\mu\frac{\partial \overline{U}_i}{\partial x_j} - \overline{\rho}\,\overline{u_i'u_j'}\right) - \overline{\rho}g_i \quad (3)$$

$$\frac{\partial}{\partial x_j}(\overline{\rho}\overline{U}_j\overline{T}) = \frac{\partial}{\partial x_j}\left(\frac{\mu}{Pr}\frac{\partial \overline{T}}{\partial x_j}\right) - \frac{\partial}{\partial x_j}\left(\overline{\rho}\,\overline{u_j't'}\right) \quad (4)$$

where U is the time-averaged velocity; p is the static pressure; μ and ρ are the dynamic viscosity and density of the transported medium, respectively; g is specific gravity; $\overline{\rho}\overline{u_i'u_j'}$ is the turbulence stress tensor; T is time averaged temperature; $\overline{\rho}\overline{u_j't'}$ is fluctuating component of velocity and temperature, and Pr is the Prandtl number defined as:

$$\mathrm{Pr} = \frac{\mu\, c_p}{\lambda} \quad (5)$$

Solving the set of equations requires that the number of equations is equal to the number of dependent variables. This approach is called the closure problem. The momentum and energy Equations (3) and (4) possess the components $-\overline{\rho}\overline{u_i'u_j'}$ and $-\overline{\rho}\overline{u_j't'}$, which are unknown and require closure. Both components can be modelled using a direct (DNS) or indirect approach. The direct method gives a high level of accuracy but is more complex and requires long computation times. Additionally, this method considers random velocity and random temperature, which is not pragmatic for most engineering applications. The indirect method gives fair accuracy in numerical prediction but is less complex and needs a much shorter computation time. In addition, the indirect method considers the time-averaged velocity and temperature, which is more relevant in engineering applications. Therefore, an indirect method was used in this study. Taking into account the Boussinesque concept, we can write [64]:

$$-\overline{\rho}\,\overline{u_i'u_j'} = \mu_t\left(\frac{\partial \overline{U}_i}{\partial x_j} + \frac{\partial \overline{U}_j}{\partial x_i}\right) - \frac{2}{3}\overline{\rho}\,k\,\delta_{i,j} \quad (6)$$

where k is the kinetic energy of the turbulence; $\delta_{i,j}$ is the Kronecker quantity; μ_t is the turbulent viscosity.

The fluctuating components in Equations (3) and (4) can be described using Equation (6) as:

$$-\overline{\rho}\,\overline{u'v'} = \mu_t\frac{\partial \overline{U}}{\partial r} \quad (7)$$

$$-\bar{\rho}\,\overline{u_j' t'} = \frac{\mu_t}{Pr_t}\frac{\partial \overline{T}}{\partial x_i} \tag{8}$$

where u′ and v′ are fluctuating parts of the velocities in the ox and or direction, respectively; U is the component of the time-averaged velocity in the ox direction, Pr_t is the turbulent Prandtl number for the boundary layer equal to $Pr_t = 0.9$ [65].

Taking into account the assumptions stated in the physical model, the final form of Equations (2)–(4), together with Equations (7) and (8), can be expressed as follows.

$$\frac{\partial}{\partial x}\left(\bar{\rho}_m \overline{U}\right) = 0 \tag{9}$$

$$\frac{1}{r}\frac{\partial}{\partial r}\left[r\,(\mu+\mu_t)\frac{\partial \overline{U}}{\partial r}\right] = \frac{\partial \overline{p}}{\partial x} + \rho_m g \tag{10}$$

$$\bar{\rho}_m \overline{U}\,\frac{\partial \overline{T}}{\partial x} = \frac{1}{r}\frac{\partial}{\partial r}\left[r\left(\frac{\mu}{Pr}+\frac{\mu_t}{Pr_t}\right)\frac{\partial \overline{T}}{\partial r}\right] \tag{11}$$

The turbulent viscosity μ_t in Equation (10) was calculated using the turbulence model of Launder and Sharma [66]. The Launder–Sharma model was successfully validated for many turbulent flows, including homogeneous slurries. Researchers emphasised that it is one of the first and most widely used models. It has been shown to agree well with measurements and DNS data for a wide range of turbulent flows and performs better than many other k-ε models [67–71]. Launder and Sharma expressed the turbulent viscosity as follows [66]:

$$\mu_t = f_\mu \frac{\overline{\rho}}{\varepsilon} k^2 \tag{12}$$

$$f_\mu = 0.09\, exp\left[\frac{-3.4}{\left(1+\frac{Re_t}{50}\right)^2}\right] \tag{13}$$

where f_m is called the damping function of the turbulence or the wall damping function. The turbulent Reynolds number (Re_t) was developed from the dimensionless analysis [66], as follows:

$$Re_t = \frac{\rho\, k^2}{\mu\, \varepsilon} \tag{14}$$

where ε is the dissipation of the kinetic energy of turbulence.

The turbulence damping Function (13) was determined for homogeneous turbulence [68]. It will be demonstrated that using Function (13) for the chosen slurries gives a higher friction than the measurements. Therefore, a specially designed turbulence-damping function was used for medium and coarse slurries. The turbulence-damping function applied in the mathematical model is the following [46]:

$$f_\mu = 0.09\left\{\frac{-3.4\left[1 + A_P^3\, d^2 (8 - 88\, A_P\, d)\, C^{0.5}\right]}{\left(1+\frac{Re_t}{50}\right)^2}\right\} \tag{15}$$

where A_P is an empirical constant ($A_P = 100$).

Function (15) was developed empirically by matching the predictions of friction loss and velocity distribution with the measurements. This function was validated for isothermal flows for medium and coarse slurries in the range of median particle diameters of 0.1 to 0.8 mm and solid concentrations of 10% to 40%, obtaining good accuracy of predictions of friction losses and velocity profiles [46,69,70,72,73].

Analysis of Equation (15) indicates that if d→0 or C→0 the turbulence-damping function approaches the standard function expressed by Equation (13). Equation (15) indicates that the median particle diameter of the solid particle plays a primary role, while the solid concentration plays a secondary role.

Considering the assumptions made in the physical model, the final forms of the equations of the kinetic energy of turbulence and its dissipation rate in the cylindrical coordinate system are as follows.

$$\frac{1}{r}\frac{\partial}{\partial r}\left[r\left(\mu+\frac{\mu_t}{\sigma_k}\right)\frac{\partial k}{\partial r}\right]+\mu_t\left(\frac{\partial \overline{U}}{\partial r}\right)^2=\rho\varepsilon+2\mu\left(\frac{\partial k^{1/2}}{\partial r}\right)^2 \tag{16}$$

$$\frac{1}{r}\frac{\partial}{\partial r}\left[r\left(\mu+\frac{\mu_t}{\sigma_\varepsilon}\right)\frac{\partial \varepsilon}{\partial r}\right]+C_1\frac{\varepsilon}{k}\mu_t\left(\frac{\partial \overline{U}}{\partial r}\right)^2=C_2\left[1-0.3\,exp\left(-Re_t^2\right)\right]\frac{\rho\varepsilon^2}{k}-2\frac{\mu}{\rho}\mu_t\left(\frac{\partial^2 \overline{U}}{\partial r^2}\right)^2 \tag{17}$$

where C_1, and C_2 are constants; σ_k and σ_ε are diffusion coefficients in the equations of k and ε, respectively.

According to the assumptions, the convective term in the energy Equation (11) requires a unique approach. To develop the convective term, consider the energy balance between the pipe, with inner diameter D and unit length $L = 1$ m, and the slurry in the pipe. The heat flux, which is transferred from the pipe to the slurry, can be expressed as

$$\dot{Q}=\alpha\, A_*\, \Delta\overline{T}_r \tag{18}$$

where α is the coefficient of convective heat transfer; $\Delta\overline{T}_r = T_w - T_b$ is temperature difference on radius r, and

$$A_*=\pi DL \tag{19}$$

The changes in the inner energy of the slurry over the pipe distance L are as follows.

$$\dot{Q}=\dot{m}\, c_P\, \Delta\overline{T}_x=\overline{\rho}_b\,\overline{U}_b\, A\, c_p\, \Delta\overline{T}_x \tag{20}$$

where $\dot{m}$ is flux of slurry mass; $\Delta\overline{T}_x$ is the slurry temperature difference over the pipe distance $\Delta x = L$; $\overline{\rho}_b$ and $\overline{U}_b$ are bulk density and velocity, respectively; the $A = \pi D^2/4$ is the cross section of the pipe.

Comparing the right-hand side of Equation (18) with the right-hand side of Equation (20) and including (19), one can write:

$$\alpha\,\pi\, D\, L\, \Delta\overline{T}_r=\overline{\rho}_b\,\overline{U}_b\,\pi\,\frac{D^2}{4}\,c_p\,\Delta\overline{T}_x \tag{21}$$

Dividing each side of Equation (21) by the unit length L = Δx, we get

$$\dot{q}=\overline{\rho}_b\,\overline{U}_b\,\pi\,\frac{D^2}{4}\,c_p\,\frac{\Delta\overline{T}_x}{\Delta x} \tag{22}$$

where q is the unit heat flux.

Finally, ∂T/∂x, which exists in the energy Equation (11), can be expressed as follows:

$$\frac{\partial\overline{T}}{\partial x}=\frac{\dot{q}}{\overline{\rho}_b\,\overline{U}_b\,\pi\, R^2 c_p}=const \tag{23}$$

where R is the radius of the pipe.

The gradient ∂T/∂x, described in Equation (23), was applied in Equation (11). The thermal properties of the carrier liquid and solid phases used in the computations are known and are listed in Tables 1 and 2.

The final form of the mathematical model of heat exchange in the turbulent flow of medium and coarse slurries in an upward vertical pipeline constitutes four partial differential equations, namely, (10), (11), (16), and (17). The model has complementary Equations (12), (14), (15), and (23). The constants in the k-ε turbulence model are the same as in the Launder and Sharma model and are the following: $C_1 = 1.44$; $C_2 = 1.92$; $\sigma_k = 1.0$; $\sigma_\varepsilon = 1.3$ [66]. The mathematical model contains five dependent variables: U(r), p(x), T(r), k(r), ε(r), and has four partial differential equations only. The closure problem was solved by assuming that the pressure gradient ∂p/∂x is given. Therefore, the mathematical model was solved for the pre-set value of ∂p/∂x. By changing ∂p/∂x, one can change the bulk velocity. This approach allows one to make the mathematical model one-dimensional and reduce the computation time.

2.3. Numerical Procedure

The following boundary conditions were applied in the numerical computations:

$$\text{pipe wall, } r = R: \quad q = 1500\ \text{W/m};\ T_w = 293.15\ \text{K};\ U = 0;\ k = 0;\ \varepsilon = 0; \tag{24}$$

$$\text{symmetry axis, } r = 0: \quad \partial T/\partial r = 0;\ \partial U/\partial r = 0,\ \partial k/\partial r = 0 \text{ and } \partial \varepsilon/\partial r = 0. \tag{25}$$

Equation (24) indicates that the wall temperature is constant and there is no slip on the pipe wall (U = k = ε = 0), while Equation (25) indicates that axially symmetric conditions were applied to the dependent variables. The set of partial differential Equations (10), (11), (16) and (17) was solved by the TDMA method with the iteration procedure and the finite-volume method [74] and using its own computer code. The finite volume was established for a pipe with a length of L = 1 m, and by rotating its radius around the symmetry axis at an angle of φ = 1 radian. The numerical computations require initial dependent variables, which were calculated as follows.

- $T(r) = T_w$;
- $U(r) = U_{max}[(R - r)/R]^{1/7}$;
- k was established on the basis of the assigned intensity of turbulence, that is, $k(r) = 1/2\ U^2(r)$;
- ε was established based on the equation: $\varepsilon(r) = 0.09^{3/4}\ k^{3/2}(r)/L$, where L was determined from the Nikuradse formula [75].

The iterative convergence criterium was defined by Equation (26).

$$\sum_j \left| \frac{\varnothing_j^n - \varnothing_j^{n-1}}{\varnothing_j^n} \right| \leq 0.001 \tag{26}$$

where $\varnothing_j^n$ is a general dependent variable ∅ = U, T, k, ε; the jth is the nodal point after the nth iteration cycle, and the $\varnothing_j^{n-1}$ is the (n − 1)th iteration cycle.

The number of grid points strongly affects the accuracy of the predictions [74,76]. Therefore, the following meshes containing 60, 70, 80, and 90 nodal points distributed on a pipe radius were applied. Additionally, the expansion coefficient has been used to achieve a nonuniform distribution of nodal points on a pipe radius, so most nodal points were located close to a pipe wall. It was found that for the expansion coefficient equal to 1.1, the iterative convergence process for 80 and 90 nodal points was similar. For this reason, the mesh with 80 nodal points was used.

The results of numerical computations allowed us to obtain the flow field of the fully developed hydrodynamically and thermally turbulent flow of medium and coarse slurries, which includes the distribution of T(r), U(r), k(r), ε(r). Additionally, the following quantities can be calculated: Nu, Pr, Re, U_b, and T_b.

2.4. Validation for Isothermal Slurry Flow

The mathematical model has been validated for isothermal slurry flows in the range of median particle diameters from d = 0.1 mm to d = 0.8 mm, and solid concentrations of C = 10% to C = 40%, and in a comprehensive range of Reynolds numbers and pipe diameters, giving reasonably good predictions of dp/dx = f(U_b) and velocity profiles [46,69,70,72,73]. To convince the reader that the chosen slurries with medium and coarse solid particles demonstrate the suppression of turbulence and that the mathematical model predicts those flows well, the results of the selected measurements and predictions will be presented.

Figures 2 and 3 show the results of the measured and predicted frictional pressure loss for four slurries, containing particles with diameters of 0.125 mm, 0.240 mm, 0.470 mm, and 0.780 mm, respectively, and for volume concentration of solids equal to 40%. The results of predictions using the modified damping Function (15) are marked as solid red lines, while predictions with the standard damping Function (13) are marked as dashed red lines.

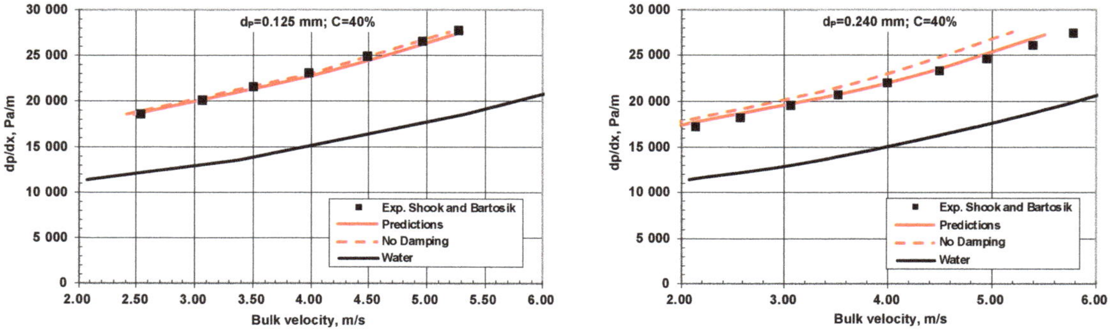

(a) (b)

Figure 2. Predictions with and without damping of turbulence and measurements [63] for glass spheres slurries: (**a**) d = 0.125 mm, *C* = 40%, D = 0.026 mm; (**b**) d = 0.240 mm, C = 40%, D = 0.026 mm.

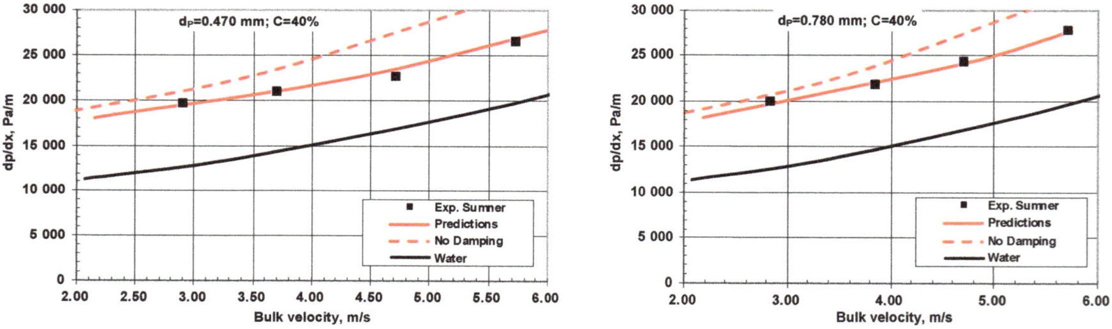

(a) (b)

Figure 3. Predictions with and without damping of turbulence and measurements [40] for sand slurries: (**a**) d = 0.470 mm, C = 40%, D = 0.026 mm; (**b**) d = 0.780 mm, C = 40%, D = 0.026 mm.

The analysis of Figures 2 and 3 shows discrepancies between the measurements and the numerical predictions if the mathematical model contains a standard wall damping function, described by Equation (13). The disparity increases with increasing particle diameter, and the highest discrepancy appears for the slurry with a solid particle diameter equal to 0.470 mm, as seen in Figure 3a. The predictions with the wall damping Function (15) match well with the measurements for all applied particles, as seen in Figures 2 and 3.

The analysis of Figures 2 and 3 indicates that dp/dx for glass sphere slurry at C = 40% should be approximately 1.57 times higher than water because this slurry's density is equal to 1574 kg/m^3. However, it is approximately 1.52 for d = 0.125 mm and 1.45 for d = 0.240 mm. If the data for the sand slurry with C = 40% (ρ_m = 1658 kg/m^3) are analysed, we expect that dp/dx should be approximately 1.66 higher compared to water, while it is only 1.49 for d = 0.470 mm and 1.51 for d = 0.780 mm. In conclusion, the experimental data presented in Figures 2 and 3 explicitly show that the friction loss of the four slurries depends on the diameter of the particle and is lower than we expected. Therefore, one can conclude that such slurries demonstrate turbulence suppression. The lowest friction loss was obtained for the slurry with d = 0.470 mm.

2.5. Validation for Non-Isothermal Slurry Flow

Due to the lack of experimental data on heat exchange in slurries used in these studies, the mathematical model was validated only for carrier liquid flow. Numerical computations were performed for turbulent carrier liquid flow in a vertical pipe with inner diameter D = 0.026 m, wall temperature T_w = 293.15 K, and unit heat flux q = 1500 W/m. The results of the predicted Nusselt number were compared with the Dittus–Boelter empirical Equation (27) [77] and are presented in Figure 4. The Dittus–Boelter equation is valid for fully developed flows for Re > 10,000 and $0.7 \leq Pr \leq 160$ [77] and is expressed as follows:

$$Nu = 0.02296\, Re^{0.8}\, Pr^{1/3} \tag{27}$$

where Reynolds and Prandtl's numbers are as follows:

$$Re = \frac{\rho\, U_b\, 2\, R}{\mu_L} \tag{28}$$

$$Pr = \frac{\mu_L\, c_p}{\lambda} \tag{29}$$

$$Relative\ Error = \frac{|Nu_{Pred} - Nu_{Exp}|}{Nu_{Exp}}\, 100\% \tag{30}$$

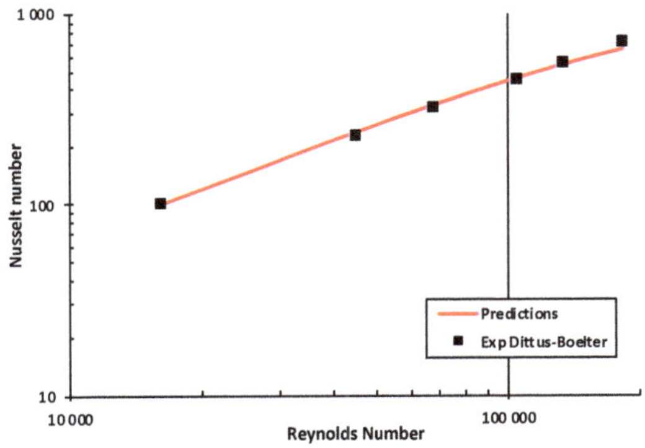

Figure 4. Validation of numerical predictions of the Nusselt number for carrier liquid flow. D = 0.026 m; T_w = 293.15 K, q = 1500 W/m.

It is seen in Figure 4 that the numerical computations of Nusselt number are close to the Dittus–Boelter data, and the averaged relative error, expressed by Equation (30), is about 2%.

3. Results of Predictions

Numerical predictions were performed based of the simplified mathematical model that contains Equations (10), (11), (16), (17), and the turbulence-damping Function (15). Predictions were made for four slurries. In each case, the slurry flow was fully developed hydrodynamically and thermally in a vertical upward pipeline with an inner diameter

D = 0.026 m. The wall temperature and the unit heat flux were constant and equal t 293.15 K and 1500 W/m, respectively. Simulations were performed for solid volume cor centrations equal to 10%, 30%, and 40% by volume, and for water. The thermal propertie of the solid and liquid phases and slurry were stated in Tables 1 and 2.

The results of the predictions of Nu = f(Re) for medium and coarse slurries at C = 10% 30%, and 40% are presented in Figures 5 and 6. The Nusselt number for the carrier liqui (water) was calculated using the empirical Dittus–Boelter Equation (27).

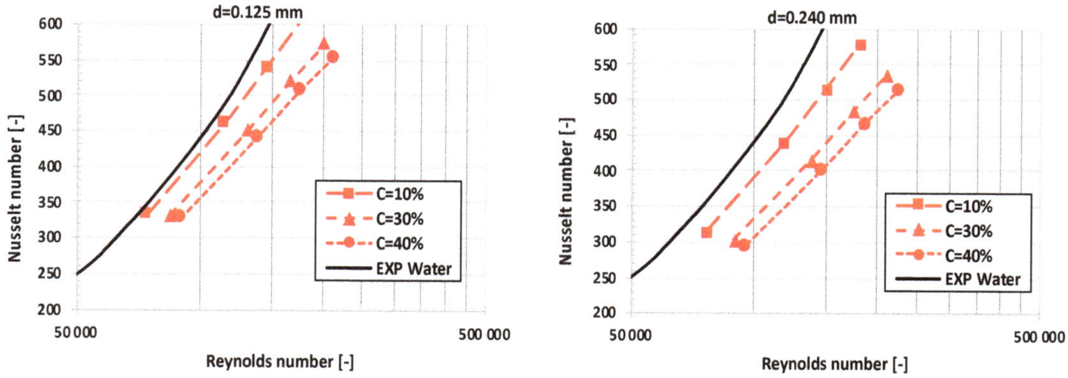

(a) (b)

Figure 5. Dependence of the Nusselt number on the Reynolds number for glass spheres slurries an water: (**a**) d = 0.125 mm and C = 10%, 30%, 40%; (**b**) d = 0.240 mm and C = 10%, 30%, 40%.

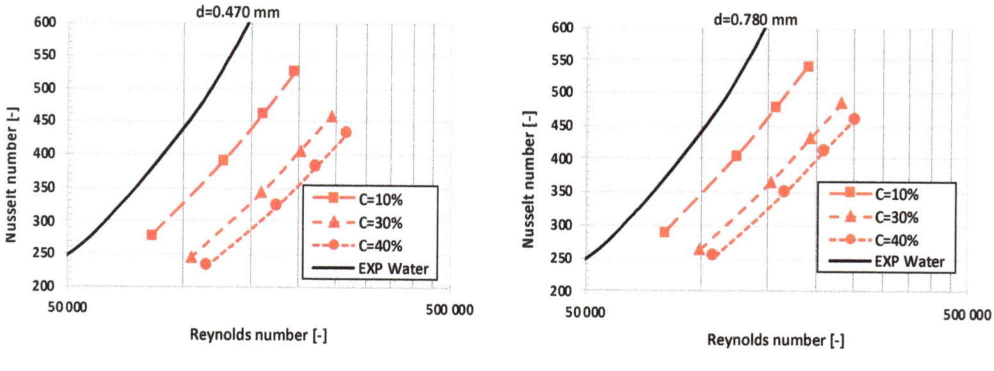

(a) (b)

Figure 6. Dependence of the Nusselt number on the Reynolds number for sand slurries and wate (**a**) d = 0.470 mm and C = 10%, 30%, 40%; (**b**) d = 0.780 mm and C = 10%, 30%, 40%.

Figures 5 and 6 demonstrate that parameters such as Re, d, and C strongly affect th Nusselt number. For all four slurries, the Nusselt number was lower compared to wate It is seen that with an increase in solid particle diameter from 0.125 mm to 0.470 mm, th Nusselt number decreases, while for d > 0.470 mm there is an increase.

The influence of particle diameter on the Nusselt number is clearly seen in Figure 7a,b Both figures present predictions for the four slurries at C = 10% and C = 40%, respectivel Analysing Figures 2 and 3 and Figure 7a,b, it is seen that there is a correlation betwee frictional pressure loss and the Nusselt number. For example, for solid concentration C = 40% it can be seen that the friction loss decreases if the diameter of the solid particles increases fro d = 0.125 mm to d = 0.470 mm (Figures 2a and 3a) and increases if d > 0.470 mm (Figure 3b). The same applies to the Nusselt number, which is seen in Figure 7b.

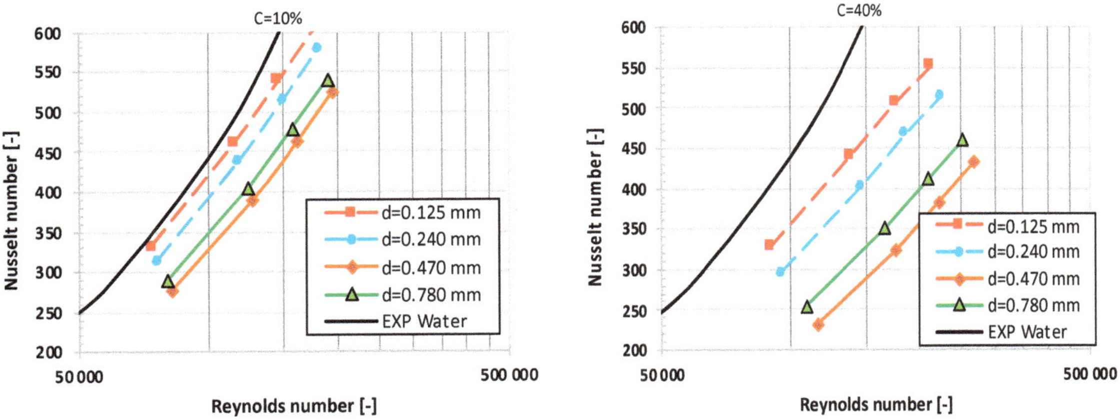

(a) (b)

Figure 7. Dependence of the Nusselt number on the Reynolds number for four slurries: (**a**) C = 10%; (**b**) C = 40%.

The relation of dp/dx = f(U_b) determines the link Nu = f(Re). Therefore, it is interesting to analyse the dependence of the Nusselt number on particle diameter for chosen solid concentrations and for chosen values of dp/dx, which is presented in Figure 8a,b and Figure 9a.

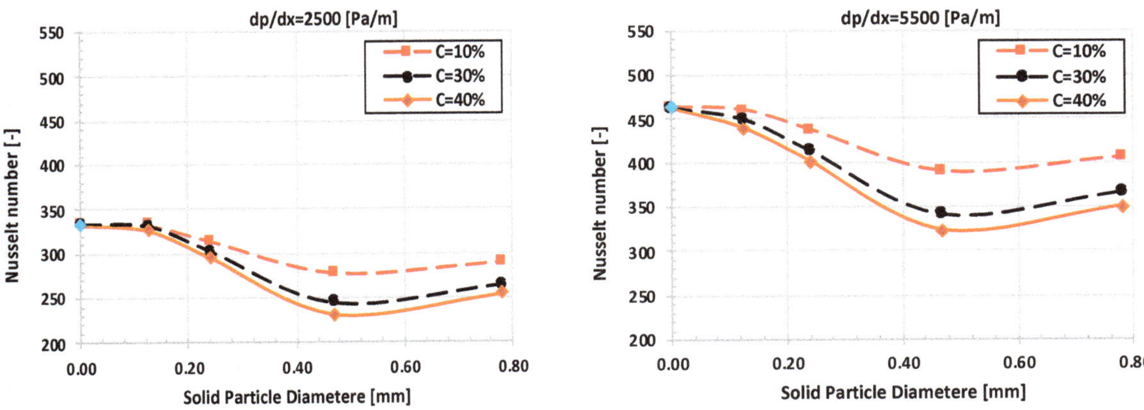

(a) (b)

Figure 8. Dependence of the Nusselt number on the particle diameter for four slurries and water: (**a**) dp/dx = 2500 Pa/m, C = 10%, 30%, 40%; (**b**) dp/dx = 5500 Pa/m, C = 10%, 30%, 40%.

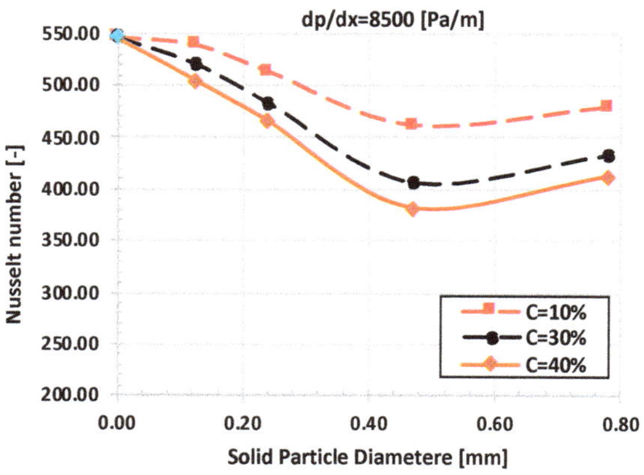

(**a**)

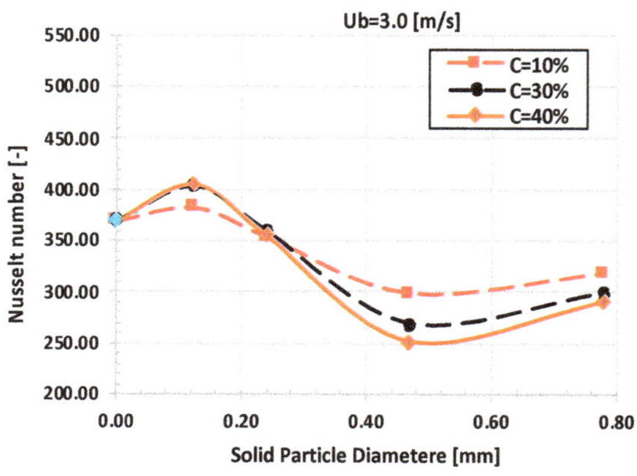

(**b**)

Figure 9. Dependence of the Nusselt number on the particle diameter for four slurries and water: (**a**) dp/dx = 8500 Pa/m, C = 10%, 30%, 40%; (**b**) U_b = const = 3.0 m/s and C = 10%, 30%, 40%.

Figure 8a,b and Figure 9a demonstrate that an increase in dp/dx causes an increase in the Nusselt number, while an increase in solid concentration causes a decrease in the Nusselt number. The first point in Figure 8a,b and Figure 9a, that is, for d = 0.0 mm, presents the data for the water, which were calculated using the Dittus–Boelter Equation (27). Analysis of Figure 8a,b and Figure 9a indicates that in the range of d = (0.125−0.780) mm, a minimum Nusselt number appears for d = 0.470 mm, regardless of solid concentration, while there is a maximum for water. This phenomenon is consistent with the behaviour of the pressure drop, which is presented in Figures 2 and 3.

It is interesting to analyse the behaviour of the Nusselt number for constant bulk velocity. Figure 9b presents the Nusselt number's dependence on the solid particle's diameter at constant bulk velocity equal to 3.0 m/s, and for C = 10%, 30%, and 40%, and for water. Again, the first point in Figure 9b shows the predictions for water (d = 0.0 mm). Figure 9b shows that with an increase in particle diameter from d = 0.0 mm (water) to d = 0.125 mm, the Nusselt number increases and reaches a maximum for d = 0.125 mm, regardless of solid concentration. If the particle diameter increases from d = 0.125 mm to d = 0.470 mm the Nusselt number decreases and reaches a minimum for d = 0.470 mm, irrespective of the solid concentration. Figure 9b shows that for constant bulk velocity (constant flow rate), the maximum and minimum of the Nusselt number exist for d = 0.125 mm and d = 0.470 mm, respectively. The maximum and minimum Nusselt numbers are consistent with the behaviour of the frictional pressure drop, which has been discussed earlier.

To illustrate why the efficiency of heat exchange between the pipe and the slurry is highest for d = 0.125 mm and the lowest for d = 0.470 mm, Figure 10 presents the temperature distributions for both slurries at C = 40% and U_b = 3.0 m/s. Taking into account Equation (18), it can be concluded that for constant heat flux, the coefficient of convective heat transfer is higher for the slurry with d = 0.125 mm compared to the slurry with d = 0.470 mm because $T_w - T_b$ is lower for the slurry with d = 0.125 mm than for d = 0.470 mm.

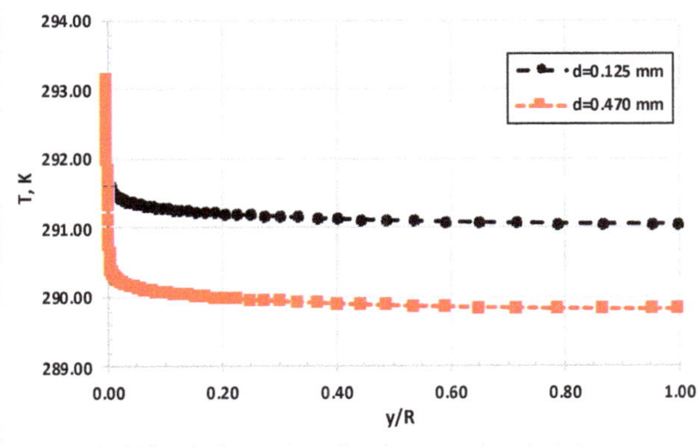

Figure 10. Temperature distributions for two slurries: with d = 0.125 mm; d = 0.470 mm; at U_b = 3.00 m/s, and C = 40%.

4. Discussion

It is well-known that fine solid particles can move freely in the viscous sublayer, and the friction process is similar to the flow of the carrier liquid. If sufficiently large particles, such as those used in this study, are considered, such particles are pushed away from the wall toward the symmetry axis, causing a decrease in the friction between the particles and the wall, and modifying the turbulence in the remaining region. Experiments have shown that, in some cases, the friction in the slurry can be lower than we expected. Such a phenomenon was observed in experiments of Charles and Charles [78] for d = 0.216 mm, Ghosh and Shook [79] for d = 0.6 mm, Matousek [36] for d = 0.37 mm, Sumner [40] for d = 0.47 mm, and d = 0.78 mm, Talmon [80] for d = (0.1–2.0) mm, and Ming-zhi et al. [81] for d = 0.125 mm and d = 0.44 mm. The process of enhancement or suppression of turbulence, caused by the presence of solid particles in the carrier liquid, is complex, especially in the viscous sublayer and buffer layer, and is still not well understood. For such complex phenomena, it is impossible to propose a universal approach to predict frictional loss or heat transfer.

This study showed that a mathematical model that uses a standard turbulence-damping function is not suitable for predicting the isothermal flow of a slurry with medium and coarse solid particles. Thus, it cannot be used to predict heat transfer. However, using the specially designed turbulence-damping function, described by Equation (15), allows one to apply a mathematical model to predict isothermal and non-isothermal slurry flow. However, it should be noted that the mathematical model assumes that the solid concentration across the pipeline is constant. In contrast, the experiments proved that the distribution is not constant, and the lowest concentration appears close to the pipe wall. The effect of non-uniform concentration distribution is compensated by the wall damping function, described by Equation (15).

Analysis of the literature indicates that the simplest way to formulate a mathematical model is to use the assumptions of constant heat flux acting on a unit of pipe length and constant wall temperature [47]. This approach was just used in this study. Looking for simplifications, the convective term in the energy equation was developed based on the heat balance between the pipe and the slurry. This approach allows one to reduce the mathematical model to one-dimensional and decrease the time of computation, causing the model to be convenient for parametric studies. The mathematical model allows for predicting the frictional pressure loss, profiles of velocity and temperature, and the Nusselt number. It is essential to emphasise that the mathematical model for heat exchange in a slurry flow should first be validated for the isothermal flow. The applied mathematical model is simple and suitable for predicting heat transfer in a medium or coarse slurry if the assumptions stated in the physical model are used. However, this model still requires validation for heat transfer if experimental data will be available. It is worth mentioning that the physical properties of slurries with glass spheres and sand particles are remarkably similar. Differences in the specific heat and thermal conductivity are minor, and the relative discrepancy is below 2.5%, while the discrepancy is about 8% for particles density.

Performed studies demonstrate that the median particle diameter of solid particles plays a substantial role in heat exchange in a slurry flow. The Nusselt number decreases in some cases; in the other cases, it increases with the change in the diameter of the solid particles. These anomalies indicate that the diameter of the solid particle affects frictional losses and is related to the physics of heat transfer and needs further study. The novelty and value of this study lies in the deeper characterisation of the influence of medium- and coarse-solid particles on heat exchange between pipe wall and slurry.

5. Conclusions

Analysis of the results of numerical predictions of heat exchange in turbulent flow of four slurries containing glass particles with d = 0.125 mm and d = 0.240 mm and sand particles with d = 0.470 mm and d = 0.780 mm in a vertical pipe directed upwards allows the following conclusions to be drawn:

- The diameter of the solid particle affects the Nusselt number and the friction loss regardless of the concentration of the solids.
- The higher the solid concentration in the slurry, the lower the Nusselt number.
- In the chosen range of median particle diameters, the Nusselt number changed sinusoidal, reaching maximum for the slurry with d = 0.125 mm, and minimum for d = 0.470 mm.
- It was found that for the same volume of transported slurry (U_b = const), the efficiency of heat transfer is the highest for the slurry with *d* = 0.125 mm, while for the slurry with d = 0.470 mm *it* is the lowest—see Figures 9b and 10.
- The influence of the particle diameter on the Nusselt number and friction losses result in the impact of the particles on the turbulence. The suppression of turbulence was determined indirectly based on frictional loss analysis.
- Solid particles with a diameter of 0.470 mm demonstrate the highest suppression of turbulence and, consequently, suppression of heat transfer.

The results obtained can serve as a reference for further study of the resistance mechanism and the enhancement or suppression of heat transfer in the transported slurry through the presence of selected solid particles. Future research should focus on a deeper understanding of the mechanism of influence of particle size, density, and shape on turbulence, especially in the region close to a pipe wall. Research on the impact of physical and thermal properties of solid particles on the transport capability of solid particulates in different Prandtl liquids to minimise pressure loss and increase or decrease heat exchange is also needed.

Funding: The research was funded by Ministry of Education and Science, grant number 01.0.00.00/1.02.001/SUBB.M 23.002.

Data Availability Statement: Data are contained within this article.

Conflicts of Interest: The author declares no conflict of interest.

Nomenclature

A	cross section of the pipe, m^2
A_P	constant in the turbulence damping function (A_P = 100)
A_*	pipe surface over distance L = 1 m, m^2
a, b, c	constants in Taylor expansion

C_i	constant in Launder and Sharma turbulence model, i = 1, 2
C	solids concentration (volume fraction of solids averaged in cross-section), %
c_p	specific heat at constant pressure, J/(kg K)
D	inner pipe diameter, m
d	median solid particle diameter, mm
f_μ	turbulence damping function
g	specific gravity, m/s^2
j	nodal point
k	kinetic energy of turbulence, m^2/s^2
L	unit pipe length / length of mixing in Nikuradse formula, m
$\dot{m}$	flux of slurry mass, kg/s
n	number of iteration cycle
Nu	Nusselt number
Pr	Prandtl number
p	static pressure, Pa
q	input power of heat per unit pipe length (unit heat flux), W/m
R	pipe radius, m
r	cylindrical coordinate, distance from symmetry axis, m
Re	Reynolds number
t′	fluctuating component of temperature, K
T	temperature, K
$U_{i,j}$	time averaged velocity components, where U = f(U, V, W), m/s
U	velocity component in the main flow direction '0x', m/s
V	velocity component in the direction '0r', m/s
W	velocity component in the direction '0φ', rad/s
u′, v′	fluctuating components of velocity U, and V, respectively, m/s
x	cylindrical coordinate, main flow direction, m
y	radial distance from a pipe wall, m
$\bar{}$	time averaged
Greek symbols	
α	convective heat transfer coefficient, W/(m^2 K)
δ_{ij}	Kronecker quantity
Δx	unit pipe length, m
ε	rate of dissipation of kinetic energy of turbulence, m^2/s^3
λ	thermal conductivity of water or slurry, W/(m K)
μ	dynamic viscosity coefficient for water, Pa·s
ν	kinematic viscosity coefficient for water, m^2/s
ρ	density, kg/m^3
σ_i	diffusion coefficients in k-ε turbulence model, i = k, ε
τ	shear stress, Pa
Φ	general dependent variable, $\Phi = \rho, c_p, \lambda$
ϕ	general dependent variable, Φ = U, k, ε
φ	cylindrical coordinate, angle around pipe symmetry axis, deg
Subscripts	
b	bulk (cross-section averaged value)
EXP	experiment
i	index, i = 1, 2
j	number of nodal points
L	liquid
m	slurry (solid–liquid mixture)
n	number of iterations
P	solid particle/solid phase
Pred	predictions
r	radius
t	turbulent
w	wall

References

1. Sape Miedema, P.A. *Slurry Transport: Fundamentals, Historical Overview and Delft Head Loss and Limit Deposit Velocity Framework*; TU Delft Open: Delft, The Netherlands, 2016; ISBN 9789461862938.
2. Zengeni, B.; Nekhavhambe, D.; Goosen, P. Sensitivity of non-Newtonian slurry viscous properties to temperature. In Proceedings of the 19th Conference on Transport and Sedimentation of Solid Particles, Cape Town, South Africa, 24–27 September 2019; pp. 285–292.
3. Dai, Y.; Zhang, Y.; Li, X. Numerical and experimental investigations on pipeline internal solid-liquid mixed fluid for deep ocean mining. *Ocean Eng.* **2021**, *220*, 108411. [CrossRef]
4. Wang, R.; Zhu, Z.; Su, X.; Mianowicz, K. Slurry pumps in deep-sea mining: A review of numerical and experimental studies. *Ocean Eng.* **2022**, *251*, 111150. [CrossRef]
5. Shook, C.A.; Roco, M.C. *Slurry Flow: Principles and Practice. Howard Brenner*; Digital Press: Clifton, NJ, USA, 2013; Available online: https://www.perlego.com/book/1884007/slurry-flow-principles-and-practice-pdf (accessed on 3 September 2023).
6. Baha, A.P.E. *Slurry Systems Handbook*, 2nd ed.; McGraw Hill: New York, NY, USA, 2021.
7. Reddy, N.V.K.; Pothal, J.K.; Barik, R.; Senapati, P.K. Pipeline slurry transportation system: An overview. *J. Pipeline Syst. Eng. Pract.* **2023**, *14*, 1–10. [CrossRef]
8. Csengeri, E. Fuel Discharge towards the Core Catcher in Sodium-Cooled Fast Reactor under Severe Accident Conditions with Mitigation Devices. Ph.D. Thesis, Université Grenoble Alpes, Grenoble, France, 2022.
9. Dong, H.; Aziz, N.A.; Shafri, H.Z.M.; Ahmad, K.A.B. Numerical study on transportation of cemented paste backfill slurry in bend pipe. *Processes* **2022**, *10*, 1454. [CrossRef]
10. Wilson, K.C.; Addie, G.R.; Sellgren, A.; Clift, R. *Slurry Transport Using Centrifugal Pumps*, 3rd ed.; Springer: Berlin/Heidelberg, Germany, 2006; Available online: https://link.springer.com/book/10.1007/b101079 (accessed on 3 September 2023).
11. Chen, Z.X.; Hu, H.X.; Guo, X.M.; Zheng, Y.G. Effect of groove depth on the slurry erosion of V-shaped grooved surfaces. *Wear* **2022**, *488–489*, 204133. [CrossRef]
12. Zhang, R.; Sun, L.; Li, Y.; Yin, J. Numerical analysis of solid particle erosion caused by slurry in dredging pipelines based on a particle separation method. *Powder Technol.* **2023**, *428*, 118826. [CrossRef]
13. Shaukat, R.; Qamar, A.; Anwar, Z.; Imran, S.; Amjad, M.; Ali, H.M. Experimental study on heat transfer performance of mPCM slurry flow in microchannels. *J. Therm. Anal. Calorim.* **2023**, 1–13. [CrossRef]
14. Rogachevskii, I. *Introduction to Turbulent Transport of Particles, Temperature and Magnetic Fields Analytical Methods for Physicists and Engineers*; Cambridge University Press: Cambridge, UK, 2021. [CrossRef]
15. Ahmed, J. *Rheology and Rheological Measurements*; Wiley: Hoboken, NJ, USA, 2021. [CrossRef]
16. Jorge, O. *Particle-Laden Turbulent Pipe Flows*; OmniScriptum GmbH Co.: Riga, Latvia, 2017.
17. Minier, J.P.; Pozorski, J. *Particles in Wall-Bounded Turbulent Flows: Deposition, Re-Suspension and Agglomeration*; Springer International Publishing: Cham, Switzerland, 2017; p. 261. [CrossRef]
18. Soo, S.L. *Instrumentation for Fluid Particle Flow*; University Press of Mississippi: Jackson, MS, USA, 2013.
19. Silva, R. Experimental characterization techniques for solid-liquid slurry flows in pipelines: A Review. *Processes* **2022**, *10*, 597. [CrossRef]
20. Mallach, M.; Gevers, M.; Gebhardt, P.; Musch, T. Fast and precise soft-field electromagnetic tomography systems for multiphase flow imaging. *Energies* **2018**, *11*, 1199. [CrossRef]
21. Ahmadi, F.; Ebrahimian, M.; Sanders, R.S.; Ghaemi, S. Particle image and tracking velocimetry of solid-liquid turbulence in a horizontal channel flow. *Int. J. Multiph. Flow* **2019**, *112*, 83–99. [CrossRef]
22. Hampel, U.; Barthel, F.; Bieberle, A.; Bieberle, M.; Boden, S.; Franz, R.; Neumann-Kipping, M.; Tas-Köhler, S. Tomographic Imaging of Two-Phase Flow. *Int. J. Adv. Nucl. React. Des. Technol.* **2020**, *2*, 86–92. [CrossRef]
23. Zou, J.; Liu, C.; Wang, H.; Wu, Z.P. Mass Flow Rate Measurement of bulk solids based on microwave tomography and microwave Doppler methods. *Powder Technol.* **2020**, *360*, 112–119. [CrossRef]
24. Yianneskis, M.; Whitelaw, J.H. Velocity characteristics of pipe and jet flows with high particle concentration. *ASME Symp. Liq. Solid Flows Eros. Wear Ind. Equip.* **1983**, *13*, 12–15.
25. Chen, R.C.; Kadambi, J.R. Discrimination between solid and liquid velocities in slurry flow using Doppler Velocimeter. *ASME Powder Technol.* **1995**, *85*, 127–134. [CrossRef]
26. Lv, Y.; Su, X.; Yang, H.; Zhang, J.; Wang, R.; Zhu, Z. Velocity slip in a deep-sea slurry pump and its effect on particle transportation. *J. Appl. Fluid Mech.* **2023**, *16*, 1654–1665. [CrossRef]
27. Nouri, J.M.; Whitelaw, J.H.; Yianneskis, M. Particle motion and turbulence in dens two-phase flows. *Int. J. Multiph. Flow* **1987**, *13*, 729–739. [CrossRef]
28. Schreck, S.; Kleis, S.J. Modification of grid-generated turbulence by solid particles. *J. Fluid Mech.* **1993**, *249*, 665–688. [CrossRef]
29. Chen, R.C.; Kadambi, J.R. Experimental and numerical studies of solid-liquid pipe flow. *ASME. Liq. Solid Flows* **1994**, *189*, 123–135. Available online: https://www.osti.gov/biblio/110079 (accessed on 3 September 2023).
30. Gai, G.; Hadjadj, A.; Kudriakov, S.; Thomine, O. Particles-induced turbulence: A critical review of physical concepts, numerical modeling and experimental investigations. *Theor. Appl. Mech. Lett.* **2020**, *10*, 241–248. [CrossRef]
31. Gore, R.A.; Crowe, C.T. Modulation of turbulence by a dispersed phase. *J. Fluid Eng.* **1991**, *113*, 304–307. [CrossRef]

2. Jianren, F.; Junmei, S.; Youqu, Z.; Kefa, C. The effect of particles on fluid turbulence in a turbulent boundary layer over a cylinder. *Acta Mech. Sin.* **1997**, *13*, 36–43. [CrossRef]
3. Eaton, J.K.; Paris, A.D.; Burton, T.M. Local distortion of turbulence by dispersed particles. *AIAA* **2012**, AIAA-99–3643. [CrossRef]
4. Fessler, J.R.; Eaton, J.K. Turbulence modification by particles in a backward-facing step flow. *J. Fluid Mech.* **1999**, *394*, 97–117. [CrossRef]
5. Li, D.; Luo, K.; Fan, J. Modulation of turbulence by dispersed solid particles in a spatially developing flat-plate boundary layer. *J. Fluid Mech.* **2016**, *802*, 359–394. [CrossRef]
6. Matousek, V. Research developments in pipeline transport of settling slurries. *Powder Technol.* **2005**, *156*, 43–51. [CrossRef]
7. Javed, K.; Vaezi, M.; Kurian, V.; Kumar, A. Frictional behaviour of wheat straw-water suspensions in vertical upward flows. *Biosyst. Eng.* **2021**, *212*, 30–45. [CrossRef]
8. Javed, K.; Kurian, V.; Kumar, A. The effect of particle size and concentration on the frictional behavior of vertical upward flows of wheat straw aqueous slurries. *Chem. Eng. Res. Des.* **2022**, *186*, 614–627. [CrossRef]
9. Nasr-El-Din, H.; Shook, C.A.; Colwell, J. The lateral variation of solids concentration in horizontal slurry pipeline flow. *Int. J. Multiph. Flow* **1987**, *13*, 661–670. [CrossRef]
0. Sumner, R.J. Concentration Variations and Their Effects in Flowing Slurries and Emulsions. Ph.D. Thesis, University of Saskatchewan, Saskatoon, SK, Canada, 1992.
1. Eaton, J.K.; Fessler, J.R. Preferential concentration of particles by turbulence. *Int. J. Multiph. Flow* **1994**, *20*, 169–209. [CrossRef]
2. Eskin, D.; Miller, M.J. A model of non-Newtonian slurry flow in a fracture. *Powder Technol.* **2008**, *182*, 313–322. [CrossRef]
3. Silva, R.; Garcia, F.A.P.; Faia, P.M.G.M.; Rasteiro, M.G. Settling suspensions flow modelling: A Review. *KONA Powder Part. J.* **2015**, *32*, 41–56. [CrossRef]
4. Cotas, C.; Asendrych, D.; Garcia, F.; Faia, P.; Rasteiro, M.G. COST Action FP1005. In Proceedings of the Final Conference EUROMECH Colloquium 566, Trondheim, Norway, 9–11 June 2015; pp. 31–33.
5. Chang, K.C.; Hsieh, W.D.; Chen, C.S. A modified low-Reynolds-number turbulence model applicable to recirculating flow in pipe expansion. *J. Fluids Eng.* **1995**, *117*, 417–423. [CrossRef]
6. Bartosik, A. Mathematical modelling of slurry flow with medium solid particles. In Proceedings of the 2nd International Conference on Mathematical Models for Engineering Science, Puerto de la Cruz, Spain, 10–12 December 2011; pp. 124–129.
7. Cieslinski, J. Numerical modelling of forced convection of nanofluids in smooth, round tubes: A review. *Energies* **2022**, *15*, 7586. [CrossRef]
8. Abidi, A.; Sajadi, S.M. Impact of fin and hybrid nanofluid on hydraulic-thermal performance and entropy generation in a solar collector using a two-phase approach. *Eng. Anal. Bound. Elem.* **2023**, *156*, 311–320. [CrossRef]
9. Bordet, A.; Poncet, S.; Poirier, M.; Galanis, N. Advanced numerical modeling of turbulent ice slurry flows in a straight pipe. *Int. J. Therm. Sci.* **2018**, *127*, 294–311. [CrossRef]
0. Hassaan, A.M. Comparing the performance of using nanofluids in two different types of heat exchangers with the same heat transfer area. *Heat Transf. Res.* **2023**, *54*, 1–16. [CrossRef]
1. Mandal, D.K.; Hazra, A.; Biswas, N. Effect of particle size on fluid flow and heat transfer in a pipe with slurry flow. *IOP Conf. Ser. Mater. Sci. Eng.* **2021**, *1080*, 012026. [CrossRef]
2. Harada, E.; Toda, M.; Kuriyama, M.; Konno, H. Heat transfer between wall and solid-water suspension flow in horizontal pipes. *J. Chem. Eng.* **1985**, *18*, 33–38. [CrossRef]
3. Iqbal, M.; Kouloulias, K.; Sergis, A.; Hardalupas, Y. Critical analysis of thermal conductivity enhancement of alumina–water nanofluids. *J. Therm. Anal. Calorim.* **2023**, *148*, 9361–9389. Available online: https://link.springer.com/article/10.1007/s10973-023-12334-7 (accessed on 14 July 2023). [CrossRef]
4. Hetsroni, G.; Rozenblit, R. Heat transfer to a solid-liquid mixture in a flume. *Int. J. Multiph. Flow* **1994**, *20*, 671–689. [CrossRef]
5. Rozenblit, R.; Simkhis, M.; Hetsroni, G.; Taitel, Y. Heat transfer in horizontal solid-liquid pipe flow. *Int. J. Multiph. Flow* **2000**, *26*, 1235–1246. [CrossRef]
6. Bhanvase, B.; Barai, D. *Nano Fluid for Heat and Mass Transfer—Fundamentals. Sustainable Manufacturing and Applications*; Academic Press in an imprint Elsevier: London, UK, 2021; p. 448. Available online: https://www.elsevier.com/books-and-journals (accessed on 30 April 2021).
7. Permanasari, A.A.; Faridi, A.F.; Sukarni, S.; Puspitasari, P.; Mohamed, N.M.; Pratama, M.M.A. Heat transfer characteristics on shell and tube heat exchanger using TiO2/water-ethylene glycol nanofluid. *AIP Conf. Proc.* **2023**, *2748*, 020048. [CrossRef]
8. Firdaus, D.R.; Permanasari, A.A.; Sukarni, S.; Puspitasari, P. Heat transfer performance of DWCNTs nanofluids in shell and tube heat exchanger with segmental baffle. *AIP Conf. Proc.* **2023**, *2687*, 040023. [CrossRef]
9. Sattar, E.A.; Ouda, A.A.; Salman, M.D. Enhancement heat transfer in shell and tube heat exchanger by used hybrid and nanofluid. *AIP Conf. Proc.* **2023**, *2787*, 030006. [CrossRef]
0. Wang, X.; Xu, X.; Choi, S.U.S. Thermal conductivity of nanoparticle—fluid mixture. *J. Thermophys. Heat Transfer.* **1999**, *13*, 474–480. [CrossRef]
1. Lotfi, R.; Saboohi, Y.; Rashidi, A.M. Numerical study of forced convective heat transfer of nanofluids: Comparison of different approaches. *Int. Commun. Heat Mass Transf.* **2010**, *37*, 74–78. [CrossRef]
2. Mokmeli, A.; Saffar-Avval, M. Prediction of nanofluid convective heat transfer using the dispersion model. *Int. J. Therm. Sci.* **2010**, *49*, 471–478. [CrossRef]

63. Shook, C.A.; Bartosik, A.S. The pseudohomogeneous flow approximation for dispersed two-phase systems. *Part. Sci. Tech.* **199** *9*, 119–134. [CrossRef]
64. Boussinesque, J. Theorie de l'ecoulement tiurbillant. *Mem. Acad. Sci.* **1877**, *23*, 46.
65. Blom, J. Experimental determination of the turbulent Prandtl number in a developing temperature boundary layer. In Proceeding of the 4th International Conference Heat Transfer, Paris, France, 31 August–5 September 1970; Volume VII.
66. Launder, B.E.; Sharma, B.I. Application of the energy-dissipation model of turbulence to the calculation of flow near a spinnin disc. *Lett. Heat Mass Transfer.* **1974**, *1*, 131–138. Available online: https://www.sciencedirect.com/science/article/abs/pii/00944 4874901507 (accessed on 7 August 2021). [CrossRef]
67. Mathur, S.; He, S. Performance and implementation of the Launder–Sharma low-Reynolds number turbulence model. *Compu Fluids* **2013**, *79*, 134–139. [CrossRef]
68. Lawn, C.J. The determination of the rate of dissipation in turbulent pipe flow. *J. Fluid. Mech.* **1971**, *48*, 477–505. [CrossRef]
69. Cotas, C.; Asendrych, D.; Garcia, F.A.P.; Fala, P.; Rasteiro, M.G. Turbulent flow of concentrated pulp suspensions in pipe—Numerical study based on a pseudo-homogeneous approach. In Proceedings of the COST Action FP1005 Final Confe ence, EUROMECH Colloquium 566, Trondheim, Norway, 9–11 June 2015; Available online: https://www.researchgate.net publication/283342328 (accessed on 7 August 2021).
70. Cotas, C.; Silva, R.; Garcia, F.; Faia, P.; Asendrych, D.; Rasteiro, M.G. Application of different low-Reynolds k-ε turbulence model to model the flow of concentrated pulp suspensions in pipes. *Procedia Eng.* **2015**, *102*, 1326–1335. [CrossRef]
71. Abir, I.A.; Emin, A.M. A comparative study of four low-Reynolds-number k-e turbulence models for periodic fully developedduc flow and heat transfer. *Numer. Heat Transf. Part B Fundam.* **2016**, *69*, 234–248.
72. Bartosik, A. Validation of friction factor predictions in vertical slurry flows with coarse particles. *J. Hydrol. Hydromech.* **2020**, *6* 119–127. [CrossRef]
73. Jones, T. Advances in Slurry Technology. In *Chapter: Numerical Modelling of Medium Slurry Flow in a Vertical Pipeline*; IntechOper London, UK, 2023; pp. 1–19. [CrossRef]
74. Roache, P.J. *Computational Fluid Dynamics*; Hermosa Publishing: Albuquerque, NM, USA, 1982.
75. Nikuradse, J. Unyersuchungen über die geshwindigkeitsverteilung in turbulenten strömungen. *Aeronaut. J.* **1927**, *31*, 117 [CrossRef]
76. Hermandes-Peres, V.; Abdulkadir, M.; Azzopardi, B.J. Grid generation issues in the CFD modelling of two-phase flow in a pip *J. Comput. Multiph. Flows.* **2011**, *3*, 1–26. [CrossRef]
77. Dittus, F.W.; Boelter, L.M.K. Heat transfer in automobile radiators of the tubular type. *Int. Comm. Heat Mass Transfer.* **1985**, *1* 3–22. [CrossRef]
78. Charles, M.E.; Charles, R.A. *Advances in Solid-Liquid Flow and Its Applications*; Pergamon Press: New York, NY, USA, 1971; Availabl online: https://www.elsevier.com/books/advances-in-solid-liquid-flow-in-pipes-and-its-application/zandi/978-0-08-015767- (accessed on 3 September 2023).
79. Ghosh, T.; Shook, C.A. Freight pipelines. In *Freight Pipelines*; Liu, H., Round, G.F., Eds.; Hemisphere: Washington, DC, USA, 199 pp. 281–288.
80. Talmon, A.M. Analytical model for pipe wall friction of pseudo-homogenous sand slurries. *Part. Sci. Tech.* **2013**, *31*, 264–27 [CrossRef]
81. Li, M.-Z.; He, Y.-P.; Liu, W.-H.; Liu, Y.-D.; Huang, C.; Jiang, R.-H. Effect of adding finer particles on the transport characteristics o coarse-particle slurries in pipelines. *Ocean Eng.* **2018**, *2020*, 108160. [CrossRef]

energies

MDPI

Article

The Application of Barocaloric Solid-State Cooling in the Cold Food Chain for Carbon Footprint Reduction

Luca Cirillo, Adriana Greco and Claudia Masselli *

Department of Industrial Engineering, University of Naples Federico II, P.le Tecchio 80, 80125 Naples, Italy; luca.cirillo2@unina.it (L.C.); adriana.greco@unina.it (A.G.)
* Correspondence: claudia.masselli@unina.it

Abstract: In this paper, the application of solid-state cooling based on the barocaloric effect in the cold food supply chain is investigated. Barocaloric solid-state technology is applied to the final links of the cold food supply chain regarding the steps of retail and domestic conservation. In this context, effective barocaloric cooling entails the refrigeration of food at 5 °C (273 K) and as such is a promising cooling technology due to its energy efficiency and environmental friendliness. The categories of food involved in this investigation are meat and fresh food products like soft cheese, yogurt, and milk. The energy performance of the barocaloric system is analyzed and compared with a commercial vapor compression refrigerator of a similar size, both operating using R600a under the same working conditions. Based on the results of this comparison, it is concluded that barocaloric cooling is a favorable technology for application in the final links of the cold food supply chain if the system operates in an ABR cycle at frequencies between 1.25 and 1.50 Hz with a regenerator comprising acetoxy silicone rubber as the solid-state refrigerant and a 50%EG–50% water mixture as the heat transfer fluid flowing at an optimal velocity of 0.15 m s^{-1}. Thus, an appropriate tradeoff between the temperature span, cooling power, and coefficient of performance is guaranteed. Under these conditions, the barocaloric system outperforms the domestic vapor compression cooler operating using R600a.

Keywords: barocaloric; food supply chain; carbon footprint reduction; solid-state cooling; energy saving

Citation: Cirillo, L.; Greco, A.; Masselli, C. The Application of Barocaloric Solid-State Cooling in the Cold Food Chain for Carbon Footprint Reduction. *Energies* **2023**, *16*, 6436. https://doi.org/10.3390/en16186436

Academic Editor: Artur Bartosik

Received: 31 July 2023
Revised: 30 August 2023
Accepted: 2 September 2023
Published: 6 September 2023

1. Introduction

In the last decade, the global community has fully grasped the risk of climate change and has started implementing countermeasures to slow its effects. Environmental sustainability has become essential for any new technological innovation; the objective is to fully realize sustainable development goals as soon as possible. One of the most significant contributors to environmental pollution is greenhouse gas emissions from air conditioning systems [1,2] in buildings and industrial warehouses, as well as coupled systems for refrigeration [3,4]. Therefore, the design and management of these systems must increasingly consider strict emission guidelines [5].

Over the years, and considering the increasing health consciousness among consumers, the food industry has evolved significantly to meet high market demand and maintain high-quality standards.

The aspect that links the refrigeration and air-conditioning sector and the food industry is the cold food chain [6]. The expression "cold chain" indicates the maintenance of fresh and frozen products at a constant temperature throughout their life cycle, from the production plant to the point of sale, through all transport and storage stages. Maintaining food storage temperature helps to ensure that the product obtained by the consumer is intact, without having undergone thermal shocks that could alter the organoleptic properties of the food. The goal is to prevent the cold chain from being interrupted in one of the supply chain stages, compromising the product's nutritional properties and posing a threat to its integrity, hygiene standards, and food safety [7].

Since a temperature that is too high or too low might negatively impact the food's quality and/or safety, each food product has a defined ideal storage temperature. Refrigeration is used at every stage of the food supply chain, from food processing to distribution, retail sales, and final consumption. Various cooling methods and strict temperature control during different processes are crucial to ensure the target production levels are reached, and the product's organoleptic qualities are kept consistent at every production stage. Reliable, efficient, sustainable, and high-quality production processes are needed to meet the requirements for reduced energy use and effective energy management. According to the findings from the High-Level Panel of Experts on Food Security and Nutrition, the concept of food systems encompasses various elements like food production, supply networks, environments, dietary habits, and consumer selection [8]. It also considers the resulting impacts on nutrition, the environment, and socioeconomic aspects. The processes and phases associated with the production, processing, distribution, marketing, consumption, and disposal of food all fall under the scope of food supply chains, which are becoming progressively intricate. The proportion of global greenhouse gas emissions attributed to the overall food system, encompassing activities beyond the retail stage like cooking and waste disposal, is projected to rise significantly if the food and agriculture sector continues to account for roughly a quarter (26%) of worldwide greenhouse gas emissions (amounting to 34% of total greenhouse gases) [9,10]. While the primary production phase undoubtedly makes the largest contribution, subsequent activities within the supply chain such as processing, transportation, packaging, and retail collectively contribute to 18% of greenhouse gas emissions related to food. Notably, emissions resulting from food waste are substantial, accounting for approximately one-quarter of emissions (equivalent to 3.3 billion tons of CO2eq) generated during food production. These emissions occur due to inefficiencies in the supply chain (contributing to 15% of food-related greenhouse gases) as well as consumer losses [11].

To ensure competitiveness in the global market while simultaneously reducing environmental impacts, the food sector must minimize its energy consumption and increase the use of ecofriendly facilities [12]. In terms of achieving all these goals, optimizing refrigeration/freezing systems represents a significant challenge [13]. European food hygiene and safety regulations provide detailed guidelines concerning the temperatures required for food preservation. Regulation EC No. 852/04 [14] states that food businesses handling perishable products must have all the necessary facilities to maintain the "cold chain" for these products. Suitable structures such as refrigerated rooms, freezers, and (or) display counters must be available for the storage and display of perishable goods. However, all these measures alone are insufficient, as different food products must be kept at specific temperatures that must be continuously monitored. Therefore, the equipment must be designed to allow for temperature control and recording.

In general, however, the legal provisions on this matter are quite elaborate and also refer to different storage periods: transportation, warehousing, and display [15,16]. Traditional food preservation devices exhibit evident limitations in terms of environmental sustainability. Specifically, most refrigeration systems currently rely on the vapor compression cycle [17,18]. Over the years, the development of this technology has faced significant challenges in terms of the environmental impact associated with the use of refrigerants, many of which have proven to be highly detrimental to the environment, both in terms of damaging the planet's stratospheric ozone layer and contributing to global warming.

In this context, research increasingly considers alternatives to traditional vapor compression systems: On the one end of the spectrum, there are renewable-energy-based technologies such as solar [19–22] or geothermal [23–26] technologies, while on the other, there are solid-state technologies based on the caloric effect [27–30]. These alternatives do not use refrigerant gases, thus ensuring a negligible contribution to the greenhouse effect [31]. They also have potentially higher performance coefficients than vapor compression systems, leading to energy savings of 50–60%, no noise during operation, and easy maintenance [32–35]. These technologies fully align with the global political context aiming to reduce greenhouse gas emissions and transition towards a low-carbon economy.

The operating principle of solid-state systems is the caloric effect: It occurs when a caloric material is subjected to an external field that induces a temperature change by varying its intensity under adiabatic conditions (ΔT_{ad}). If the field varies under isothermal conditions, an entropy change (ΔS_{iso}) can be observed in the material [36–38]. Depending on the nature of the applied external field and the thermophysical properties of the caloric material, specific caloric effects can be achieved: magnetocaloric (magnetic field) [39], electrocaloric (electric field) [40,41], and mechanocaloric (mechanical stress) [42–46]. The caloric effect is divided into two types depending on the applied external field. If the external field is generated due to mechanical stress, such as compression [47], tension [48], or torsion [49], it is called the elastocaloric effect [50,51]. This effect occurs in specific materials known as shape memory alloys [52–54]. On the other hand, if the applied field is the result of hydrostatic pressure, it is referred to as the barocaloric effect [55,56].

Recently, there has been a growing interest in the solid-state cooling effect known as barocaloric [57]. As a result, there are relatively fewer scientific publications with respect to other techniques related to caloric effects [58]. The interest in barocaloric effects emerged when studies suggested that magnetocaloric materials undergoing first-order phase transitions might experience a temperature change under adiabatic pressure variation [59].

Various investigations were conducted, and some years later, it was experimentally confirmed that many materials exhibiting magnetocaloric properties also show the possibility of detecting the barocaloric effect. Moreover, it was found that the barocaloric effect could be coupled with other caloric effects, leading to multicaloric effects in the same material. Another promising class of barocaloric materials includes elastomer natural rubbers.

Researchers like Moya et al. have provided comprehensive reviews [58,59] on barocaloric materials. However, the number of barocaloric materials that have been developed and investigated for cooling applications continuously increases. Besides materials, the research on the development of devices for cooling applications also lacks experimental results derived from the operation of barocaloric prototype devices. To the best of our knowledge, in 2014, Czernuszewicz et al. proposed the concept of a magnetobarocaloric test stand for cooling applications [60], which, similar to their previously developed magnetocaloric test stand [61], allows for assessing the application potential of materials that exhibit temperature changes under the influence of magnetization and applied pressure.

In 2018, Aprea et al. conducted a comprehensive study on caloric materials used in an active caloric cooler, including some of the most promising barocaloric refrigerants [62]. This investigation provided an up-to-date overview of their energy performance, establishing a general framework for these materials.

One significant outcome of the study above is the confirmation that barocaloric cooling holds great promise and is comparable to vapor compression (VC) cooling. Selecting promising barocaloric materials for cooling applications is crucial to developing efficient active barocaloric coolers. Additionally, the use of more efficient auxiliary fluids is necessary to ensure that this technique is environmentally friendly. Typically, water or a water–glycol mixture is used as the auxiliary fluid, depending on whether the caloric system operates above or below zero degrees Celsius.

Thus, the aim of this investigation is to study the application of barocaloric solid-state technology to the final links of the cold food supply chain, for retail and domestic food storage. In this context, we tested a barocaloric cooler for keeping food at 5 °C (273 K), to evaluate the potential of this method as a promising, energy-efficient, and environmentally friendly cooling technology. The food categories of interest were meat and fresh food products (soft cheese, yogurt, and milk).

In summary, the main objectives of this research are as follows:

- We aim to examine the energy efficiency of a solid-state barocaloric cooling system designed to function as a refrigeration machine for selling cold products, which need to be kept at 278 K. The scenario is assumed to be a hot indoor environment t examine the potentialities of this technology in extreme summer conditions (313 K Our purpose is to compare its performance with the performance of an existing vapo compression refrigerator commonly used in industrial settings. The solid-state system involves active barocaloric regeneration, using a heat-transfer fluid consisting of a 50% ethylene glycol (EG) and 50% water mixture, while the solid-state refrigerant i acetoxy silicone rubber (ASR) [63–66].

- We aim to assess and compare the energy efficiency of the active barocaloric cycle wit that of a commercial vapor compression refrigerator of a similar size, both operating under the same working conditions.

2. The Project "PNRR on Foods" and the Cold Chain

The research investigation presented in this paper is part of the maxi-project PNR "ONFOODS" research activities. The concept of sustainability has only recently entere everyday use. Sustainability refers to patterns of production and consumption that respec natural resources and their usual rhythms, focusing on long-term resilience and avoiding the depletion of resources and environmental degradation.

The 2030 Agenda for Sustainable Development, adopted by all United Nations Mem ber States in 2015, describes 17 Sustainable Development Goals (SDGs), ranging from ending poverty to improving health and education, reducing inequality, spurring economi growth, tackling climate change, and working to preserve our oceans and forests.

The project ONFOODS (Research and Innovation Network on Food and Nutritio Sustainability, Safety, and Security—Working on Foods) is financed by the European Com munity and is among the activities of the Italian PNRR, i.e., the National Recovery an Resilience Plan (PNRR). This plan is part of the Next Generation EU (NGEU) program, EUR 750 billion package, accounting for about half of the grants agreed upon by the Euro pean Union in response to the pandemic crisis of 2020. Within this framework, ONFOOD is an action plan with a comprehensive approach, joining together and synergizing the strengths and competencies of several different disciplines, ranging from social and jurid cal sciences to agricultural economics, food chemistry, food technology and engineering logistics, microbiology, marketing, human nutrition, and many disciplines of medicine Only through a holistic approach can we preserve the environment, develop strategie for the effective implementation of our food systems, and, at the same time, improve the well-being of individuals and extend their healthy life span by reducing avoidable an premature mortality and inequity and putting people at the center of a more effective, saf sustainable, and fair social security system.

ONFOODS faces this challenge through a coordinated action involving seven spoke each one focusing on a very specific, although broad, aspect of food production, trans formation, and effect. All these spokes will be deeply intertwined in a complex an well-developed network of connections and complementation, where ideas that originat in one spoke will become innovation only after being validated in the work of researcher collaborating in other relevant spokes within the program.

The identified objectives of ONFOODS are as follows:

- To promote the sustainability of food production;
- To increase the adherence to more sustainable dietary patterns;
- To promote the sustainability of food distribution;
- To increase the quality of foods and diets;
- To develop smart innovative technologies for sustainable food production and con sumption;
- To guarantee food safety and food security at the whole population level and i specific vulnerable targets of the population.

The research group of the University of Naples Federico II, to which the authors belong is focused on the ONFOODS activities to "promote the sustainability of food distribution" i.e., to potentiate the sustainability of the cold chain in all its steps.

The cold chain begins with production; it continues through the storage phase in refrigerated platforms, followed by the transport phase up to the sales warehouses and refrigerated counters of the points of sale, and ends with the final consumers' coolers. Every step is essential for the success of the cold chain process. More details are provided in Figure 1, which illustrates the phases of the food cold supply chain.

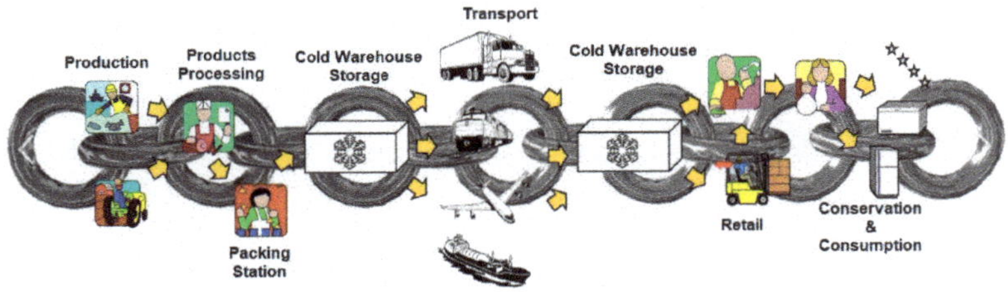

Figure 1. The phases of the food cold supply chain.

I. Production and Product Processing: Upon leaving the packaging lines, measuring the core temperature of the product is necessary to determine the time required to reach the temperature indicated on the packages. After physical–chemical and bacteriological assessments, thermosensitive food products—fresh or frozen—are managed under refrigeration and stored in temperature-controlled warehouses.
II. Cold Warehouse Storage: The storage of food products in cold rooms guarantees an overall reduction in their core temperature, and the picking and order preparation activities can also be carried out in the same environment. Based on the type of product, it must also be ensured that the ideal temperatures and hygienic conditions of containers and packaging are maintained; loading areas must also be under a controlled temperature and possibly adjacent to the logistics cell.
III. Transport: This phase is of crucial importance for the cold chain. In fact, vehicles must be equipped with refrigerators already at the required temperature, which should be maintained during transport to another facility for cold storage or the points of sale of large-scale retail trade. This occurs through the use of tools that allow operators to keep the air temperature in the compartment under control.
IV. Retail: Products are displayed in retail outlets on special refrigerated shelves and are available for purchase by the consumer.

Conservation and Consumption: After purchasing food products, the consumer keeps them in their refrigerators until they are used for consumption.

During all these steps of the cold chain, in order to avoid any contamination, operators are required to meticulously implement and monitor the following features:

- The cleanliness and hygienic conditions of warehouses, equipment, and transport;
- The pre-established critical limits in various stages of the cold chain;
- The temperatures of the different actors of the cold chain.

In this context, Spoke 3 of ONFOODS (Research and Innovation Network on Food and Nutrition Sustainability, Safety, and Security—Working on Foods) has the following objectives:

- To report on best storage and operational solutions for food waste control and reduction in energy consumption;
- To focus on the use of phase change materials (PCMs) for the steps of the cold chain process related to (II) cold warehouse storage and (III) transport;
- To apply ecofriendly technologies as an alternative to vapor compression for refrigeration systems used in the steps related to (IV) retail and (V) conservation and consumption.

Temperature-sensitive products must therefore follow precise hygienic measures and comply with the temperatures required by law. As suggested by Italian Legislative Decree 27 January 1992, n.110 [67], frozen foods must maintain a constant temperature of −18° Celsius; however, during transport, brief upward swings not exceeding 3 °C are allowed. As far as fresh food is concerned, however, each one has its own reference temperature:

- Meats: +5 °C to +7 °C;
- Fish: +2 °C;
- Fresh products: +5 °C;
- Ice creams: −22 °C;
- Chocolate: +18 °C to +20 °C.

Following the description of Spoke 3 of PNRR ONFOODS, this investigation aims to introduce barocaloric solid-state technology as an alternative to vapor compression for the links of the cold food supply chain related to retail and domestic conservation. The food categories of interest in this investigation are meat and fresh food products (soft cheese, yogurt, and milk), kept at 5 °C (273 K) using an energy-efficient and environmentally friendly barocaloric cooler. The system is tested while operating in a hot indoor environment to examine the potentialities of the technology in extreme summer conditions (313 K). The method is also compared with the vapor compression technology.

3. The Thermodynamic Cycle Based on the Barocaloric Effect

Active barocaloric refrigeration involves an active barocaloric regenerative cycle in which a barocaloric material serves as the refrigerant. The system consists of a barocaloric regenerator housed in a pressure cell to enable variations in the working pressure. This regenerator interacts with cold and hot heat exchangers, and a heat transfer fluid is used to facilitate the heat exchange while passing through the barocaloric material. The regenerator can be designed in various ways, such as using porous media, parallel plates, or wires of the barocaloric material, creating channels through which the auxiliary fluid flows, among other possibilities.

Figure 2 illustrates an example of the active barocaloric regenerative refrigeration cycle using a parallel plate design. It is essential to clarify that the purpose of Figure 2 is to conceptually depict the four processes of the active barocaloric regenerative refrigeration cycle from a thermodynamic standpoint rather than presenting a detailed engineering design. The ABR cycle comprises two isobaric and two adiabatic processes, as evident in Figure 2a.

With reference to Figure 2b, during the first process (Step I), the barocaloric regenerator undergoes adiabatic loading, increasing hydrostatic pressure from p_0 to p_1. This results in a temperature rise in the barocaloric material due to the BCE (transformation 1–2 of Figure 2a). Once the pressure reaches p_1, the loading process ends, and the system exchanges heat with the hot heat exchanger through the heat transfer fluid (HTF) flow (Step II of Figure 2b). Consequently, as depicted in the transformation 2–3 of Figure 2a, the ABR starts to decrease in temperature. Subsequently, Step III (Figure 2b) involves adiabatic unloading, with the pressure decreasing from p_1 to p_0, which causes the further cooling of the ABR, as shown in transformation 3–4 in Figure 2a.

Lastly, as seen in Step IV (Figure 2b), the HTF flowing in the ABR results in heat being subtracted from the cold heat exchanger, thus achieving the desired cooling effect. Consequently, the regenerator undergoes an increase in temperature (transformation 4-1 of Figure 2a).

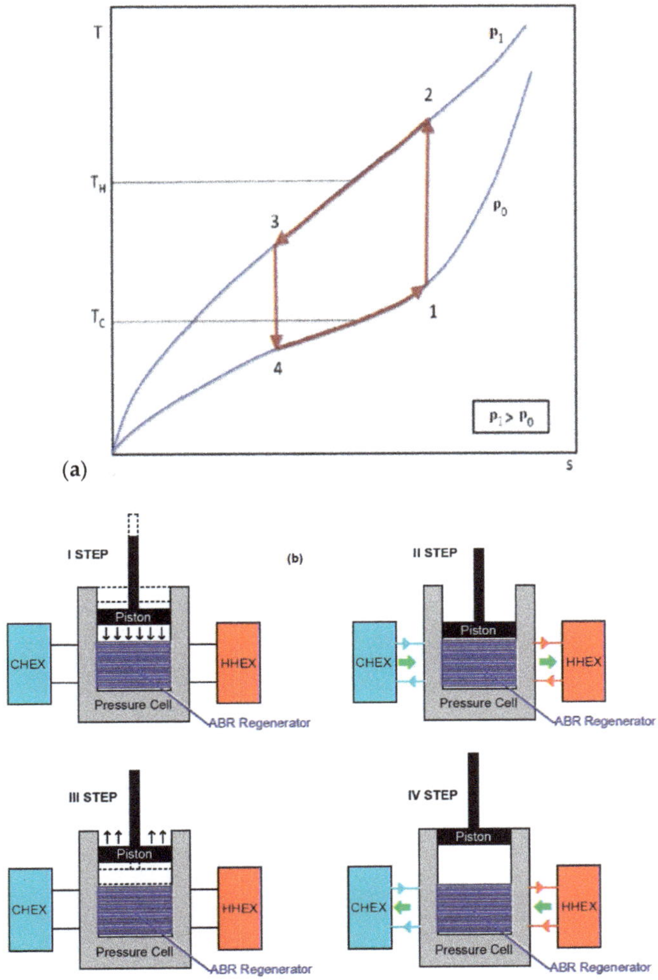

Figure 2. The Brayton-based ABR cycle: (**a**) on T–s diagram; (**b**) applied to the cooling system.

3.1. The Barocaloric Material

The selection of an appropriate barocaloric substance for cooling applications is influenced by the range of the desired operating temperature and the specific intended use. To be considered a promising refrigerant for caloric cooling, a material must meet several essential criteria, such as affordability, ease of synthesis, and non-toxicity [68]. Moreover, the material must exhibit significant adiabatic temperature changes (ΔT_{ad}) within the designated temperature range. With these objectives in mind, we opted to utilize acetoxy silicone rubber (ASR) [63] in our investigation, as it is a widely available, cost-effective, and environmentally friendly elastomer. ASR is a type of elastomer that belongs to the vulcanizing rubber category, and its thermophysical properties are reported in Table 1.

Table 1. Thermophysical properties of acetoxy silicone rubber.

Material	Thermal Conductivity ($W\,m^{-1}\,K^{-1}$)	Density ($kg\,m^{-3}$)	Adiabatic Temperature Change (K) at Δp = 0.390 GPa
ASR	1.48	960	41.1

One of the key characteristics of this group of materials is their remarkable elasticity: These vulcanizing rubber materials can be repeatedly stretched to their original length when subjected to relatively low pressures. This ability makes them noteworthy as potential mechanocaloric materials, as they combine their elastomeric properties with promising structural transitions. Acetoxy silicone rubber, composed of polydimethylsiloxane, additives, and illers, demonstrates an extraordinary barocaloric effect, displaying a significant ΔT_{ad} peak during crystalline (41.1 K) to amorphous transitions, which leads to rearrangements of the polymer chains. As shown in Figure 3, Imamura et al. [63] generated a diagram for the change in the adiabatic temperature of acetoxy silicone rubber during the decompression process at a pressure of Δp = 0.390 GPa, plotted against the working temperature of the material at the start of the transformation. The complete details of the measurements and the methodology utilized can be found in a previous study [63]. The peak barocaloric effect occurs at approximately 41.1 K, near 298 K. Considering a suitable working temperature range for domestic refrigeration purposes (278 ÷ 313 K), the observed values of ΔT_{ad} remain above 30 K, affirming the applicability of ASR for this type of cooling application. ASR is classified as a vulcanizing rubber material, and there is a significant concern regarding its fatigue behavior. Researchers such as Sebald et al. [68] investigated the impact of fatigue behavior on the degradation of barocaloric properties, while Woo et al. [69] developed a tool for predicting fatigue life. The findings by Sebald et al. [68] indicate that the barocaloric effect (BCE) temperature degradation is much lower than the magnitude of stress degradation. On the other hand, Woo et al. [69] observed that the fatigue failure of such materials typically occurs after approximately 6×10^5 cycles (in the order of magnitude).

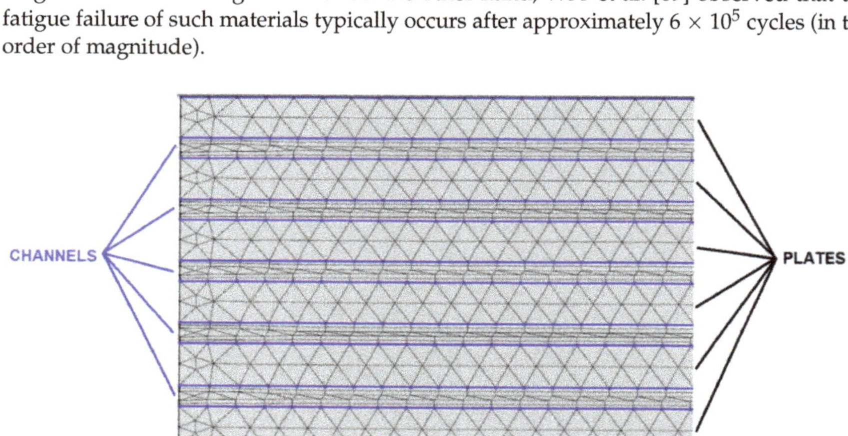

Figure 3. An image of the regenerator's triangular mesh is utilized to solve the model.

3.2. Mathematical Modeling

This analysis was conducted utilizing a 2D model of an ABR refrigerator. The regenerator (measuring 20 mm × 45 mm) was constructed with 54 parallel plates made of barocaloric substances, enabling HTF (heat transfer fluid) flow for exchanging heat with the hot and cold heat exchangers. The thickness of every plate was 0.25 mm, whereas the distance stacked between two consecutive plates, i.e., the height of the channel where the auxiliary fluid flows, was 0.125 mm.

Before introducing the model, the following assumptions should be noted:

- Laminar is the regime of motion of the auxiliary fluid.
- The auxiliary fluid is considered incompressible.
- The viscous dissipation is neglected since the mass flow is small.
- The fluid direction is modeled through a velocity that has a positive sign if the fluid flows from the cold to the hot heat exchanger and a negative sign for fluid flowing from the hot to the cold heat exchanger.

- The presence of the cold and hot heat exchangers is ensured by the Dirichlet boundary conditions driving the model during the fluid flow steps (temperature of the cold heat exchanger if the fluid flows from cold to hot; temperature of the hot heat exchanger if the fluid flows from hot to cold side).

The mathematical framework governing the four processes of the ABR cycle is outlined as follows (Navier–Stokes equations and the equation of the energy for the fluid and the solid [70]):

$$\begin{cases} \frac{\partial u}{\partial x} + \frac{\partial v}{\partial y} = 0 \\ \frac{\partial u}{\partial t} + u\frac{\partial u}{\partial x} + v\frac{\partial u}{\partial y} = -\frac{1}{\rho_f}\frac{\partial p}{\partial x} + v\left(\frac{\partial^2 u}{\partial x^2} + \frac{\partial^2 u}{\partial y^2}\right) \\ \frac{\partial v}{\partial t} + u\frac{\partial v}{\partial x} + v\frac{\partial v}{\partial y} = -\frac{1}{\rho_f}\frac{\partial p}{\partial y} + v\left(\frac{\partial^2 v}{\partial x^2} + \frac{\partial^2 v}{\partial y^2}\right) \\ \frac{\partial T_f}{\partial t} + u\frac{\partial T_f}{\partial x} + v\frac{\partial T_f}{\partial y} = \frac{k_f}{\rho_f C_f}\left(\frac{\partial^2 T_f}{\partial x^2} + \frac{\partial^2 T_f}{\partial y^2}\right) \\ \rho_s C\frac{\partial T_s}{\partial t} = k_s\left(\frac{\partial^2 T_s}{\partial x^2} + \frac{\partial^2 T_s}{\partial y^2}\right) + Q \end{cases} \quad (1)$$

where u and v are the components of the fluid velocity vector along the x and y axes; p is the fluid pressure; and T_f and T_s are the fluid and solid temperatures, respectively. k, ρ, and C are the thermal conductivity, density, and heat capacity of the fluid (if they are coupled with the f subscript) or of the solid (if they are followed by the s subscript).

In Steps I and III of the ABR cycle, the fluid is not moving, and the elastocaloric effect occurs through +Q and $-Q$, respectively; thus, Equation (1) becomes

$$\begin{cases} \rho_f C_f\frac{\partial T_f}{\partial t} = k_f\left(\frac{\partial^2 T_f}{\partial x^2} + \frac{\partial^2 T_f}{\partial y^2}\right) \\ \rho_s C\frac{\partial T_s}{\partial t} = k_s\left(\frac{\partial^2 T_s}{\partial x^2} + \frac{\partial^2 T_s}{\partial y^2}\right) + Q \end{cases} \quad (2)$$

with

$$Q = Q(field, T_S) = \frac{\rho_s\, C(field, T_s)\Delta T_{ad}(field\,, T_s)}{\tau} \quad (3)$$

The Q functions were derived using specialized software that identifies the closest mathematical representation to the input table function. These functions were generated based on experimental data previously published in the literature. Table 2 contains the mathematical expressions corresponding to the various investigated pressure loads. To solve the equations of the model, the finite element method was employed, and the regenerator underwent meshing using the free triangular meshing approach (depicted in Figure 3).

Table 2. $Q\,{*}\tau$ functions, related to the field increasing process, obtained starting from the experimental data.

Material	**Δp (GPa)**	**Mathematical Expression**
ASR	0.390	$Q_{load} * \tau = 10^6 *$ $[(1.33 * T + 0.000211 * (T^2 * sin(4.33 + sin(-0.0636 * T) * sin(4.12 + sin(0.0726 * T) - 165 - 0.00216 * T)]$

The concept was previously presented in one of our earlier investigations and experimentally validated through accurate comparison with experimental data obtained using a rotary permanent magnet magnetocaloric cooler prototype, a collaborative project developed by the University of Naples Federico II and the University of Salerno [39,62]. In fact, one of the key characteristics of our model is its ability to accurately simulate the behavior of an active regenerator with the use of any solid-state refrigerant, regardless of the particular caloric impact displayed.

The finite element method (FEM) is employed to solve the model, while the ABR cycle is an iterative process, repeating multiple times until reaching a steady-state condition that fulfills the cutoff criterion at every point of the ABR regenerator:

$$\delta = max\{T(x,y,0+n\theta) - T(x,y,4\tau+n\theta)\} < \bar{\varepsilon} \tag{4}$$

4. Thermal Performance Analysis

The ABR regenerator utilizes acetoxy silicone rubber as the barocaloric refrigerant, with Δp = 0.390 GPa during the adiabatic compression–decompression process. This study was conducted with specific parameters set at a temperature range of 278 ÷ 313 K, where the lower and upper limits represent the cold and hot heat exchanger temperatures, respectively. Each ABR cycle had a total duration of 0.8 s, with each step lasting 0.2 s. The Heat transfer fluid employed in the system was a mixture of 50% ethylene glycol (EG) and 50% water, with a whole freezing point of 236 K.

Table 3 summarizes the parameters and ranges used in this investigation.

Table 3. Operating conditions of the numerical investigation.

Parameter	Unit	Value
$T_{set\ point}$	K	278
f	Hz	1.25
Δp	GPa	0.390
v	m s^{-1}	0.04–0.2

The most significant parameters characterizing the energy performances were determined as follows:

$$\Delta T_{span} = T_H - \frac{1}{\tau}\int_{3\tau+q\theta}^{4\tau+q\theta} T_f(0,y,t)dt \tag{5}$$

where ΔT_{span} is the temperature span measured across the regenerator during the last process of the ABR cycle in steady-state conditions.

$$\dot{Q}_{ref} = \frac{1}{\theta}\int_{3\tau+q\theta}^{4\tau+q\theta} \dot{m}_f C_f\left(T_C - T_f(0,y,t)\right)dt \tag{6}$$

where $\dot{Q}_{ref}$ is the proper refrigeration power of the ABR refrigerator.

$$COP = \frac{\dot{Q}_{ref}}{\dot{W}_{TOT}} = \frac{\dot{Q}_{ref}}{\dot{W}_{field} + \dot{W}_{pump}} \tag{7}$$

where the COP (coefficient of performance) measures the cooling power of the refrigerator relative to the combined expenses involved in its operation. W_{TOT} represents the contribution of changes in the external field (W_{field}) and the mechanical power needed to move the heat transfer fluid (W_{pump}).

To evaluate the energy efficiency of the barocaloric refrigerator, its energy performance was compared with the experimental results obtained from a standard food cooler operating under identical conditions. This allowed for a reference point to gauge the level of performance attained by the solid-state system.

5. Results

In this section, the energy performance of the elastocaloric device is evaluated when functioning in the cooling mode. The graph in Figure 4 illustrates that the variation in the temperature span, represented as ΔT_{span}, consistently exceeds 35 K. This is an essential parameter for evaluating the performance of a thermal device, such as the elastocaloric device introduced in the previous section. A wide temperature span indicates a higher capacity to transfer heat and provide cooling or heating effects. At a speed of 0.10 m s^{-1} and a frequency of 1.5 Hz, the maximum ΔT_{span} value reaches 39.2 K. It is interesting to note that there are two distinct regions in the graph. In the low-speed zone, the temperature increases, while in the high-speed zone, the temperature decreases. This effect is mainly due to the exchange time between fluid and barocaloric material; as the speed increases, the exchange time between fluid and material decreases, not allowing the thermal flow to fully undergo the exchange process. Therefore, there is an optimum air velocity for each frequency beyond which the temperature span values decrease.

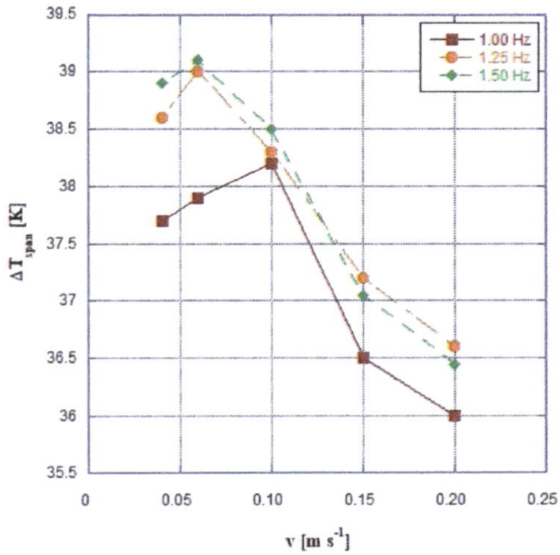

Figure 4. Temperature span as a function of the fluid velocity for different frequencies of the ABR cycle.

The specific cooling capacity of the barocaloric system is shown in Figure 5 as a function of fluid flow velocity. We compared the proposed system with a vapor compression cooler, which is a Samsung RB38T705CB1 VC system using 62.5 g of R600a.

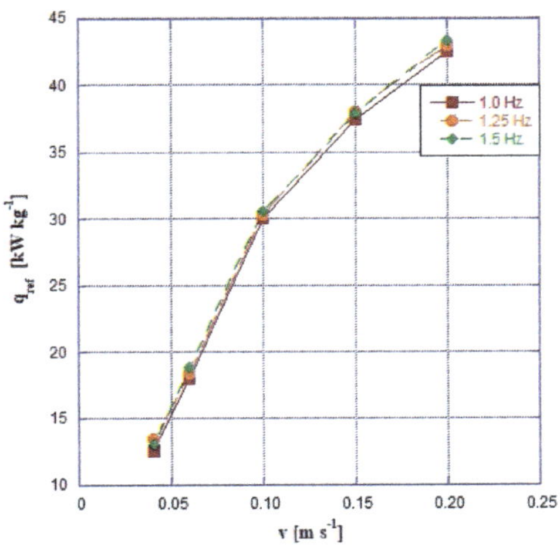

Figure 5. Specific cooling power as a function of the fluid velocity for different frequencies of the ABR cycle.

The VC system has a 1 kW cooling capacity and is intended for household use. The system's cooling capacity per kilogram is 16 kW kg^{-1}. Figure 5 shows that the barocaloric system outperforms the vapor compression system for fluid velocities greater than 0.06 m s^{-1}. The specific cooling power of the barocaloric system is always greater than the conventional system for HTF velocities greater than 0.05 m s^{-1}. At a frequency of 1.5 Hz and a speed of 0.20 m s^{-1}, it reaches its maximum value of 44 kW kg^{-1}.

The fluctuations in the coefficient of performance (COP) with HTF velocity at different frequencies are shown in Figure 6. The most favorable COP readings are observed at a frequency of 1.25 Hz and an HTF velocity of 0.20 m s^{-1}, reaching a peak value of approximately 4.5. Even the standard vapor-compression-based system achieves highly satisfactory *COP* values.

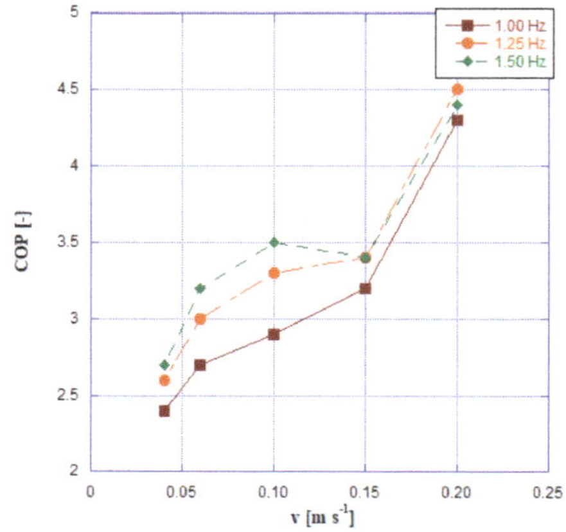

Figure 6. Coefficient of performance as a function of the fluid velocity for different frequencies of the ABR cycle.

Figure 7 demonstrates the results of the *COP* comparison between a residential refrigerator based on vapor compression and a barocaloric cooler. The following equation was used to obtain the comparison data:

$$\frac{\Delta COP}{COP} = \frac{COP_{ABR} - COP_{VC}}{COP_{VC}} \quad (8)$$

As shown in Figure 7, the velocity values of the HTF equal to or greater than 0.15 m s^{-1} always ensure energy savings if the domestic cooler is based on the ABR cycle. If the velocity is 0.10 m s^{-1}, the frequencies of 1.25 and 1.50 ensure a positive gain in terms of *COP*. The results demonstrate that the ABR cycle consistently outperforms the traditional vapor compression refrigerator, with up to 40% improvement. For low-velocity values and all frequencies, the barocaloric system does not achieve the same *COP* values as the traditional system. However, for velocities higher than 0.06 m s^{-1}, the barocaloric device manages to achieve improvements of over 40% for all frequencies.

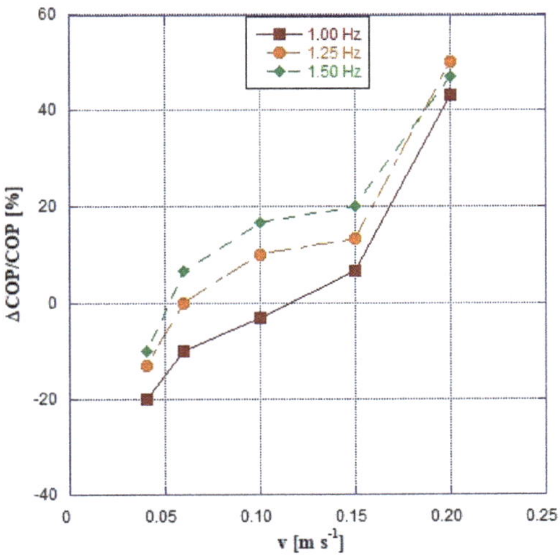

Figure 7. Coefficient of performance of the barocaloric cooler compared with that of a domestic vapor compression cooler.

6. Conclusions

In this research, the energy efficiency analysis of an active barocaloric refrigerator intended to serve cold food storage is presented. Using a 2D tool based on the finite element approach, we assessed its energy performance, considering factors like temperature span, cooling power, mechanical power, and coefficient of performance. The system was tested while operating in a hot indoor environment to examine the potentialities of the technology in extreme summer conditions (313 K). The food categories of interest in this investigation were meat and fresh food products (soft cheese, yogurt, and milk), kept at 5 °C (278 K) using a barocaloric cooler, which is considered an energy-efficient and environmentally friendly system. The features of Spoke 3 of PNRR ONFOODS were also used to compare the proposed method with vapor compression technology. In this investigation, vapor compression was replaced with barocaloric solid-state technology for the steps of the cold food supply chain associated with retail and household food preservation. The barocaloric system runs on an active barocaloric regenerative refrigeration cycle, with acetoxy silicone rubber serving as the solid-state refrigerant and ethylene glycol serving as HTF. From this investigation, the following conclusions can be drawn:

- The temperature span always exceeds 35 K, and it reaches a peak of 39.2 at a speed of 0.10 m s^{-1} and a frequency of 1.5 Hz.
- For fluid velocity greater than 0.06 m s^{-1}, the barocaloric system outperforms the vapor compression system in terms of cooling power. The peak of 44 kW kg^{-1} is reached at a speed of 0.20 m s^{-1} and a frequency of 1.5 Hz.
- The most favorable COP readings are observed at a frequency of 1.25 Hz and an HTF velocity of 0.20 m s^{-1}, reaching a peak value of approximately 4.5.
- Velocity values of the HTF equal to or greater than 0.15 m s^{-1} always ensure energy savings if the domestic cooler is based on the ABR cycle.

Based on the following considerations, it is concluded that barocaloric cooling is a favorable technology to employ in the final links of the cold food supply chain if the system operates with the ABR cycle at frequencies between 1.25 and 1.50 Hz with a regenerator comprising acetoxy silicone rubber as the solid-state refrigerant and 50%EG–50%water mixture as the heat transfer fluid flowing at the optimal velocity of 0.15 m s^{-1}. This solution guarantees an appropriate tradeoff between the temperature span, the cooling power, and the coefficient of performance. Under these conditions, the barocaloric system outperforms the domestic vapor compression cooler operating with R600a.

Author Contributions: Conceptualization, L.C., A.G. and C.M.; methodology, L.C., A.G. and C.M.; validation, L.C., A.G. and C.M.; formal analysis, L.C., A.G. and C.M.; investigation, L.C., A.G. and C.M.; data curation, L.C., A.G. and C.M.; writing—original draft preparation, L.C., A.G. and C.M.; writing—review and editing, L.C., A.G. and C.M.; visualization, L.C., A.G. and C.M.; supervision, L.C., A.G. and C.M.; project administration, A.G. All authors have read and agreed to the published version of the manuscript.

Funding: This research was funded by the University Ministry of Research of Italy within the scope of the project "PNRR ONFOODS Spoke 3", one of the spokes of the maxi project ONFOODS (Research and Innovation Network on Food and Nutrition Sustainability, Safety, and Security—Working on Foods) financed by the European Community and among the activities of the Italian PNRR, i.e., the National Recovery and Resilience Plan (PNRR), which is part of the Next Generation EU (NGEU) program established by the European Union in response to the pandemic crisis of 2020.

Data Availability Statement: Data are available upon request.

Acknowledgments: The authors acknowledge the University Ministry of Research of Italy for financing within the scope of the project "PNRR ONFOODS Spoke 3", one of the spokes of the maxi project ONFOODS (Research and Innovation Network on Food and Nutrition Sustainability, Safety, and Security—Working on Foods) financed by the European Community and among the activities of the Italian PNRR, i.e., the National Recovery and Resilience Plan (PNRR), which is part of the Next Generation EU (NGEU) program established by the European Union in response to the pandemic crisis of 2020.

Conflicts of Interest: The authors declare no conflict of interest.

Nomenclature

Roman symbols	
c	Specific heat capacity, J kg^{-1} K^{-1}
COP	Coefficient of performance, -
f	Frequency, Hz
k	Thermal conductivity, W m^{-1} K^{-1}
L	The whole length of the system, mm
n	Number of times
p	Pressure, Pa
$\dot{Q}$	Power density due to caloric effect, W m^{-3}
q	Thermal power density, W m^{-2}
S	Entropy, J kg K
T	Temperature, K
t	Time, s
u	Longitudinal velocity coordinate, m s^{-1}
V	Volume, m^3
v	Orthogonal velocity coordinate, m s^{-1}
x	Longitudinal spatial coordinate, m
X	Conjugate field
y	Orthogonal spatial coordinate, m
Y	Driving field
Greek symbols	
Δ	Finite difference
δ	Infinitesimal difference
ρ	Density, kg m^{-3}
Subscripts	
ad	Adiabatic
C	Cold heat exchanger
cycle	Cycle
f	Fluid
H	Hot heat exchanger

p	Constant pressure
ref	Refrigeration
s	Solid

References

1. Benhadid-Dib, S.; Benzaoui, A. Refrigerants and their environmental impact Substitution of hydro chlorofluorocarbon HCFC and HFC hydro fluorocarbon. Search for an adequate refrigerant. *Energy Procedia* **2012**, *18*, 807–816. [CrossRef]
2. Aprea, C.; Greco, A. An exergetic analysis of R22 substitution. *Appl. Therm. Eng.* **2002**, *22*, 1455–1469.
3. Aprea, C.; Greco, A.; Rosato, A. Comparison of R407C and R417A heat transfer coefficients and pressure drops during flow boiling in a horizontal smooth tube. *Energy Convers. Manag.* **2008**, *49*, 1629–1636.
4. Llopis, R.; Torrella, E.; Cabello, R.; Sánchez, D. HCFC-22 replacement with drop-in and retrofit HFC refrigerants in a two-stage refrigeration plant for low temperature. *Int. J. Refrig.* **2012**, *35*, 810–816.
5. United Nation Environment Program. *Montreal Protocol on Substances that Deplete the Ozone Layer*; United Nation Environment Program (UN): New York, NY, USA, 1987.
6. James, S.J.; James, C.J.F.R.I. The food cold-chain and climate change. *Food Res. Int.* **2010**, *43*, 1944–1956.
7. Mercier, S.; Villeneuve, S.; Mondor, M.; Uysal, I. Time–temperature management along the food cold chain: A review of recent developments. *Compr. Rev. Food Sci. Food Saf.* **2017**, *16*, 647–667. [PubMed]
8. Audsley, E.; Brander, M.; Chatterton, J.C.; Murphy-Bokern, D.; Webster, C.; Williams, A.G. How low can we go? An assessment of greenhouse gas emissions from the UK food system and the scope reduction by 2050. In *Report for the WWF and Food Climate Research Network*; WWF-UK: London, UK, 2010.
9. Garnett, T. Where are the best opportunities for reducing greenhouse gas emissions in the food system (including the food chain)? *Food Policy* **2011**, *36*, S23–S32.
10. Oenema, O.; Velthof, G.; Kuikman, P. Technical and policy aspects of strategies to decrease greenhouse gas emissions from agriculture. *Nutr. Cycl. Agroecosyst.* **2001**, *60*, 301–315.
11. Camanzi, L.; Alikadic, A.; Compagnoni, L.; Merloni, E. The impact of greenhouse gas emissions in the EU food chain: A quantitative and economic assessment using an environmentally extended input-output approach. *J. Clean. Prod.* **2017**, *157*, 168–176.
12. Cascetta, F.; Di Lorenzo, R.; Nardini, S.; Cirillo, L. A Trnsys Simulation of a Solar-Driven Air Refrigerating System for a Low-Temperature Room of an Agro-Industry site in the Southern part of Italy. *Energy Procedia* **2017**, *126*, 329–336. [CrossRef]
13. Aprea, C.; de Rossi, F.; Greco, A.; Renno, C. Refrigeration plant exergetic analysis varying the compressor ca-pacity. *Int. J. Energy Res.* **2003**, *27*, 653–669. [CrossRef]
14. Regulation (EC) No 852/2004 of The European Parliament and of the Council. 2004.
15. Zhang, Y.; Xu, Y.; Lu, R.; Zhang, S.; Hai, A.M.; Tang, B. Form-stable cold storage phase change materials with durable cold insulation for cold chain logistics of food. *Postharvest Biol. Technol.* **2023**, *203*, 112409. [CrossRef]
16. Bozorgi, M.; Roy, P.; Siddique, A.R.M.; Venkateshwar, K.; Tasnim, S.; Mahmud, S. Experimental investigation and life cycle assessment of a phase change material (PCM) based thermoelectric (TE) refrigerator. *Int. J. Thermofluids* **2023**, *19*, 100394. [CrossRef]
17. Aprea, C.; Greco, A.; Vanoli, G. Condensation heat transfer coefficients for R22 and R407C in gravity driven flow regime within a smooth horizontal tube. *Int. J. Refrig.* **2003**, *26*, 393–401.
18. Aprea, C.; Rossi, F.; Greco, A. Experimental evaluation of R22 and R407C evaporative heat transfer coefficients in a vapour compression plant. *Int. J. Refrig.* **2000**, *23*, 366–377. [CrossRef]
19. Jiao, C.; Li, Z. An Updated Review of Solar Cooling Systems Driven by Photovoltaic–Thermal Collectors. *Energies* **2023**, *16*, 5331.
20. Alahmer, A.; Ajib, S. Solar cooling technologies: State of art and perspectives. *Energy Convers. Manag.* **2020**, *214*, 112896.
21. Otanicar, T.; Taylor, R.A.; Phelan, P.E. Prospects for solar cooling–An economic and environmental assessment. *Sol. Energy* **2012**, *86*, 1287–1299.
22. Hwang, Y.; Radermacher, R.; Al Alili, A.; Kubo, I. Review of solar cooling technologies. *HvacR Res.* **2008**, *14*, 507–528. [CrossRef]
23. D'agostino, D.; Esposito, F.; Greco, A.; Masselli, C.; Minichiello, F. Parametric analysis on an earth-to-air heat exchanger employed in an air conditioning system. *Energies* **2020**, *13*, 2925. [CrossRef]
24. D'Agostino, D.; Esposito, F.; Greco, A.; Masselli, C.; Minichiello, F. The energy performances of a ground-to-air heat exchanger: A comparison among köppen climatic areas. *Energies* **2020**, *13*, 2895. [CrossRef]
25. Greco, A.; Masselli, C. The optimization of the thermal performances of an earth to air heat exchanger for an air conditioning system: A numerical study. *Energies* **2020**, *13*, 6414. [CrossRef]
26. Ketfi, O.; Abdi, H.; Lounici, B.; Bourouis, M. Performance Analysis of Low-Capacity Water–LiBr Absorption–Cooling Systems Using Geothermal Heat-Sinks in Hot Climates. *Energies* **2023**, *16*, 809. [CrossRef]
27. Fähler, S.; Rößler, U.K.; Kastner, O.; Eckert, J.; Eggeler, G.; Emmerich, H.; Entel, P.; Müller, S.; Quandt, E.; Albe, K. Caloric effects in ferroic materials: New concepts for cooling. *Adv. Eng. Mater.* **2012**, *14*, 10–19. [CrossRef]
28. Vopson, M.M.; Fetisov, Y.K.; Hepburn, I. Solid-state heating using the multicaloric effect in multiferroics. *Magnetochemistry* **2021**, *7*, 154. [CrossRef]

29. Garcia-Ben, J.; Delgado-Ferreiro, I.; Salgado-Beceiro, J.; Bermudez-Garcia, J.M. Simple and low-cost footstep energy-recove barocaloric heating and cooling device. *Materials* **2021**, *14*, 5947. [CrossRef] [PubMed]
30. Kabirifar, P.; Žerovnik, A.; Ahčin, Ž.; Porenta, L.; Brojan, M.; Tušek, J. Elastocaloric cooling: State-of-the-art and future challenge in designing regenerative elastocaloric devices. *Stroj. Vestn./J. Mech. Eng.* **2019**, *65*, 615–630. [CrossRef]
31. Aprea, C.; Greco, A.; Maiorino, A.; Masselli, C. The employment of caloric-effect materials for solid-state heat pumping. *Int. Refrig.* **2020**, *109*, 1–11. [CrossRef]
32. Alahmer, A.; Al-Amayreh, M.; Mostafa, A.O.; Al-Dabbas, M.; Rezk, H. Magnetic refrigeration design technologies: State of the ar and general perspectives. *Energies* **2021**, *14*, 4662. [CrossRef]
33. Silva, D.J.; Ventura, J.; Araújo, J.P. Caloric devices: A review on numerical modeling and optimization strategies. *Int. J. Energy Re* **2021**, *45*, 18498–18539. [CrossRef]
34. Takeuchi, I.; Sandeman, K. Solid-state cooling with caloric materials. *Phys. Today* **2015**, *68*, 48–54. [CrossRef]
35. Bachmann, N.; Schwarz, D.; Bach, D.; Schäfer-Welsen, O.; Koch, T.; Bartholomé, K. Modeling of an Elastocaloric Cooling Systen for Determining Efficiency. *Energies* **2022**, *15*, 5089. [CrossRef]
36. Fähler, S.; Pecharsky, V.K. Caloric effects in ferroic materials. *MRS Bull.* **2018**, *43*, 264–268. [CrossRef]
37. Kitanovski, A.; Plaznik, U.; Tomc, U.; Poredoš, A. Present and future caloric refrigeration and heat-pump technologies. *Int. Refrig.* **2015**, *57*, 288–298. [CrossRef]
38. Moya, X.; Mathur, N.D. Caloric materials for cooling and heating. *Science* **2020**, *370*, 797–803. [CrossRef] [PubMed]
39. Aprea, C.; Greco, A.; Maiorino, A.; Masselli, C. Analyzing the energetic performances of AMR regenerator working with differen magnetocaloric materials: Investigations and viewpoints. *Int. J. Heat Technol.* **2017**, *35*, S383–S390. [CrossRef]
40. Greco, A.; Masselli, C. Electrocaloric cooling: A review of the thermodynamic cycles, materials, models, and devices. *Magnetc chemistry* **2020**, *6*, 67. [CrossRef]
41. Li, Q.; Shi, J.; Han, D.; Du, F.; Chen, J.; Qian, X. Concept design and numerical evaluation of a highly efficient rotary electrocalori refrigeration device. *Appl. Therm. Eng.* **2021**, *190*, 116806. [CrossRef]
42. Engelbrecht, K. Future prospects for elastocaloric devices. *J. Phys. Energy* **2019**, *1*, 021001. [CrossRef]
43. Ismail, M.; Yebiyo, M.; Chaer, I. A review of recent advances in emerging alternative heating and cooling technologies. *Energie* **2021**, *14*, 502. [CrossRef]
44. Chen, J.; Lei, L.; Fang, G. Elastocaloric cooling of shape memory alloys: A review. *Mater. Today Commun.* **2021**, *28*, 10270 [CrossRef]
45. Porenta, L.; Trojer, J.; Brojan, M.; Tušek, J. Experimental investigation of buckling stability of superelastic Ni-Ti tubes under cycli compressive loading: Towards defining functionally stable tubes for elastocaloric cooling. *Int. J. Solids Struct.* **2022**, *256*, 11194 [CrossRef]
46. Lloveras, P.; Tamarit, J.-L. Advances and obstacles in pressure-driven solid-state cooling: A review of barocaloric materials. *MR Energy Sustain.* **2021**, *8*, 3–15. [CrossRef]
47. Chen, J.; Zhang, K.; Kan, Q.; Yin, H.; Sun, Q. Ul-tra-high fatigue life of NiTi cylinders for compression-based elastocaloric coolin *Appl. Phys. Lett.* **2019**, *115*, 093902. [CrossRef]
48. Tušek, J.; Engelbrecht, K.; Mikkelsen, L.P.; Pryds, N. Elastocaloric effect of Ni-Ti wire for application in a cooling device. *J. App Phys.* **2015**, *117*, 124901. [CrossRef]
49. Wang, R.; Fang, S.; Xiao, Y.; Gao, E.; Jiang, N.; Li, Y.; Mou, L.; Shen, Y.; Zhao, W.; Li, S.; et al. Torsional Refrigeration by Twiste Coiled, and Supercoiled Fibers. Available online: http://science.sciencemag.org/ (accessed on 30 July 2023).
50. Greibich, F.; Schwödiauer, R.; Mao, G.; Wirthl, D.; Drack, M.; Baumgartner, R.; Kogler, A.; Stadlbauer, J.; Bauer, S.; Arnold, N et al. Elastocaloric heat pump with specific cooling power of 20.9 W g–1 exploiting snap-through instability and strain-induce crystallization. *Nat. Energy* **2021**, *6*, 260–267. [CrossRef]
51. Cirillo, L.; Farina, A.R.; Greco, A.; Masselli, C. Numerical optimization of a single bunch of niti wires to be placed in a elastocaloric experimental device: Preliminary results. *Magnetochemistry* **2021**, *7*, 67. [CrossRef]
52. Ahčin, Ž.; Kabirifar, P.; Porenta, L.; Brojan, M.; Tušek, J. Numerical modeling of shell-and-tube-like elastocaloric regenerato *Energies* **2022**, *15*, 9253. [CrossRef]
53. Ma, S.; Zhang, X.; Zheng, G.; Qian, M.; Geng, L. Toughening of Ni-Mn-Based Polycrystalline Ferromagnetic Shape Memor Alloys. *Materials* **2023**, *16*, 5725. [CrossRef]
54. Zhu, Y.; Tsuruta, R.; Gupta, R.; Nam, T. Feasibility Investigation of Attitude Control with Shape Memory Alloy Actuator on Tethered Wing. *Energies* **2023**, *16*, 5691. [CrossRef]
55. Cirillo, L.; Greco, A.; Masselli, C. Cooling through barocaloric effect: A review of the state of the art up to 2022. *Therm. Sci. En Prog.* **2022**, *33*, 101380. [CrossRef]
56. Aprea, C.; Greco, A.; Maiorino, A.; Masselli, C. The use of barocaloric effect for energy saving in a domestic refrigerator wit ethylene-glycol based nanofluids: A numerical analysis and a comparison with a vapor com-pression cooler. *Energy* **2020**, *19* 116404. [CrossRef]
57. Liu, J.; Li, Z.; Liu, H.; Yu, L.; Zhang, Y.; Cao, Y.; Xu, K.; Liu, Y. Martensitic Transformation and Barocaloric Effect in Co-V-Ga-F Paramagnetic Heusler Alloy. *Metals* **2022**, *12*, 516. [CrossRef]
58. Moya, X.; Defay, E.; Heine, V.; Mathur, N.D. Too cool to work. *Nat. Phys.* **2015**, *11*, 202–205. [CrossRef]

59. Moya, X.; Kar-Narayan, S.; Mathur, N.D. Caloric materials near ferroic phase transitions. *Nat. Mater.* **2014**, *13*, 439–450. [CrossRef] [PubMed]
60. Czernuszewicz, A.; Kaleta, J.; Lewandowski, D.; Przybylski, M. An idea of the test stand for studies of mag-netobarocaloric materials properties and possibilities of their application. *Phys. Status Solidi C Curr. Top. Solid State Phys.* **2014**, *11*, 995–999. [CrossRef]
61. Czernuszewicz, A.; Kaleta, J.; Królewicz, M.; Lewandowski, D.; Mech, R.; Wiewiórski, P. A test stand to study the possibility of using magnetocaloric materials for refrigerators. *Int. J. Refrig.* **2014**, *37*, 72–77. [CrossRef]
62. Aprea, C.; Greco, A.; Maiorino, A.; Masselli, C. Solid-state refrigeration: A comparison of the energy perfor-mances of caloric materials operating in an active caloric regenerator. *Energy* **2018**, *165*, 439–455. [CrossRef]
63. Imamura, W.; Usuda, O.; Paixão, L.S.; Bom, N.M.; Gomes, A.M.; Carvalho, A.M.G. Supergiant Barocaloric Effects in Acetoxy Silicone Rubber over a Wide Temperature Range: Great Potential for Solid-state Cooling. *Chin. J. Polym. Sci.* **2020**, *38*, 999–1005. [CrossRef]
64. Shit, S.C.; Shah, P. A review on silicone rubber. *Natl. Acad. Sci. Lett.* **2013**, *36*, 355–365. [CrossRef]
65. Guo, Y. Investigation of Silicone Rubber Blends and Their Shape Memory Properties. Ph.D. Thesis, University of Akron, Akron, OH, USA, 2018.
66. Halim, Z.A.A.; Ahmad, N.; Yajid, M.A.M.; Hamdan, H. Thermal insulation performance of silicone rubber/silica aerogel composite. *Mater. Chem. Phys.* **2022**, *276*, 125359. [CrossRef]
67. DECRETO LEGISLATIVO 27 gennaio 1992, n. 110. Attuazione Della Direttiva n. 89/108/CEE in Materia di Alimenti Surgelati destinati All'alimentazione Umana. Decreto del Presidente Della Repubblica. Available online: https://www.gazzettaufficiale.it/atto/serie_generale/caricaDettaglioAtto/originario?atto.dataPubblicazioneGazzetta=1992-02-17&atto.codiceRedazionale=092G0147 (accessed on 30 July 2023).
68. Sebald, G.; Xie, Z.; Guyomar, D. Fatigue effect of elastocaloric properties in natural rubber. *Philos. Trans. R. Soc. A Math. Phys. Eng. Sci.* **2016**, *374*, 20150302. [CrossRef] [PubMed]
69. Woo, C.-S.; Kim, W.-D.; Lee, S.-H.; Choi, B.-I.; Park, H.-S. Fatigue life prediction of vulcanized natural rubber subjected to heat-aging. *Procedia Eng.* **2009**, *1*, 9–12. [CrossRef]
70. Boyer, F.; Fabrie, P. *Mathematical Tools for the Study of the Incompressible Navier-Stokes Equations and Related Models*; Springer Science & Business Media: Berlin/Heidelberg, Germany, 2012; Volume 183.

energies

Article

Heat Production Capacity Simulation and Parameter Sensitivity Analysis in the Process of Thermal Reservoir Development

Yi Yang [1], Guoqiang Fu [2,*], Jingtao Zhao [1] and Lei Gu [3]

1 College of Geoscience and Survey Engineering, China University of Mining and Technology, Beijing 100083, China; yangyi19961103@163.com (Y.Y.); diffzjt@cumtb.edu.cn (J.Z.)
2 School of Resources and Geosciences, China University of Mining and Technology, Xuzhou 221116, China
3 Jining Energy Development Group Co., Ltd., Jinqiao Coal Mine, Jining 272000, China; gl18454129297@163.com
* Correspondence: fuguoqiang@cumt.edu.cn

Abstract: The development of a geothermal system involves changes in the temperature field (T), seepage field (H), stress field (M), and chemical field (C) and the influence among them and injecting the heat extraction working fluid into the injection well that flows (migrating) through natural fractures and exchanges heat with the geothermal high-temperature rock. At the same time, the injection of low-temperature working fluid will induce thermal stress, resulting in changes in the reservoir temperature field and stress field. To study the influence factors and influence degree of heat production performance and mining life under multi-field coupling in the process of thermal reservoir development, based on THMC multi-field coupling numerical simulation software, this paper deeply studies the control differential equations and boundary coupling conditions of rock mass (fracture) deformation, seepage, heat exchange, the chemical reaction, and other processes based on the numerical solution method of the discrete fracture network model, simulating heat production capacity during the deep geothermal resource extraction process. The reservoir geological model analysis and generalization, parameter setting, boundary conditions, initial condition settings, mesh generation, and other steps were carried out in turn. Two different heat extraction working fluids, water, and CO_2 were selected for numerical simulation in the mining process. The changes in the thermal reservoir temperature, net heat extraction rate, and SiO_2 concentration during the thirty years of systematic mining were compared. The results show that CO_2 has a better heat extraction effect. Finally, the reservoir thermal conductivity, heat capacity, well spacing, injection temperature, fracture spacing, fracture permeability, fracture number, fracture length, and other parameters were set, respectively. The parameter variation range was set, and the parameter sensitivity analysis was carried out. The numerical simulation results show that the engineering production conditions (injection temperature, well spacing) have little effect on the thermal efficiency and mining life, and the properties of fractures (fracture permeability, fracture number, fracture length) have a great influence.

Keywords: thermal storage development; THMC multi-field coupling; discrete fracture network; numerical simulation; sensitivity analysis

Citation: Yang, Y.; Fu, G.; Zhao, J.; Gu, L. Heat Production Capacity Simulation and Parameter Sensitivity Analysis in the Process of Thermal Reservoir Development. *Energies* **2023**, *16*, 7258. https://doi.org/10.3390/en16217258

Academic Editor: Artur Bartosik

Received: 16 September 2023
Revised: 21 October 2023
Accepted: 23 October 2023
Published: 25 October 2023

1. Introduction

The development and utilization of geothermal energy play an important role in alleviating the global energy crisis, reducing carbon emissions, protecting the ecological environment, and improving energy security. In recent years, geothermal energy development has been extensively studied [1]. However, geothermal wells are required to produce high flow rates and sustained power output to increase the efficiency of heat production in geothermal reservoirs. The exploitation of geothermal resources is extremely complex due to the dynamics of spatio-temporal evolution, which involves multi-field coupling (THMC), such as the temperature field, seepage field, mechanical field, and chemical field.

In recent years, with the continuous development of computer technology and simulation algorithms, numerical modeling methods have been essential for solving complex multi-field coupled problems in the geothermal field [2].

The numerical simulation of geothermal energy development is mainly based on the finite element method (finite difference, finite volume) under a dual medium and porous medium. Many scholars at home and abroad have developed various numerical simulation software for geothermal resource development and have carried out quantitative numerical simulation research, including ABAQUS, ANSYS, FEFLOW, COMSOL Multiphysics, FEFLOW, Visual MODFLOW, and so on [3–7]. Li Xinxin [8] used the T-H coupling equivalent simulation method in three-dimensional fractured rock mass, combined with COMSOL Multiphysics software for secondary development, to solve the fully coupled mathematical and physical model of hydrothermal coupling and applied this method to the numerical calculation of large-scale geothermal energy well system development at the level of a complex fractured site. Horne and Juliusson [9] optimized the injection flow rate of a multi-well system based on FEFLOW simulation software. Domestic and foreign scholars also use a variety of computer code programming to solve underground complex multi-field coupling geothermal energy development and EGS simulation problems, including TOUGH2, OpenGeoSys, ROCMAS, FRACture, and so on [10–13]. Li and Zhang et al. [14,15] used TOUGH2 software to numerically calculate the hydrothermal output of EGS double vertical wells and single fracture mining in this area and to evaluate and determine the area's dynamic reserves and space–time evolution law. Guo and Zhang et al. [16] considered the two working conditions of the shielding layer and the non-shielding layer of the geothermal reservoir and numerically simulated the hydrothermal production of the single-fractured horizontal well in the Xujiaweizi area of the Songliao Basin. Lei and Zhang et al. [17,18] used TOUGH2 to establish vertical and horizontal well models for geothermal reservoirs in the Qiabuqia area of the Gonghe Basin, Qinghai Province, for numerical simulation of heat production. Wu et al. [19] proposed a CMPSF-enhanced THMC simulator by integrating the rock damage model and CO_2 physical property model into the coupled COMSOL-MATLAB-PHREEQC (CMP) framework via MATLAB, which was combined with numerical and experimental example validation to analyze the THMC field and fracture evolution. Mohammad [20] used the thermoporous elastic displacement discontinuity (DD) method combined with different forms of finite element methods in conjunction with the fracture intrinsic relationship to consider the deformation properties of the fracture and applied advanced numerical methods to simulate and analyze the effect of the coupling process injecting cold water into the fracture/matrix system on the temporal and spatial variations of the matrix–fracture stresses, temperatures, and pore pressures.

The heat production capacity is a decisive criterion for evaluating the effect of high-temperature thermal reservoir reconstruction. Therefore, an accurate evaluation of the heat production capacity of the reservoir has important guiding significance for the dynamic mining and operation life of the EGS system. To comprehensively analyze the influence of EGS parameters on the heat recovery performance of the system and the influence of the parameters on each other, the Tengchong Rehai geothermal field in Yunnan, the Qiabuqia geothermal field in the northeast of the Qinghai–Tibet Plateau, and the Yangyi geothermal field in Tibet were taken as the geological background, respectively. The orthogonal design idea was used to analyze the changes in well spacing, injection flow, injection temperature, reservoir permeability, and other factors with the numerical simulation method [21–23]. The study shows that the factor that has less influence on the thermal performance of EGS extraction is the injection temperature, and the greater the mining flow, the shorter the operating life of EGS. Zhang et al. [24], based on the borehole geological data of the Qiabuqia geothermal area in the northeastern Qinghai–Tibet Plateau, established a three-dimensional hydrothermal coupled fractured reservoir model using the finite element method to evaluate its heat production potential. N. Tenma et al. [25] estimated the parameters of a typical underground system using a two-well model based on the FEHM code. Six different heat extraction schemes have been studied by changing the system parameters to explore the best scheme. Xu et al. [26] established an oil–water two-phase THMC temperature coupling prediction model based on discrete fractures to study the effect of different hydraulic fracturing parameters on formation temperature after injecting low-temperature fracturing fluid for the coupling process of THMC in high-temperature reservoirs. The effects of the fracturing fluid temperature, reservoir temperature, dimensionless conductivity, Young's modulus, injection rate, cluster spacing, and proportion of the branch fracture area on the reservoir and the fracture temperatures were investigated. Aliyu [27] analyzed the sensitivity of deep geothermal reservoirs by studying the influence of various reservoir parameters on the production temperature, showing that the fluid injection temperature is the most influential parameter for the experiment. Liu [28] took two groups of middle and deep geothermal wells in the Xiwenzhuang geothermal field of the Taiyuan Basin as the research object. Based on the summary and analysis of the distribution characteristics of the geothermal reservoirs, the geothermal characteristics, hydrogeological characteristics, and the drilling data in the Xiwenzhuang area, numerical simulation technology was used to simulate the geothermal energy mining process, and the heat recovery efficiency of the geothermal wells under different conditions was studied.

In summary, it can be seen that the current multi-field coupling research for geothermal systems is mainly two-field (TH) or three-field (THM, THC) coupling, heat flow curing (THMC), and a four-field coupling mutual influence effect injected into the heat extraction working fluid where there is less comparative research on water and CO_2 geothermal heat production capacity [29]. In the geothermal system mining multi-field coupling process respectively, the mining conditions and natural fracture parameter changes in the heat extraction efficiency of the research parameters are not comprehensive. Therefore, in this paper, the numerical simulation method of the fractured rock body based on the discrete fracture network model simulates the THMC coupling calculation model under the natural fracture of geothermal system hot dry rock and analyzes and compares the reservoir temperature and net heat extraction rate of the two types of work fluid, the water, and the CO_2 in the same reservoir conditions and the operating parameters of the heat extraction system in the extraction over thirty years to evaluate the performance of two types of work masses for the extraction of heat in the geothermal system. Under the condition that water is used as the heat extraction working fluid, the influence of each parameter on the heat extraction performance of the thermal storage was analyzed separately; i.e., the parameter sensitivity analysis and the results of the numerical simulation using multi-field coupling could be used to determine the important influencing parameters at the design stage of the geothermal resource development.

2. Mathematical Model

2.1. Basic Assumption

When establishing the THMC coupling mathematical model of fractured media, the following assumptions are introduced:

(1) The rock mass is composed of matrix rock blocks and fractures in the dual media of the pores and fractures.
(2) The rock matrix can be simplified as a quasi-continuous medium model and homogeneous isotropic elastic body, and the fracture can be simplified as a fracture medium model.
(3) Fracture seepage obeys Darcy's law.
(4) The effective stress law of fractures is:

$$\sigma' = \sigma - \alpha_f P \quad (1)$$

In this formula, σ'—effective stress; σ—initial stress; α_f—the ratio of the connected area to the total area in the fracture; P—pressure.

(1) The thermoelastic constitutive law of matrix rock is:

$$\sigma' = \lambda\delta_{ij}\varepsilon_{kk} + \frac{E}{1+v}\varepsilon_{ij} - \frac{E}{1-2v}\alpha_T\Delta T\delta_{ij} \tag{2}$$

In this formula, σ' and ε—stress and strain; λ—Lame constant; δ_{ij}—Kronecker symbol; E and v—the elastic modulus and Poisson's ratio, respectively; α_T—thermal expansion coefficient; ΔT—temperature increment.

(2) The heat in a rock mass can be transferred via conduction, convection, and radiation. In many cases, the radiation heat can be ignored.
(3) Heat exchange and heat transfer occur in fractured water through convection and conduction. To simplify the calculation, it is considered that the rock mass and fractures are always in the elastic state based on the assumption of a small deformation [30].

2.2. Seepage Process

The seepage field of the thermal reservoir satisfies the mass conservation equation:

$$\frac{\partial m}{\partial t} + \nabla \cdot (\rho_w q_w) = Q_s \tag{3}$$

In this formula, $m = \rho_w\phi, \rho_w$ (kg/m^3)—fluid density; ϕ—porosity; t(s)—time; q_w(m/s)—Darcy speed; Q_s (kg/m^3 s)—source of fluid mass.

$$q_w = -\frac{k}{u}\nabla p_w \tag{4}$$

In this formula, u(Pa s)—fluid viscosity; k(m^2)—permeability; p_w(Pa)—pore water pressure.

Substitute (4) into (3) to obtain:

$$\frac{\partial \rho_w \phi}{\partial t} - \nabla \cdot \left(\frac{k}{u}\rho_w \nabla p_w\right) = Q_s \tag{5}$$

2.3. Heat Transfer Process

Generally, the temperature distribution inside the rock mass depends on the heat exchange inside the rock and the heat exchange between the rock and the external medium, which is related to time. Heat exchange within the rock mass is described using the Fourier heat transfer equation:

$$\rho c\frac{\partial T}{\partial t} = \frac{\partial}{\partial x}\left(\lambda_x\frac{\partial T}{\partial x}\right) + \frac{\partial}{\partial y}\left(\lambda_y\frac{\partial T}{\partial y}\right) + \frac{\partial}{\partial z}\left(\lambda_z\frac{\partial T}{\partial z}\right) + \overline{Q} \tag{6}$$

In this formula, ρ(kg/m^3)—rock mass density; c(J/(kg K))—specific heat capacity of the rock mass; $\lambda_x, \lambda_y, \lambda_z$(W/(m K))—thermal conductivity of the rock mass; T(K)—rock temperature; t(s)—time; $\overline{Q}$ (W/m^3)—heat source density in the rock mass.

The hot dry rock (HDR) mass geothermal system's thermal energy comes from the cyclic heating of the injected low-temperature heat extraction working fluid through the underground HDR mass. Driven by the temperature difference between the thermal reservoir rock and the fracture water flow, there is heat conduction inside the thermal reservoir rock, and there is heat conduction and heat convection between the wall surface of the fracture and the internal fluid. Heat transfer is a local heat imbalance process. It is assumed that the heat exchange model of the natural fracture water flow in the reservoir rock is shown in Figure 1.

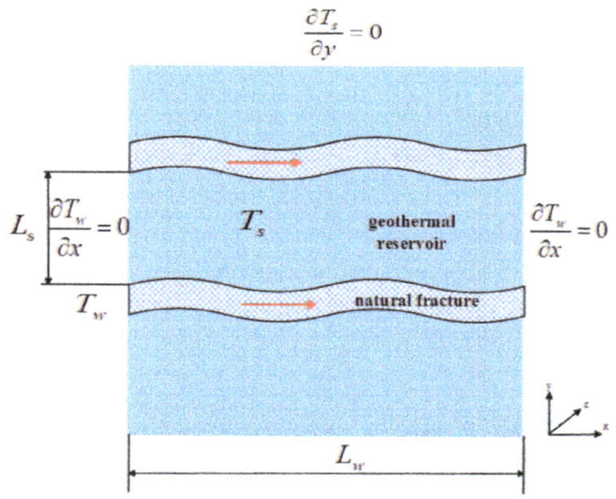

Figure 1. Heat exchange model between matrix system and natural fracture in thermal reservoir [31].

(1) The heat conduction equation in the reservoir rock skeleton is:

$$\frac{\partial[\rho_s c_s(1-\phi)\Delta T_s]}{\partial t}=T_s\alpha_T(1-\phi)K\frac{\partial\varepsilon_V}{\partial t}+\nabla\cdot[\lambda_s(1-\phi)\nabla T_s]+(1-\phi)Q_{Ts} \tag{7}$$

In this formula, $\rho_s(\text{kg/m}^3)$—rock density; $c_s(\text{J/(kg K)})$—rock-specific heat capacity; ϕ—rock porosity; $T_s(\text{K})$—thermal reservoir rock temperature; $\alpha_T(1/\text{K})$—thermal expansion coefficient; ε_V—volumetric strain; K(Pa)—rock bulk modulus; $\lambda_s(\text{W/(m K)})$—coefficient of thermal conductivity; Q_{Ts}—heat source term.

If the porosity of a rock is far less than 1, approximately taken as 0, then Formula (7) can be simplified as:

$$\frac{\partial(\rho_s c_s\Delta T_s)}{\partial t}=T_s\alpha_T K\frac{\partial\varepsilon_V}{\partial t}+\nabla\cdot(\lambda_s\nabla T)+Q_{Ts} \tag{8}$$

(2) The fluid heat transfer equation in the fracture is:

$$\frac{\partial(\rho_w c_w\phi\Delta T_w)}{\partial t}=-\nabla\cdot(\rho_w c_w q_w T_w)+\nabla\cdot(\lambda_w\phi\nabla T_w)+\phi Q_{Tw} \tag{9}$$

In this formula, $\rho_w(\text{kg/m}^3)$—fluid density; $c_w(\text{J/(kg K)})$—specific heat capacity of the fluid; ϕ—porosity; $T_w(\text{K})$—fluid temperature in the fracture; $q_w(\text{m/s})$—Darcy velocity; $\lambda_w(\text{W/(m K)})$—thermal conductivity of the fluid; ϕQ_{Tw}—heat source term.

Assuming that the solid and the fluid are always in a thermal equilibrium state, the heat transfer control equation of the fluid in the reservoir is obtained by adding Formulas (8) and (9):

$$\frac{\partial(\rho_s c_s\Delta T+\rho_w c_w\phi\Delta T)}{\partial t}=T\alpha_T K\frac{\partial\varepsilon_V}{\partial t}-\nabla\cdot(\rho_w c_w\mathbf{q}_w T)+\nabla\cdot[(\lambda_s+\lambda_w\phi)\nabla T]+Q_T \tag{10}$$

2.4. Rock Mechanics Process

The numerical model of the solid mechanical field in the multi-field coupling simulation of the geothermal mining system is based on BIOT consolidation theory, which considers:

(1) The interaction between the groundwater and rock mass is the result of the change in the pore water pressure and rock mass deformation.
(2) Assuming that the thermal reservoir rock is an isotropic homogeneous porous medium, its deformation is small, the temperature change causes the deformation of the solid skeleton, and the thermal strain generated is only positive strain. Therefore, the rock strain can be expressed as the sum of the thermal strain due to the temperature change and the effective stress-induced strain:

$$Gu_{i,kk} + \frac{G}{1-2v}u_{k,ki} - \alpha_T KT_{,i} - \alpha_{p,i} + f_{,i} = 0 \quad (11)$$

In this formula, G(Pa)—shear modulus; u_i—offset component; v—Poisson's ratio; $\alpha_T(1/\text{K})$—thermal expansion coefficient; K(Pa)—bulk modulus; T(K)—temperature; $\alpha_{p,i}$—the influence of fluid seepage on rock mass deformation; $-\alpha_T KT_{,i}$—the influence of temperature change on rock mass deformation; $f_{,i}$—volume force component.

2.5. Chemical Reaction Process

Most studies on the thermal efficiency of deep thermal storage consider the coupling of two or three fields and often ignore the influence of the chemical field. However, the interaction between the heat extraction working fluid (water, CO_2) and rocks is a key issue in developing geothermal resources. As the temperature and pressure of the heat extraction working fluid change, the reservoirs, wellbores, and pipelines in the geothermal system generally undergo chemical reactions, resulting in dissolution and precipitation. In addition, due to the mismatch between the temperature and chemical composition of the injected working fluid and the original geothermal water in the reservoir, the evolution of the temperature field and seepage field in the thermal reservoir changes the initial chemical equilibrium and reaction rate, which in turn affects the flow and heat transfer process.

Taking water injection as an example, chemical field reaction mainly includes two processes: solute transport and water chemical reaction [32]. Most of the HDR thermal storage rocks are dominated by granite. The main components of the hot reservoir rocks in the study area are quartz (SiO_2), potassium feldspar ($KAlSi_3O_8$), albite ($NaAlSi_3O_8$), muscovite ($KAl_2(AlSi_3O_{10})(OH)_2$), and calcite ($CaCO_3$).

(1) Solute transport

The dissolution and precipitation reactions of quartz (SiO_2) are generally considered the most important chemical factors in developing HDR resources. The solute transfer equation of quartz minerals in the rock matrix and fracture is described as:

$$\begin{cases} c'_m = c_m - c_m^{eq} \\ c'_f = c_f - c_f^{eq} \end{cases}$$
$$\begin{cases} \frac{\partial c'_m}{\partial t} = D_m \frac{\partial^2 c'_m}{\partial y^2} - \frac{k_m}{\theta_m} c'_m \\ \frac{\partial c'_f}{\partial t} = -v_f \frac{\partial c'_f}{\partial x} - k'_f c'_f + \frac{\theta_m D_m}{b} \frac{\partial c'_m}{\partial y} |y = n \end{cases} \quad (12)$$

In this formula, c_m(ppm)—the total dissolved concentration of solute in the matrix; eq(ppm)—the superscript representing the solubility of the solute in different media with the temperature; c_f(ppm)—the total concentration of solute dissolved in the fracture; k_m(m/s)—the dissolution rate of the solute in the rock matrix with the temperature change; θ_m—the porosity of the rock matrix; k'_f(m/s)—the dissolution rate of the solute in the fracture with the temperature change.

In Formula (10), the coefficients in the solute transfer equation in the natural fractures and matrix are related to the temperature environment of the medium and need to be solved by coupling with the fluid–matrix heat transfer equation in fractures.

(2) Chemical reaction

In simulating geothermal mining, this paper only considers water–rock interactions due to the deep circulation of the geothermal fluids in the reservoir in fractures and the long-term effects of high-temperature environments. Geothermal water usually has the characteristics of high soluble salt. The governing equation for the reaction of quartz with water reacts on the surface to form a silicic acid monomer or to form a dissolved silica. The reaction equation is:

$$SiO_2 + H_2O \rightarrow H_4SiO_4 \quad (13)$$

When the injection well injects a large amount of cold water through the fracture t heat the production well, the material reaction equation between the water and the granit is Formula (13). Assuming that the concentration of the reaction substance (quartz) is lowe than that of the solvent liquid (water), the Fickian diffusion law can be used to describe th diffusion term in mass transport. Therefore, the convection–diffusion equation can be use to model the mass transfer of SiO_2, H_2O, and H_4SiO_4.

$$u \cdot \nabla c_i = \nabla \cdot (D_i \nabla c_i) + R_i \tag{14}$$

In this formula, u—flow velocity; c_i (mol/m^3)—concentration; D_i (m^2/s)—diffusivit R_i (mol/(m^3 s)—material reaction rate.

$$R = -k_f \cdot c_{i1} \cdot c_{i2} \tag{15}$$

In this formula, k_f—reaction rate constant.

In the process of geothermal exploitation, when the heat extraction working fluid i CO_2, carbonic acid is produced by the reaction between the reservoir and the water in i Then the chemical reaction with the components of the rock is as follows:

$$\begin{aligned}
&CO_2 + H_2O \rightarrow H_2CO_3 \\
&\mathrm{3KAlSi_3O_8 + 2H^+ + H_2O \rightleftharpoons KAl_3Si_3O_{10}(OH)_2 + 6SiO_2 + 2K^+ + H_2O} \\
&NaAlSi_3O_8 + H_2O + CO_2 \rightarrow Al_2Si_2O_5(OH)_4 + 2\mathrm{Na^+} + 2\mathrm{HCO_3^-} + 4\mathrm{H_4SiO_4} \\
&KAl_2(AlSi_3O_{10})(OH)_2 + 10H^+ \rightarrow K^+ + 3Al^{3+} + 3SiO_2(aq) + 6H_2O \\
&CaCO_3 + H^+ \rightarrow Ca^{2+} + \mathrm{HCO_3^-}
\end{aligned} \tag{16}$$

According to the study by Kamali Asl et al. [33] and the material reaction equatio it is found that the products of the diagenetic minerals of granite contain SiO_2 after th water–rock reaction. Therefore, only the concentration of SiO_2 in the effluent near th production well is used as the main evaluation index of the degree of water–rock respons and the concentrations of Ca^{2+}, $\mathrm{Na^+}$, Al^{3+}, and K^+ are used as the auxiliary index i this study.

3. Construction of Numerical Model

3.1. Construction of Numerical Model

The deep geothermal reservoir conditions and mining operation data at home an abroad were collected. Combined with the real-time high-temperature environment, th actual triaxial hydraulic fracturing physical simulation experiment in the study area, an the complex fracture network initiation and expansion evolution experiment under th coupling of heat flow solidification, the heat production capacity of the geothermal deve opment is controlled by the aspects of reservoir physical properties, well location, hea extraction working fluid, thermal conductivity, injection–production pressure differenc and flow rate. The key parameters of the rock matrix (500 m × 500 m × 500 m) and natura fracture (200 m × 40 m × 0.01 m) in the model are set and shown in Tables 1 and 2. Th fractures in the model are regarded as a three-dimensional simple plane random discret distribution in the geothermal reservoir. The average size is set to 200 m × 40 m × 0.01 m and its spatial position is evenly distributed in the center of the reservoir.

Table 1. The characteristics of the solid medium used in the simulation.

Parameters		Units	Values
geothermal reservoir	density	kg/m^3	2650
	porosity	-	1.0×10^{-5}
	permeability	m^2	5×10^{-17}
	thermal conductivity	W/(m K)	2.9
	heat capacity	J/(kg K)	850
	thermal expansion coefficient	1/K	1×10^{-5}
	thickness	m	500
fracture	porosity	-	0.1
	permeability	m^2	2.73×10^{-13}
	width	m	0.01
	length	m	200
	distance	m	20
	height	m	40
overlying layer thickness		m	230
underlying layer thickness		m	230
productivity index		m^3	5×10^{-12}

Table 2. Production conditions in numerical simulation.

Parameters	Units	Values
Injection well/production well length	m	140
Injection well/production well diameter	m	0.1
Horizontal well distance	m	200
Injection/production speed	kg/s	2.5
Injection temperatures	K	293.15
Injection pressure	MPa	60
Production pressure	MPa	5

3.2. Initial Conditions and Boundary Conditions

Boundary conditions impose additional conditions on the solution and some modeling domains (such as surfaces, edges, or points). The same model can use a variety of different boundary conditions. The numerical models studied in this paper involve temperature, fluid, mechanical, and chemical boundaries. The model ignores the mass transfer and heat transfer process of the medium outside the reservoir, the water insulation boundary.

The temperature field, in addition to the heat conduction equation, calculates the temperature distribution inside the object and also needs to specify the initial conditions and boundary conditions. According to the static temperature and pressure measurement results of geothermal logging in the simulated study area, the bottom of the rock matrix is within the depth of 4500 m to 5000 m underground, and its temperature change is ignored. In this model, the ground temperature gradient is set to 0.04 K/m, and the ground temperature is 293.15 K. Therefore, the initial temperature of the rock matrix at 5000 m can be calculated by the Formula (17):

$$T|_{t=0} = T_0(x, y, z) = 293.15(\mathrm{K}) + 0.04(\mathrm{K/m}) \times 5000(\mathrm{m}) = 493.15(\mathrm{K}) \quad (17)$$

The initial reservoir temperature of the model is 220 °C; the upper and lower boundaries are regarded as the thermal insulation boundary; the surrounding boundary is the constant temperature boundary; and the initial injection temperature is set to 293.15 K (20 °C).

In the seepage field, the upper and lower boundaries are impermeable, the thermal reservoir is large enough to ignore the influence of the boundary, and the surrounding boundary is regarded as the constant pressure boundary. Since the surrounding rock of the deep thermal reservoir in the study area is a granite-dense rock layer, no fracturing is performed. It is assumed that it is highly closed and that the permeability is extremely low. Therefore, there is no fluid loss in the rock outside the boundary of the fractured reservoir.

In the stress field, modeling thermal storage at 5000 m depth, the surface pressure is 0.1 MPa, and the ground pressure gradient is set to 8500 Pa/m. The initial pressure of the rock matrix can be calculated using Formula (18). The initial reservoir pressure of the model is 42.6 MPa, the injection well pressure is 60 MPa, and the production well pressure is 5 MPa.

$$P|_{t=0} = P_0(x, y, z) = 100000(\text{Pa}) + 8500(\text{Pa/m}) \times 5000(\text{m}) = 42.6(\text{MPa}) \quad (18)$$

In the chemical field, the initial injection of different working fluids (water or CO_2) produces chemical stimulation to the reservoir, reacts with the reservoir rock to produce dissolution and precipitation, and finally reaches a chemical equilibrium state.

3.3. Model Construction

The Comsol Multiphysics modeling software simulates the natural fracture development strata with a 4500 m~5000 m depth and an average temperature of 200 °C as the geothermal mining section. The model size is a 500 m × 500 m × 500 m cube.

The main consideration when laying out wells is the fracture pattern, which is affected by the ground stress. When the minimum principal stress is horizontal, the fracture is vertical, with a length of 200 m, a height of 40 m, and a width of 0.01 m. The numerical model is set up with eight vertical fractures spaced 20 m apart in the center of the rock mass. Due to the uniform distribution of fractures in the simulated study area, this paper adopts the mining mode of one injection and one mining. Injection wells are set at 140 m long and 0.1 m in diameter. The well spacing is set at 200 m. Cold water is injected from the injection wells, flows through rocks and fractures to be heated, and is withdrawn from the production well. The geometric conceptual model is shown in Figure 2.

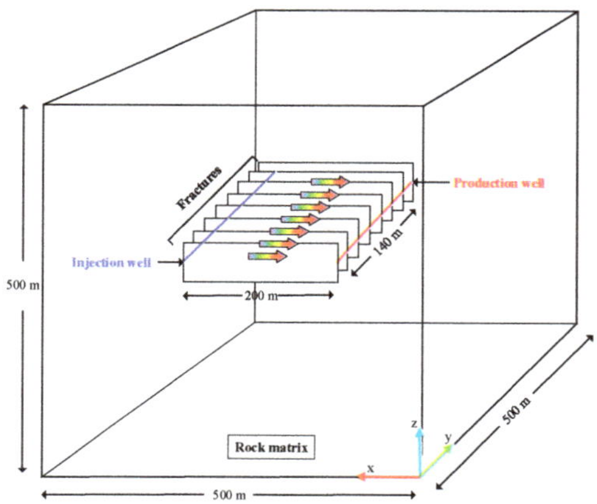

Figure 2. Geometric conceptual model diagram.

3.4. Grid Generation

Mesh generation is an important part of the established physical model in the calculation process. The size and shape of the mesh have a direct impact on the calculation results. To reduce the influence of boundary effects in the finite element solution of numerical simulation software, the method of combining coarse and fine grids is adopted in the calculation, and the numerical simulation of the three-dimensional system of geothermal development in high-temperature rock mass is successfully realized. In principle, the denser the mesh, the higher the calculation accuracy, but the greater the workload and the amount of calculation. Considering the concentrated development area of fractures and the vicinity of injection–production wells as the key research areas, the calculation method of first coarse and then fine is adopted. That is, the whole geothermal system is calculated by using the coarse calculation model. Then the geothermal system near the injection–production wells and fracture areas is calculated using the fine calculation model. The value obtained by the coarse calculation method is used as the boundary condition of the partition calculation to reduce the influence of the boundary effect on the system [34]. According to the spatial structure characteristics of the simulated deep geothermal body geological model, the free triangular mesh is used for mesh generation, and the simulation mesh generation method and mesh generation density are determined. The mesh is refined in the injection–production well and the fracture development area. The complete mesh contains 84,307 domain units and 770 side units, as shown in Figure 3.

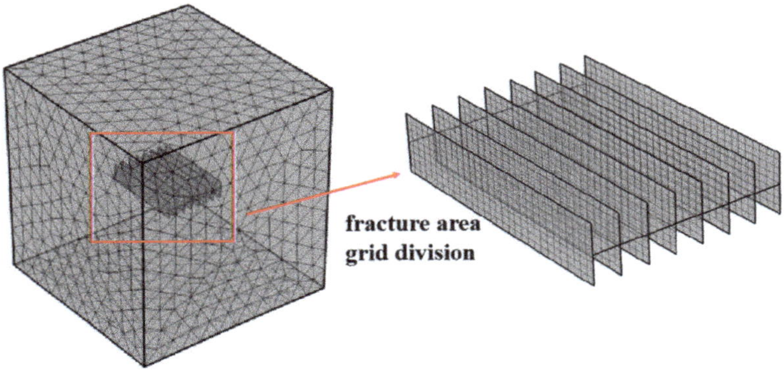

Figure 3. Gridding model.

3.5. THMC Coupled Model Validation

To verify the correctness of the THMC coupling model, this chapter establishes a horizontal well injection and extraction model that simulates 30 years of geothermal production, and the geometric model is shown in Figure 2. In this section, the feasibility and correctness of the above THMC coupling model are verified by simulating the temperature evolution experiment of rock samples during real-time high-temperature and high-pressure geothermal extraction experiments under THMC coupling. By simulating the construction conditions in the study area with a reservoir temperature of 220 °C and one injection and one extraction, the model parameters and engineering conditions are referenced in Tables 1 and 2. Under the coupling effect, the temperature field, seepage field, stress field, and chemical field interact with each other. To highlight the influence of the temperature field on other fields, the model focuses on the influence of the variation of each parameter on the reservoir temperature and the system heat extraction efficiency. By comparing the experimental solution with the numerical solution in this study, the simulation results are in good agreement with the experimental solution, as shown in Figure 4. The correctness of the numerical solution in this study is proved.

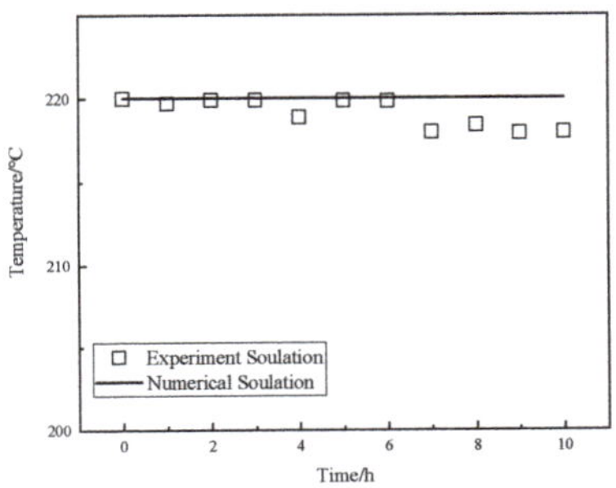

Figure 4. Numerical and experimental solutions for reservoir temperature.

4. Multi-Field Couplings Numerical Model Solution

Using the method of numerical simulation, the injection–production method of on injection and one production is adopted. The two working fluids of water and CO are selected to compare the changes in reservoir temperature and net heat recovery rat under the same reservoir conditions and operating parameters of the heat recovery syster for 30 years to evaluate the heat recovery performance of the two working fluids in th geothermal system.

4.1. Simulation of the Thermal Capacity of Water Injection

In the selection of heat extraction working fluid, due to the abundant water reserve and convenient utilization, the working fluid selected in the previous operation project was water, and the temperature was mainly between 0 °C and 100 °C. Based on the flov model of water injection in the natural fracture, this section couples the heat transfer mode of the fluid and rock matrix in fracture, the dilatancy opening change model caused b pressure change in the fracture, and the chemical corrosion opening model caused b solute exchange between the fluid and rock mass in long-term injection. Finally, a THM coupling model for analyzing geothermal system mining is established. The parameter of water as the heat extraction working fluid are shown in Table 3, and other parameter remain unchanged.

Table 3. Simulate the parameters of injected water in the wellbore.

Parameters	Units	Values
density	kg/m^3	1000
heat capacity	J/(kg K)	4200
thermal conductivity	W/(m K)	0.62
viscosity	Pa s	0.001

(1) The change in temperature

During the exploitation of geothermal systems, the injected low-temperature wate continues to take away the heat of the rock fracture or pore surface, resulting in changes i the reservoir temperature field. Figure 5 shows the temperature changes of the therma reservoir in 0, 10, 20, and 30 years, respectively, in which purple represents the low ten perature and dark red represents the high temperature. As can be seen from Figure 5, i the 10th year of system production, the temperature of the rock mass closer to the fractur surface gradually decreased due to heat transfer, forming a low-temperature zone, wit the effects of mining spreading as far as 88.72 m from the vertical distance of the injectio well. In the 20th year of production, the reservoir temperature around the injection well i decreasing, and the influence of mining is farthest spread to the range of a 119.31 m vertica distance from the injection well. At this time, the rock mass temperature at the productio well remained essentially unchanged, and the system can ensure continuous heat outpu With the continuous increase in mining time (the 30th year), the low-temperature area c the reservoir is continuously expanded along the injection well to the periphery of th fracture with the influence of geothermal extraction, and the low-temperature area in th reservoir is also expanded and gradually extended to the production well, the farthes spread to the range of a 130.45 m vertical distance from the injection well.

(1) Heat production rate analysis

The energy efficiency of the geothermal system can be defined by the net heat extra tion rate (W_h) [35], and the calculation formula is Formula (19).

$$W_h = Q_{out}h_{out} - Q_{inj}h_{inj} \quad (19$$

Q_{out} and Q_{inj} are production and injection rate, respectively, kg/s; h_{out} and h_{inj} ar the enthalpy of production and the injection fluid, respectively, kJ/kg. The model sets th injection/production rate as 2.5 kg/s, and the enthalpy value (H) at each temperature (T) corresponding to the injection of water is calculated with Formula (20).

$$H = 4.187 \times T + 0.7025 \tag{20}$$

The reservoir temperature change and net heat recovery rate (W_h) when the injected working fluid is water during the 30 years of simulated geothermal mining are shown in Figure 6. In the process of numerical simulation, the reservoir temperature change shows a stable downward trend. The reservoir temperature begins to decrease from the initial reservoir temperature, and the cold water reaches the production well from the injection well through the fracture, resulting in a decrease in the temperature around the injection well. It can be seen from Figure 7 that in three decades, the stable stage from the beginning of the heat production lasted for about three years, during which the temperature was maintained at 210 °C, and the net heat recovery rate was maintained at 2007 KJ/s. The system began to decline in the third year, and the temperature decreased from 210 °C to 185.84 °C, which decreased by about 11.5%. The net heat recovery rate decreased from 2006.31 KJ/s to 1751.19 KJ/s, a decrease of about 12.72%.

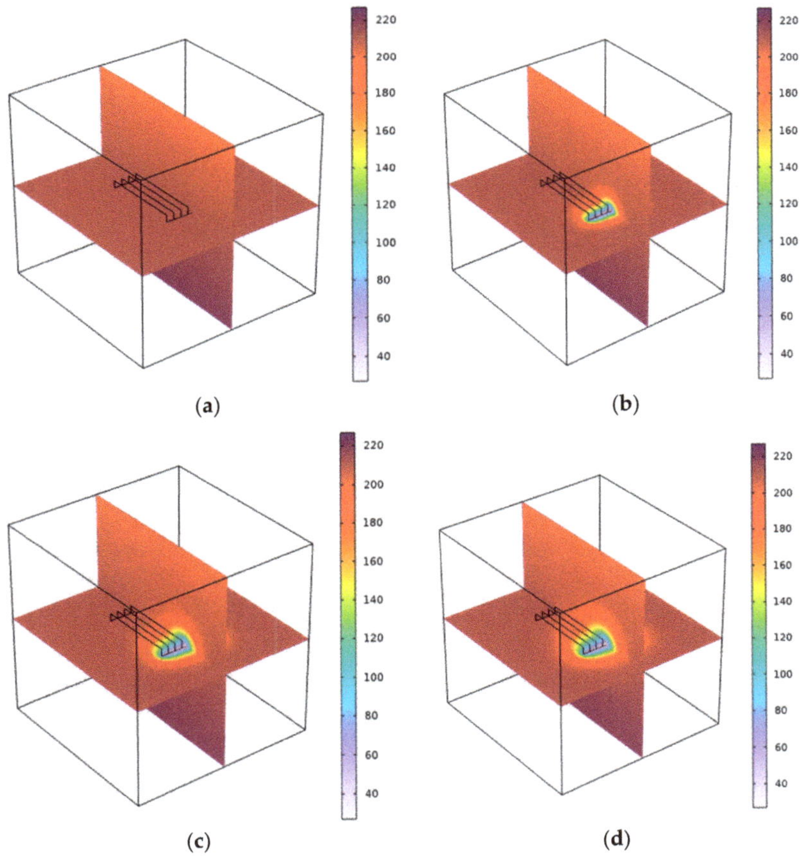

Figure 5. Temperature(°C) change in the reservoir when the heat extraction working fluid is water during mining. (**a**) The 0th year of mining; (**b**) the 10th year of mining; (**c**) the 20th year of mining; (**d**) the 30th year of mining.

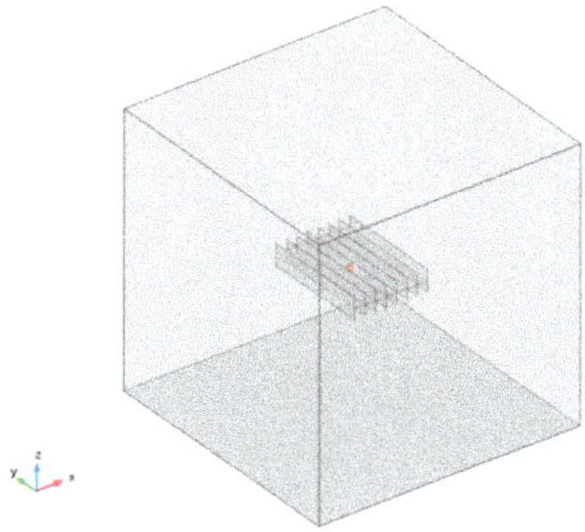

Figure 6. 'Reservoir temperature' center point probe.

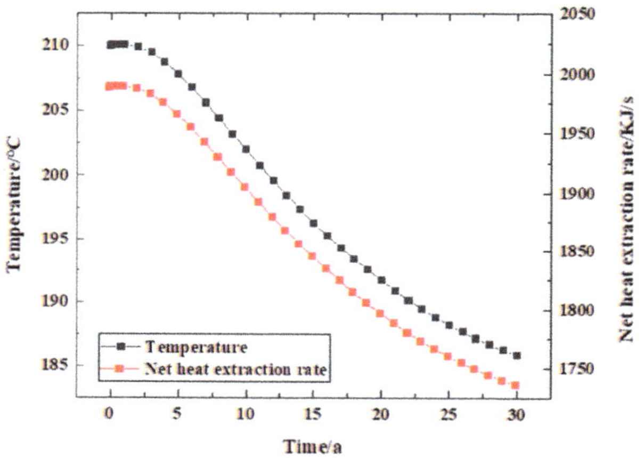

Figure 7. Changes in reservoir temperature and net heat recovery when the injected heat extraction working fluid is water during mining.

Changes in reservoir temperature and net heat recovery rate during mining.

The value of the 'reservoir temperature' in Figure 7 is the value of the point probe set at the center of the model, which is used to represent the change in reservoir temperature during the numerical simulation process. The position is shown in Figure 6.

(2) Chemical reaction

The natural fractures of HDR are continuously injected with cold water, and the fluid reacts with quartz before and after heat exchange to produce new substances. The concentration change and distribution of SiO_2 near the production well are compared, as shown in Figure 8.

During the early stages of mining, a large amount of cold fluid is injected into the vicinity of the injection wells, which leads to geochemical reactions in the reservoir due to changes in temperature, especially near the location of the injection wells, so the concentration of H_4SiO_4 around the injection well increases gradually. Subsequently, the liquid reacts with the reservoir rock in the fracture, and the chemical reaction and solute transport occur continuously. After the reaction with the aqueous solution, SiO_2 reaches the location of the production well with the cold fluid passing through the fracture, and the concentration gradually diffuses and flows out of the production well.

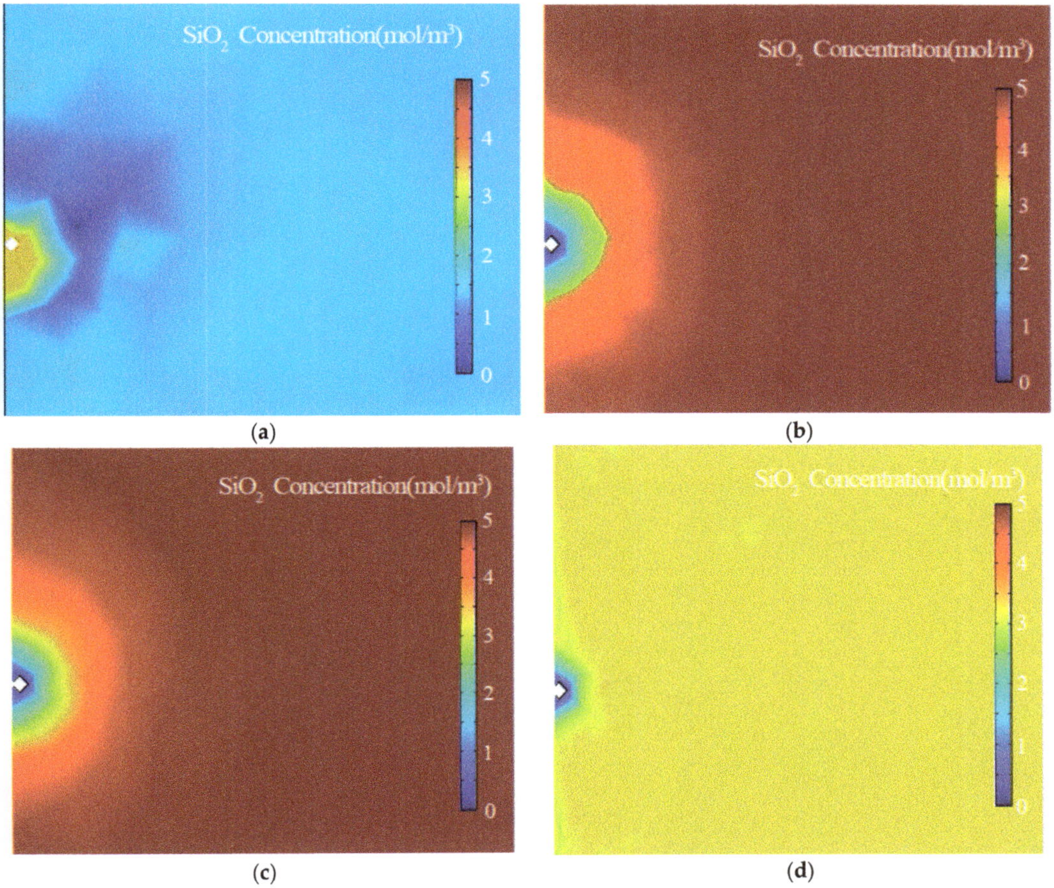

Figure 8. Distribution of SiO_2 concentration near the production wells during mining when the injected heat extraction working fluid is water.(**a**) The 0th year of mining; (**b**) the 10th year of mining; (**c**) the 20th year of mining; (**d**) the 30th year of mining.

As shown in Figures 8 and 9, the change in SiO_2 concentration shows two stages, and the overall trend is to decrease rapidly and then gradually stabilize.

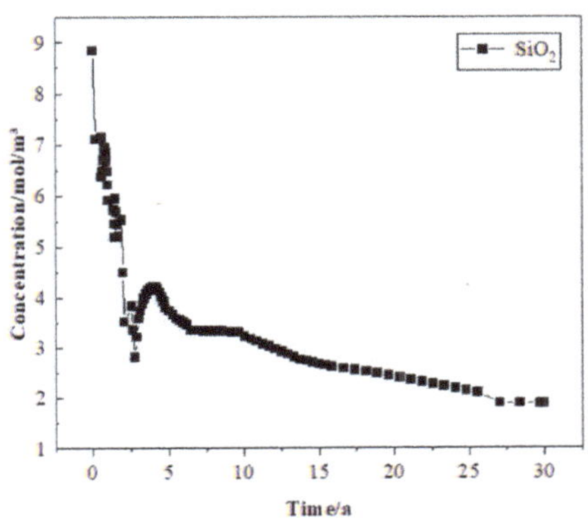

Figure 9. The variation of SiO_2 concentration near the production well when the injected heat extraction working fluid is water during mining.

4.2. Simulation of the Thermal Capacity of CO_2 Injection

In the exploitation of geothermal resources, water is usually used as the heat extractio working fluid. It is rare to study carbon dioxide with the advantages of a low viscosity an density, the small flow resistance of the system, the low solubility of rock minerals, an carbon fixation [36]. In recent years, because CO_2 has superior physical properties and ca be used for geological burial to reduce atmospheric carbon emissions, the use of CO_2 as working circulating fluid has gradually become a hot spot in geothermal development [37 In addition, CO_2 is not easy to react with the wellbore, while water is prone to corrosio resulting in wellbore scaling. At the same time, industrial waste gas is rich in a larg amount of CO_2, which can be used as a geothermal development medium to achieve CC geological storage, reduce atmospheric carbon emissions, and contribute to slowing glob warming and promoting environmental protection.

However, at the same time, CO_2 injection into the reservoir will affect the mechanic properties of a rock mass, such as the strength and elastic modulus, and cause changes i the reservoir stress field and temperature field. These changes may significantly affect th seepage characteristics of the reservoir, thereby affecting the net heat recovery rate [38 The parameters of CO_2 as the heat extraction working fluid are shown in Table 4, and othe parameters remain unchanged.

Table 4. Simulate the parameters of CO_2 injection in the wellbore.

Parameters	Units	Values
density	kg/m^3	1560
heat capacity	J/(kg K)	1.2×10^3
thermal conductivity	W/(m K)	0.0137
viscosity	Pa s	0.64×10^{-4}

(1) The change in temperature

Figure 10 shows the distribution of temperature in the thermal reservoir at 0, 1 20, and 30 years of mining, in which purple represents low temperature, and dark re represents high temperature. The temperature of the rock mass near the distribution are of the fracture channel changes faster. This is because these penetrating fractures constitut the main seepage zone, the flow velocity is higher, and the thermal convection of the hea transfer material in the fracture is obvious. The changing trend of reservoir temperature i consistent with that of water injection mining. In the 10th and 20th years of productio the mining influence spreads to the farthest range of 111.60 m and 152.97 m from th injection well, respectively. In the 30th year of production, the 'cold front' has affecte the temperature near the production well in the reservoir, and the farthest has spread t a range of 197.84 m from the injection well, and the temperature near the injection we has reduced to 42.55 °C. The analysis results show that in addition to natural fractures i the geothermal system, the proper use of hydraulic fracturing to generate fractures wi enhance the conductivity of the rock mass, thereby improving the heat recovery efficienc

(2) Heat production rate analysis

In the process of numerical simulation, according to the variation characteristics c reservoir temperature, the heat production process shows a stepwise downward tren (Figure 11). In the first 1–2 years of system mining, the initial reservoir temperature i maintained, and then the initial reservoir temperature begins to decline. CO_2 reaches th production well through the injection well, and the cold front breaks through the reservoi The temperature decreases to 147.84 °C, a decrease of about 29.6%, and the net heat recover rate decreases from 10,876.29 KJ/s to 7594.39 KJ/s, a decrease of about 30.17%.

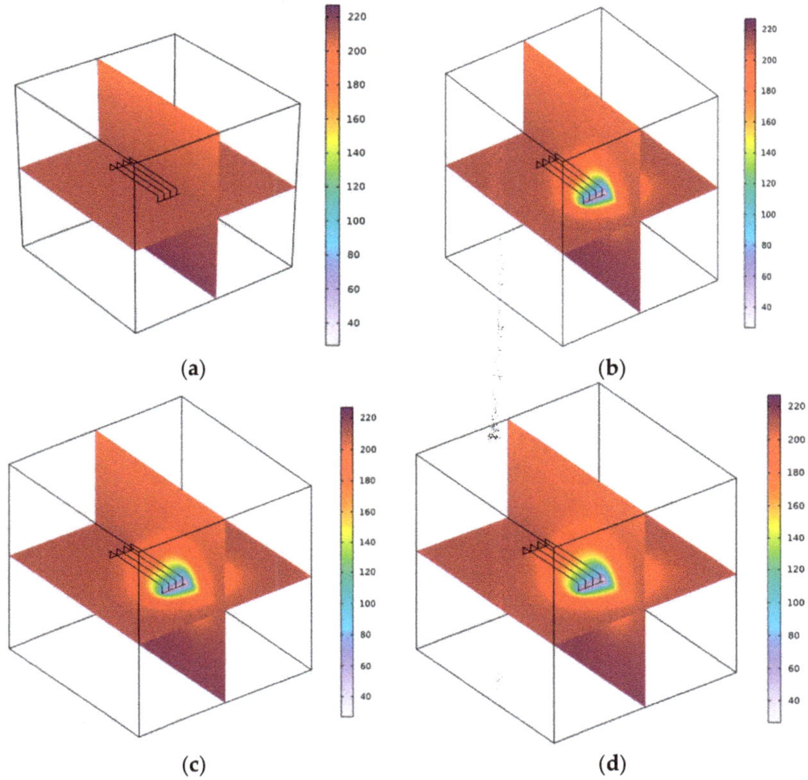

Figure 10. Temperature(°C) change in the reservoir when the heat extraction working fluid is CO_2 during mining. (**a**) The 0th year of mining; (**b**) the 10th year of mining; (**c**) the 20th year of mining; (**d**) the 30th year of mining.

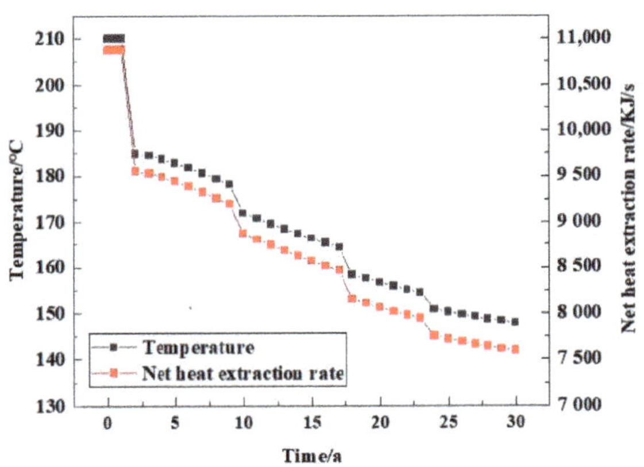

Figure 11. Changes in reservoir temperature and net heat recovery when the injected heat extraction working fluid is CO_2 during mining.

(3) Chemical reaction

When the injection well is injected with CO_2, CO_2 will react with the material in the rock matrix when it reaches the production well through the fracture heat, and the material reaction equation is Formula (14). The concentration change and distribution of SiO_2 near the production well are compared, as shown in Figure 12.

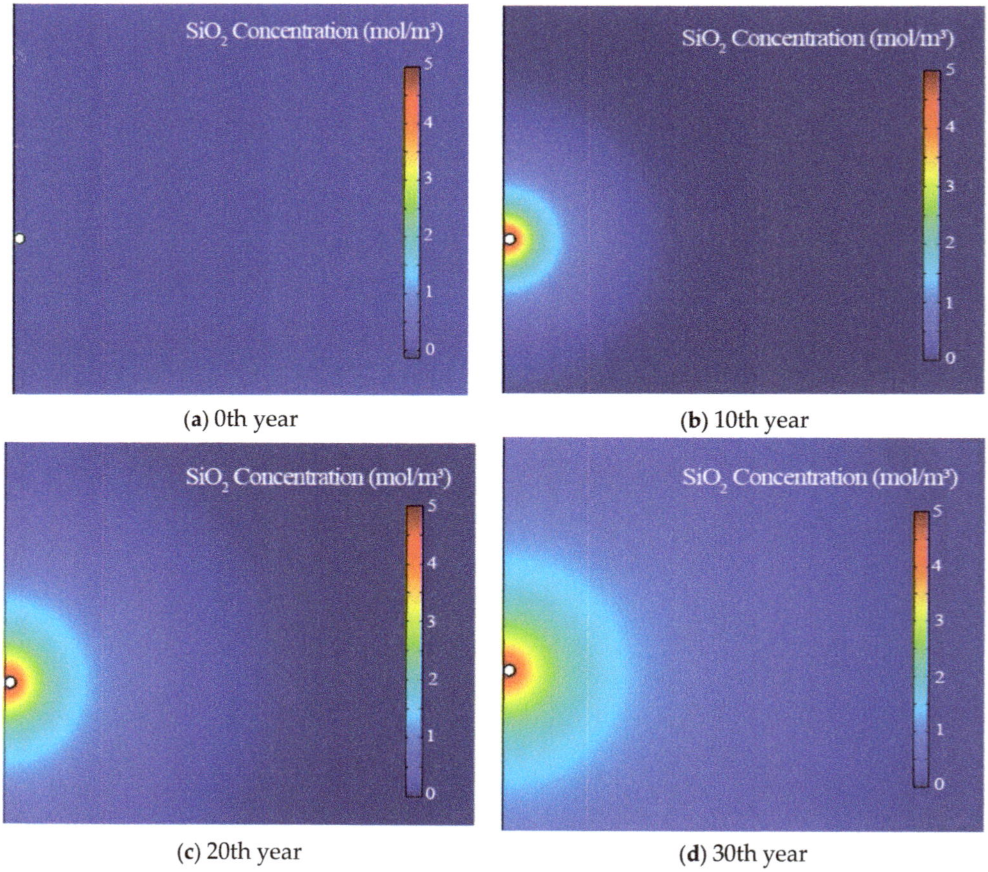

Figure 12. Distribution of SiO_2 concentration near the production wells during mining when the injected heat extraction working fluid is CO_2. (**a**) The 0th year of mining; (**b**) the 10th year of mining; (**c**) the 20th year of mining; (**d**) the 30th year of mining.

As shown in Figures 12 and 13, initially, the injection of CO_2 first reacts with the water in the rock fissures to form carbonic acid. And then, more complex reactions occur between the rock surface and geothermal fluids in the geothermal reservoir's high-temperature and high-pressure environment. CO_2 and fluid circulation promote the dissolution of minerals such as calcite, sodic feldspar, and muscovite. Under the action of high confining pressure, the minerals on the surface of the rock are broken, and the structure is destroyed, which leads to the separation of the minerals from the rock surface into the fluid, resulting in a large amount of precipitation insoluble in water. Finally, the fluid flowing in the fracture is discharged from the flow outlet. Therefore, the concentration of SiO_2 around the production well increased rapidly (the first six years), and then, with the reaction equilibrium, the concentration of SiO_2 gradually stabilized, and the concentration distribution range gradually diffused from the injection well to the production well.

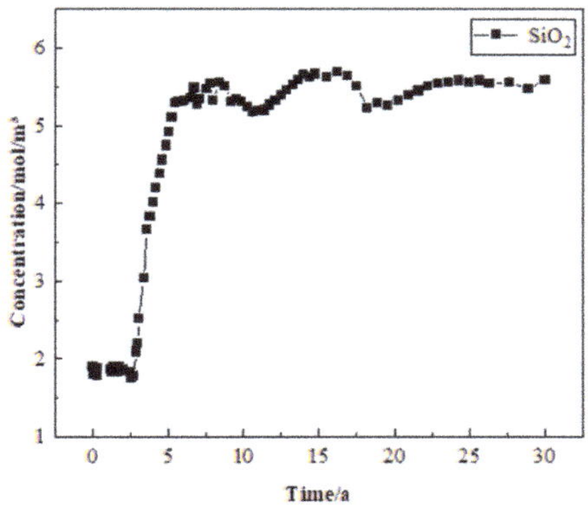

Figure 13. The variation of SiO_2 concentration near the production well when the injected heat extraction working fluid is CO_2 during mining.

4.3. Analysis of Comparative Results

The numerical simulation of geothermal development is carried out for the injection of water and CO_2, respectively. In Figure 14, CO_2 has low viscosity and superfluidity in the well. Although the heat value is lower than that of water, it is easy to form a thermal breakthrough prematurely under the condition of constant injection flow production. Its output flow rate is large, the development time is short, and the thermal efficiency is high. The overall heat recovery effect is nearly five times that of water injection.

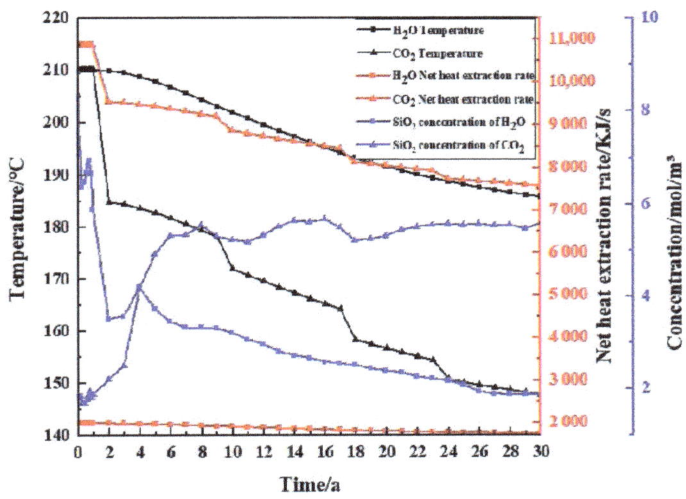

Figure 14. Reservoir temperature, SiO_2 concentration, and net heat extraction rate under different working fluid injection conditions.

5. Parameter Sensitivity Analysis

To explore the influencing factors and degree of geothermal exploitation, based on the established numerical model, water is used as the heat extraction working fluid for numerical simulation. The parameter variation range of each influencing factor is set, and the parameter sensitivity analysis is carried out on parameters such as reservoir thermal conductivity, specific heat capacity, well spacing between the production well and injection well, water injection temperature, fracture spacing, permeability, number, and fracture length. According to the simulation results, the main controlling factors of thermal efficiency and mining life are determined.

5.1. Analysis of Influencing Factors in Heat Recovery Performance

5.1.1. Properties of Reservoirs

(1) Thermal conductivity

Under the same other conditions, when the thermal conductivity of the reservo is 2.4 W/(m K), 2.9 W/(m K), and 3.4 W/(m K), respectively, the reservoir temperatur change in geothermal exploitation for 30 years is analyzed. When the thermal conductivit of reservoir rock increases from 2.4 W/(m K) to 3.4W/(m K), the change and rate c temperature are consistent, which proves that the change in rock thermal conductivity ha little effect on the change in the reservoir temperature. However, the higher the therma conductivity of the rock, the faster the heat transfer from the internal rock to the fractur wall and circulating water will be, so the larger the thermal conductivity is, the faste the reservoir temperature decreases in the later stage. However, based on the numeric simulation process, this has a negligible impact on production performance and efficienc (Figure 15).

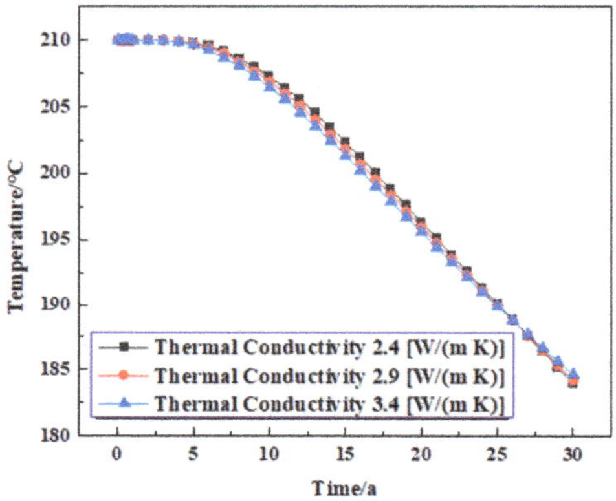

Figure 15. The variation law of reservoir temperature under different rock thermal conductivity condition

(2) Specific heat capacity

Under the same conditions, the temperature changes in the reservoir during the thre decades of geothermal exploitation were analyzed when the specific heat capacities of th reservoir were 800 J/(kg K), 850 J/(kg K), and 900 J/(kg K), respectively. It can be see that different heat capacities affect the change in the reservoir temperature. When the roc specific heat capacity decreases from 900 J/(kg K) to 800 J/(kg K), the reservoir temperatur change trend during the three decades of geothermal exploitation is the same, but th reservoir temperature drops even lower when the specific heat capacity is 800 J/(kg K This is because the decrease in specific heat capacity leads to the deterioration of the hea storage performance of the rock, so the temperature decreases faster. In general, the specifi heat capacity of the rock has little effect on geothermal production performance (Figure 16

5.1.2. Project Production Conditions

(1) Well spacing

Under the same conditions, the temperature changes in the reservoir in the thre decades of geothermal exploitation were analyzed when the well spacing was 100 m, 200 n and 300 m, respectively. In the process of simulated geothermal mining, when the we spacing is 300 m, the geothermal extraction is in a stable production stage for a long tim and the temperature is stable at the initial reservoir temperature and continues to declin slightly for about 20 years. When the well spacing is 200 m, the reservoir temperature show a stable downward trend, and the reservoir temperature reaches 184.24 °C in the 30th yea When the well spacing is 100 m, because the well spacing is too close, the temperature c the thermal reservoir decreases sharply from mining to 125.03 °C and then remains stable. Therefore, when the well spacing between the injection well and the production well is 200 m, the temperature drop is more stable, and the heat extraction effect is better, which ensures the long-term exploitation of geothermal resources (Figure 17).

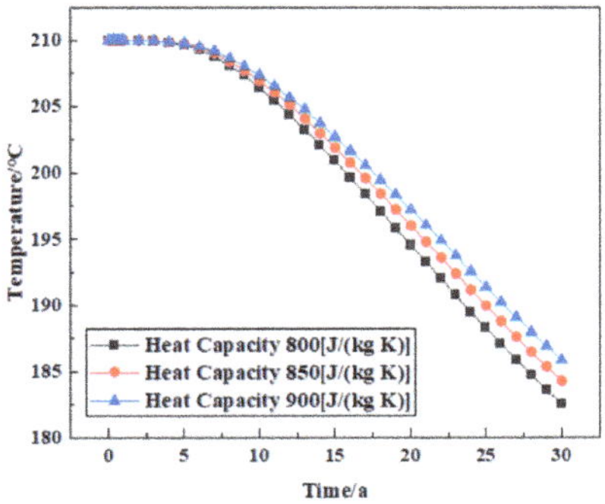

Figure 16. The variation law of reservoir temperature under different rock-specific heat capacity conditions.

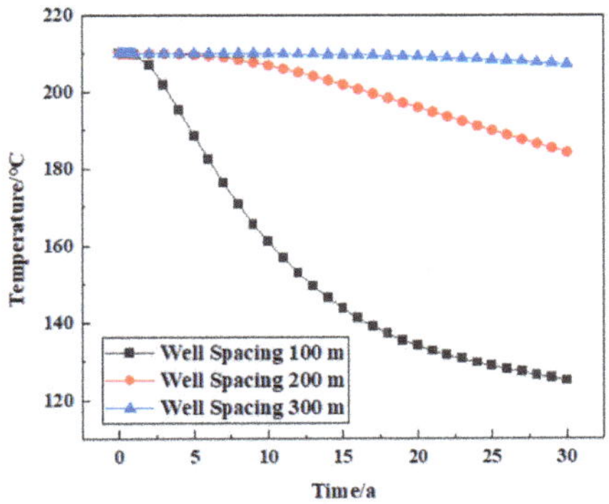

Figure 17. The variation law of reservoir temperature under different well spacing conditions.

(2) Injection temperature

In the case of the same other conditions, the temperature changes in the reservoir in the three decades of geothermal exploitation were analyzed when the injection temperatures were 20 °C, 40 °C, and 60 °C, respectively. The change law of reservoir temperature is consistent at different injection temperatures and divided into two stages: In the first stage, the reservoir temperature does not decrease significantly for about six years and remains at the initial temperature of the reservoir. This stage is usually called the stable production stage of geothermal extraction. Then, with the continuous exploitation of geothermal energy, it enters the second stage, during which the reservoir temperature decreases. Under different injection temperatures, the decline rate of the reservoir temperature is different, which shows that the lower the injection temperature, the faster the reservoir temperature decreases. In the thirtieth year of mining, when the injection temperature is 20 °C, the reservoir temperature is 184.24 °C; when the injection temperature is 60 °C, the reservoir temperature is 208.52 °C. This shows that increasing the injection temperature appropriately can improve the heat recovery temperature and operation life of the system. The lower the temperature of the injected fluid, the greater the temperature difference with the initial reservoir, the stronger the thermal convection, the faster the heat loss, and the lower the heat storage efficiency of the system (Figure 18).

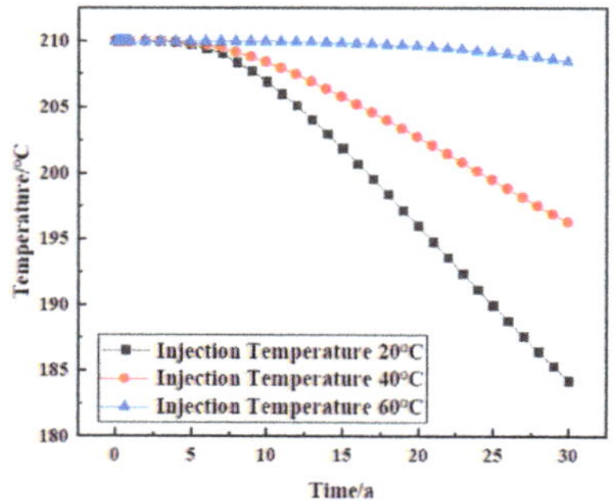

Figure 18. The change rule of reservoir temperature under different injection temperature conditions.

5.1.3. Properties of Fracture

(1) Fracture spacing

Under the same conditions, the temperature changes in the reservoir in the three decades of geothermal exploitation were analyzed when the fracture spacing was 20 m, 40 m, and 60 m, respectively. The heat exchange area of the 20 m fracture spacing increases, and the reservoir temperature decreases at the lowest and the fastest, compared to the 40 m fracture spacing. Increasing the fracture spacing (60 m) leads to a decrease in the heat exchange area between the fracture surface and the rock matrix. Therefore, at the same mining time scale, more heat will be stored in the thermal reservoir with a large fracture spacing (the rock matrix temperature is relatively higher), and the heat energy of the rock matrix taken away by the circulating fluid is reduced. This is mainly because in the case of a certain volume fraction of geothermal reservoir fractures, the smaller the fracture spacing (20 m), the larger the contact area between the fracture and the reservoir matrix, and the smaller the fracture spacing of the geothermal reservoir, which leads to an increase in the heat transfer area between the fluid and the matrix when the fluid migrates in the fracture, and the higher the heat extraction rate, so it is more conducive to the exploitation of the geothermal system (Figure 19).

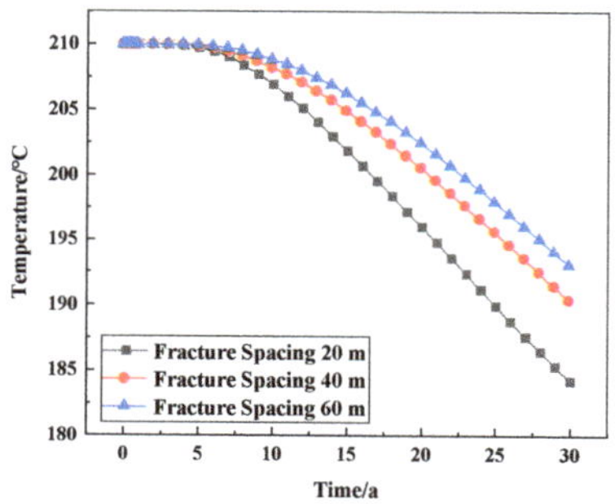

Figure 19. Variation of reservoir temperature under different fracture spacing conditions.

(2) Fracture permeability

The model sets the permeability of the rock matrix to be 5×10^{-17} m^2. Under the same other conditions, the temperature changes of the reservoir in the three decades of geothermal exploitation were analyzed when the fracture permeability was 2.73×10^{-14} m^2, 2.73×10^{-13} m^2, and 2.73×10^{-12} m^2, respectively. In the case of high fracture permeability, lower injection pressure will cause the bottom hole temperature and reservoir temperature to drop sharply after 30 years of mining because high permeability brings a high-quality mining rate, and the net heat extraction rate also reaches the highest. Therefore, from the perspective of the geothermal engineering life cycle and optimization, the fracture permeability of 10^{-13} m^2 and above is more appropriate. If the permeability is too small, the heat extraction rate is low, and if the permeability is too large, the heat extraction is too fast, which shortens the engineering life cycle (Figure 20).

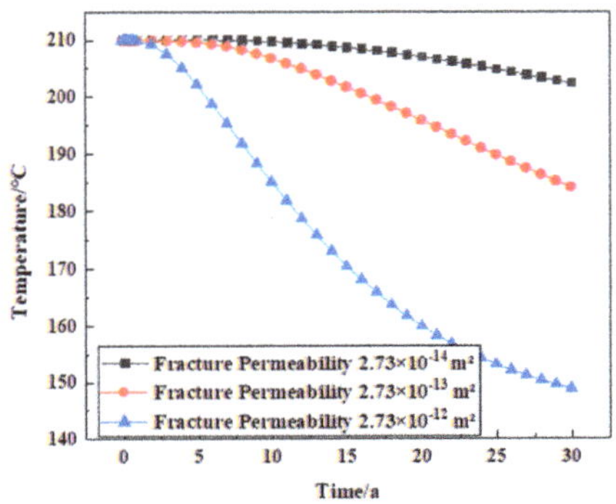

Figure 20. The change law of reservoir temperature under different fracture permeability conditions.

(3) Number of fractures

To analyze the relationship between the number of fractures and geothermal mining, the fractures are evenly arranged between the injection well and the mining well. In the case of the same other conditions, the temperature changes in the reservoir in the three decades of geothermal exploitation were analyzed when the number of fractures was two, four, and eight, respectively. The reservoir temperature after geothermal mining under different fracture numbers was studied. With the increase in fracture number, the reservoir temperature decreases faster. This is because the increase in fractures and the contact area of the injected fluid increase, and the thermal convection with rock increases, resulting in a faster temperature drop. Therefore, in some HDR geothermal resources, hydraulic fracturing is usually used to increase the number of fractures to achieve the result of system stimulation (Figure 21).

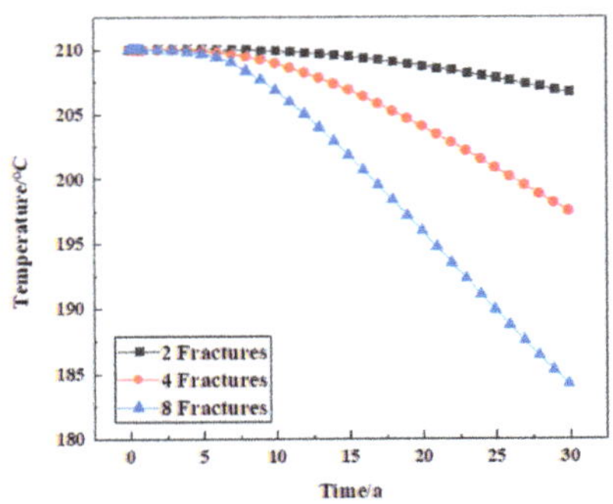

Figure 21. The variation law of reservoir temperature under different fracture numbers.

(4) Fracture length

Under the same conditions, the temperature changes in the reservoir in the three decades of geothermal exploitation were analyzed when the fracture lengths were 100 m, 200 m, and 300 m, respectively. When the fracture length is 100 m, the cold fluid flow distance becomes shorter and reaches the production well in a short time, and the reservoir temperature drops rapidly to 125.03 °C. When the fracture length is 300 m, the water and the hot rock mass have enough time to exchange heat, and the temperature remains at a high level. The results show that the local temperature of the reservoir decreases rapidly when the surface fracture length is small, which reflects that it has a great influence on the heat recovery performance of the geothermal system (Figure 22).

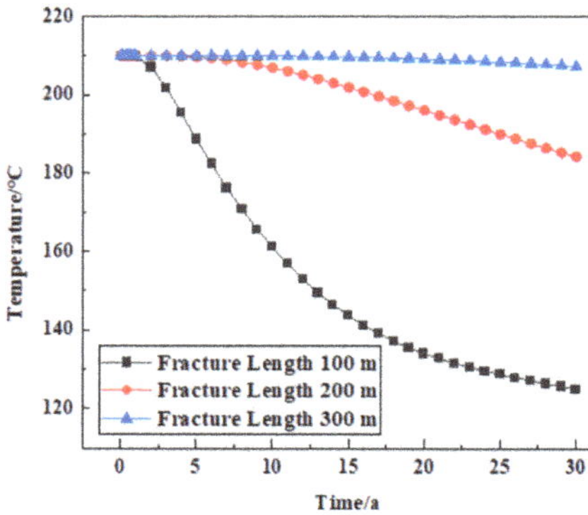

Figure 22. The variation law of reservoir temperature under different fracture length conditions.

5.2. Analysis of Influencing Factors in Mining Life

The life of thermal reservoir mining is affected by many factors. According to the methods of the literature review, indoor physical tests, and numerical simulations, this paper summarizes the following main influencing factors:

(1) The reserves and temperature of thermal reservoir resources are the most important factors affecting the life of thermal reservoir exploitation. The greater the reserves of resources, the higher the temperature and the longer the life of thermal reservoir exploitation.
(2) The physical properties of thermal storage rock mass, such as rock permeability, fracture distribution, etc., will affect the effect and life of thermal storage mining. For example, rocks with poor permeability make it difficult to allow hot water to flow through, resulting in reduced hot water penetration and heat energy transfer efficiency, thus affecting the mining life of thermal reservoirs.
(3) The advancement and applicability of thermal reservoir mining technology will directly affect the life of thermal reservoir mining. For example, the use of efficient drilling technology and a perfect water injection system can improve the efficiency and life of thermal storage.
(4) Environmental factors, such as climate, geological conditions, etc., will also affect the life of thermal reservoir exploitation. For example, natural disasters such as earthquakes may destroy the physical structure of the thermal reservoir rock mass and affect the mining effect and life of the thermal reservoir.
(5) Thermal storage mining methods: Different thermal storage mining methods will also have an impact on life. For example, flash power generation is more efficient than direct heat utilization but also consumes thermal storage resources faster.
(6) The time of thermal reservoir exploitation will also affect its life. Overexploitation may lead to the depletion of thermal reservoir resources, thus affecting the life of thermal reservoir exploitation. Therefore, it is necessary to scientifically evaluate and plan thermal storage resources to control the mining intensity and time and ensure the sustainable utilization of thermal storage resources.
(7) Geological stress is also one of the factors affecting the life of thermal reservoir mining. In the process of mining, the change in underground rock strata may lead to a change in geological stress, which will affect the physical structure and stability of the thermal reservoir rock mass, thus affecting the mining life of a thermal reservoir.
(8) Water is needed to transfer heat energy in thermal storage mining, and water quality problems will also affect the life of thermal storage mining. For example, water may contain corrosive substances, and long-term use will lead to the corrosion of heat exchange equipment, thus affecting the life of thermal reservoir exploitation.

In general, the mining life of a thermal reservoir is a comprehensive result affected by many factors. To improve the life of thermal storage mining, it is necessary to scientifically evaluate thermal storage resources, formulate reasonable mining plans, select appropriate mining technologies and management methods, and pay attention to the impact of environmental and economic factors to achieve sustainable utilization of thermal storage resources.

6. Conclusions

In this paper, the corresponding mathematical model was established by analyzing the multi-physical field process and its coupling effect in the process of geothermal exploitation. According to the literature research, the real-time high-temperature environment true triaxial hydraulic fracturing physical simulation experiment in the study area, and the complex fracture network initiation and propagation evolution experiment under the coupling of heat flow and solidification, the numerical simulation parameters, initial conditions, and boundary conditions were determined. The discrete network model was constructed, and the mesh was refined in the fracture and the well. Finally, the reservoir temperature, net heat recovery rate, and SiO_2 concentration after 30 years of geothermal system exploitation under two different heat extraction working fluids of water and CO_2 were simulated and evaluated with a numerical simulation. The sensitivity of the parameters related to the thermal performance and mining life of the deep thermal reservoir was analyzed, which provided a theoretical basis for clarifying the multi-field temporal and spatial evolution law and for optimizing the flow control method in the process of geothermal mining. The conclusions are as follows:

(1) For a 500 m $\times$ 500 m $\times$ 500 m geothermal reservoir, the numerical simulation software compares and analyzes the heat storage temperature, net heat recovery rate, and SiO_2 concentration of water and CO_2 under the same conditions. The results show that the heat of the reservoir near the injection well is first taken away, and the temperature is reduced during the seepage of the injected heat extraction working fluid to the mining well. With time, the low-temperature area gradually expands to the mining well. The net heat extraction rate of CO_2 is about five times that of water, but CO_2 will also undergo water–rock–gas interaction during heat extraction, which easily causes salt precipitation, increases SiO_2 concentration, blocks pores, and affects the efficiency and stability of heat extraction.

(2) In the analysis of different reservoir rock thermal conductivity, specific heat capacity, well spacing, injection temperature, fracture spacing, fracture permeability, fracture number, fracture length, and other parameters on the influence of the reservoir temperature, the results show that the rock thermal conductivity and specific heat capacity of geothermal system heat recovery performance have little effect; the smaller the well spacing is, the lower the injection temperature is, the faster the reservoir temperature decreases, and the higher the geothermal mining efficiency is. The smaller the fracture spacing, the larger the heat exchange area, and the higher the net heat extraction rate. Fracture permeability in the order of magnitude of 10–13 m^2 and above is more suitable for a 30-year development cycle; the more fractures, the shorter the length, the larger the heat exchange area with the reservoir, the faster the temperature of the thermal reservoir decreases during the mining process, and the higher the heat extraction rate.

This paper can be a reference for both the theoretical THMC coupling model and practical geothermal production. However, the limitations of this study are that the simulated formation depth has some difficulties in actual geothermal system exploitation; the considered fractures are natural fractures, and no operations, such as hydraulic fracturing or acid fracturing, have been performed, and several parameters can be integrated together in the parameter sensitivity analysis. Therefore, the above key constraints will be considered in subsequent studies to provide a better theoretical, experimental, and simulation research basis for geothermal development.

Author Contributions: Conceptualization, G.F.; Methodology, G.F.; Software, Y.Y.; Model validation, Y.Y.; Formal analysis, L.G.; Investigation, L.G.; Resources, J.Z.; Writing—original draft preparation, Y.Y.; Writing—review and editing, Y.Y.; Funding acquisition, G.F. All authors have read and agreed to the published version of the manuscript.

Funding: This research was funded by the Jiangsu Province Carbon Peak Carbon Neutral Technology Innovation Project in China (BE2022034).

Data Availability Statement: Not applicable.

Conflicts of Interest: The authors declare no conflict of interest.

Nomenclature

C	the chemical field
H	the seepage field
HDR	the hot dry rock
M	the stress field
T	the temperature field
THMC	the coupling of the temperature field, seepage field, stress field, and chemical field.
c_f	the total concentration of solute dissolved in the fracture, ppm
c_i	concentration, mol/m^3
c_m	the total dissolved concentration of solute in the matrix, ppm
c	specific heat capacity of the rock mass, J/(kg K)
c_s	specific heat capacity of the rock mass, J/(kg K)
c_w	specific heat capacity of the fluid, J/(kg K)
D_i	diffusivity, m^2/s
eq	the superscript representing the solubility of the solute in different media with temperature, ppm
E	elastic modulus
$f_{,i}$	volume force component
G	shear modulus, Pa
h_{out}	the enthalpy of the production fluid, kJ/kg
h_{inj}	the enthalpy of the injection fluid, kJ/kg
k	permeability, m^2
K	rock bulk modulus, Pa
k'_f	the dissolution rate of the solute in the fracture with the temperature change, m/s
k_f	reaction rate constant
k_m	the dissolution rate of the solute in the rock matrix with the temperature change, m/s
m	mass, kg
P	pressure, Pa
p_w	pore water pressure, Pa
q_w	Darcy speed, m/s
Q_s	source of fluid mass, kg/m^3 s
$\overline{Q}$	heat source density in the rock mass, W/m^3
Q_{Ts}	heat source term
Q_{out}	production rate, kg/s
Q_{inj}	injection rate, kg/s
R_i	Material reaction rate, mol/(m^3 s)
t	time, s
T	thermal reservoir rock temperature, K
T_s	thermal reservoir rock temperature, K
T_w	fluid temperature in the fracture, K
u	fluid viscosity, Pa s
u_i	offset component
v	Poisson's ratio
σ'	effective stress
σ	initial stress

α_f	the ratio of the connected area to the total area in the fracture
ε,ε_{kk},ε_{ij}	strain
ε_V	volumetric strain
λ	Lame constant
δ_{ij}	Kronecker symbol
α_T	thermal expansion coefficient, 1/K
ΔT	temperature increment
ρ_w	fluid density, kg/m^3
ϕ	porosity
ρ,ρ_s	rock mass density, kg/m^3
$\lambda_x, \lambda_y, \lambda_z$	thermal conductivity of rock mass, W/(m K)
λ_s	coefficient of thermal conductivity, W/(m K)
λ_w	thermal conductivity of fluid, W/(m K)
ϕQ_{Tw}	heat source term
Q_T	add Q_{Ts} and ϕQ_{Tw}
$\alpha_{p,i}$	the influence of fluid seepage on rock mass deformation
$-\alpha_T KT_{,i}$	the influence of temperature change on rock mass deformation
θ_m	the porosity of the rock matrix
x	x-direction
y	y-direction
z	z-direction
∇	gradient operator (total differential in all directions of space)
Δ	denotes the change in a physical quantity

eferences

Melikoglu, M. Geothermal energy in Turkey and around the World: A review of the literature and an analysis based on Turkey's Vision 2023 energy targets. *Renew. Sustain. Energy Rev.* **2017**, *76*, 485–492. [CrossRef]

Zeng, Y.J. Technical Progress and Thinking for Development of Hot Dry Rock(HDR)Geothermal Resources. *Pet. Drill. Tech* **1900**, *43*, 1–7.

Chen, B.G. Study on Numerical Methods for Coupled Fluid Flow and Heat Transfer in Fractured Rocks of Doublet System. Ph.D. Thesis, Tsinghua University, Beijing, China, 2014.

Pruess, K.; Oldenburg, C.M.; Moridis, G.J. *TOUGH2 User's Guide Version 2*; Lawrence Berkeley National Lab (LBNL): Berkeley, CA, USA, 1999.

Zhao, Y.S.; Feng, Z.J.; Feng, Z.C.; Yang, D.; Liang, W.G. THM (Thermo-hydro-mechanical) coupled mathematical model of fractured media and numerical simulation of a 3D enhanced geothermal system at 573 K and buried depth 6000–7000 m. *Energy* **2015**, *82*, 193–205. [CrossRef]

Zhao, Z.H.; Liu, G.H.; Wang, J.C.; Xu, H.R. Coupled multi-field effect on sustainable development of deep geothermal energy in cities. *J. China Coal Soc.* **2023**, *48*, 1126–1138.

Abdul, R.S.; Sheik, S.R.; Nam, N.H.; Thanh, T. Numerical Simulation of Fluid-Rock Coupling Heat Transfer in Naturally Fractured Geothermal System. *Appl. Therm. Eng.* **2011**, *31*, 1600–1606.

Li, X.X.; Li, D.Q.; Xu, Y. Equivalent Simulation Method of Three-Dimension Asleep Age and Heat Transfer Coupling in Fractured Rockmass of Geothermal-Borehole System. *Eng. Mech.* **2019**, *36*, 238–247.

Egill, J.; Roland, N.H. Optimization of injection scheduling in fractured geothermal reservoirs. *Geothermics* **2013**, *48*, 80–92.

Yu, Z.W. Research on Multiphase—Multicomponent THCM Coupling Mechanism and Its Application. Ph.D. Thesis, Jilin University, Changchun, China, 2013.

Zhang, J.N. Experimental Research on Physical-mechanical Properties and Productivity Evaluation of Geothermal Reservoir in Guide, Qinghai. Ph.D. Thesis, Jilin University, Changchun, China, 2018.

2. Taron, J.; Elsworth, D.; Min, K.B. Numerical simulation of thermal-hydrologic-mechanical-chemical processes in deformable, fractured porous media. *Int. J. Rock Mech. Min. Sci.* **2009**, *46*, 842–854. [CrossRef]

3. Zeng, Y.C.; Tang, L.S.; Wu, N.Y.; Cao, Y.F. Analysis of influencing factors of production performance of enhanced geothermal system: A case study at Yangbajing geothermal field. *Energy* **2017**, *127*, 218–235. [CrossRef]

4. Zhang, Y.J.; Li, Z.W.; Guo, L.L.; Gao, P.; Jin, X.P.; Xu, T.F. Electricity generation from enhanced geothermal systems by oilfield produced water circulating through reservoir stimulated by staged fracturing technology for horizontal wells: A case study in Xujiaweizi area in Daqing Oilfield, China. *Energy* **2014**, *78*, 788–805. [CrossRef]

5. Zhang, Y.J.; Li, Z.W.; Yu, Z.W.; Guo, L.L.; Jin, X.P.; Xu, T.F. Evaluation of developing an enhanced geothermal heating system in northeast China: Field hydraulic stimulation and heat production forecast. *Energy Build.* **2015**, *88*, 1–14. [CrossRef]

5. Zhang, Y.J.; Guo, L.L.; Li, Z.W.; Yu, Z.W.; Xu, T.F.; Lan, C.Y. Electricity generation and heating potential from enhanced geothermal system in Songliao Basin, China: Different reservoir stimulation strategies for tight rock and naturally fractured formations. *Energy* **2015**, *93*, 1860–1885. [CrossRef]

17. Lei, Z.H.; Zhang, Y.J.; Yu, Z.W.; Hu, Z.J.; Li, L.Z.; Zhang, S.Q.; Fu, L.; Zhou, L.; Xie, Y.Y. Exploratory research into the enhance geothermal system power generation project: The Qiabuqia geothermal field, Northwest China. *Renew. Energy* **2019**, *139*, 52–7 [CrossRef]
18. Lei, Z.H.; Zhang, Y.J.; Zhang, S.Q.; Fu, L.; Hu, Z.J.; Yu, Z.W.; Li, L.Z.; Zhou, J. Electricity generation from a three-horizontal-we enhanced geothermal system in the Qiabuqia geothermal field, China: Slickwater fracturing treatments for different reservo scenarios. *Renew. Energy* **2020**, *145*, 65–83. [CrossRef]
19. Wu, L.; Hou, Z.; Xie, Y.; Luo, Z.; Xiong, Y.; Cheng, L.; Wu, X.; Chen, Q.; Huang, L. Fracture initiation and propagation supercritical carbon dioxide fracturing in calcite-rich shale: A coupled thermal-hydraulic-mechanical-chemical simulation. *Int. Rock Mech. Min. Sci.* **2023**, *167*, 105389. [CrossRef]
20. Ebrahimi, M.; Ameri, M.J.; Vaghasloo, Y.A.; Sabah, M. Fully coupled thermohydro-mechanical approach to model fractu response to injection process in enhanced geothermal systems using displacement discontinuity and finite element method. *J. P Sci. Eng.* **2022**, *208*, 109240. [CrossRef]
21. Duan, Y.X.; Yang, H. Analysis of Influencing Factors on Heat Extraction Performance of Enhanced Geothermal System. *J. Jil Univ. (Earth Sci. Ed.)* **2020**, *50*, 1161–1172.
22. Xu, T.F.; Yuan, Y.L.; Jia, X.F.; Lei, Y.D.; Li, S.T.; Feng, B.; Hou, Z.Y.; Jiang, Z.J. Prospects of power generation from an enhance geothermal system by water circulation through two horizontal wells: A case study in the Gonghe Basin, Qinghai Provinc China. *Energy* **2018**, *148*, 196–207. [CrossRef]
23. Ling, L.L.; Su, Z.; Zhai, H.Z.; Wu, N.Y. During EGS Exploitation, Yangyi of Tibet Numerical Simulation Study of the Paramete Effect on Temperature Distribution and Mining Life. *Adv. New Renew. Energy* **2015**, *3*, 367–374.
24. Zhang, C.; Jiang, G.Z.; Jia, X.F.; Li, S.T.; Zhang, S.S.; Hu, D.; Hu, S.B.; Wang, Y.B. Parametric study of the production performan of an enhanced geothermal system: A case study at the Qiabuqia geothermal area, northeast Tibetan plateau. *Renew. Energy* **201** *132*, 959–978. [CrossRef]
25. Norio, T.; Kasumi, Y.; George, Z. Model study of the thermal storage system by FEHM code. *Geothermics* **2003**, *32*, 603–607.
26. Xu, R.; Guo, T.; Qu, Z.; Chen, H.; Chen, M.; Xu, J.; Li, H. Numerical simulation of THMC coupling temperature prediction f fractured horizontal wells in shale oil reservoir. *J. Pet. Sci. Eng.* **2022**, *217*, 110782. [CrossRef]
27. Aliyu, M.D.; Chen, H. Sensitivity analysis of deep geothermal reservoir: Effect of reservoir parameters on production temperatu *Energy* **2017**, *129*, 101–113. [CrossRef]
28. Liu, J. Numerical Simulation Research on Multi-field Coupling of Middle-Deep Geothermal Development Process in Xiwenzhua Taiyuan. Master's Thesis, China University of Mining and Technology, Xuzhou, China, 2021.
29. Zhao, Y.A. Analysis of Heat Extraction Performance of Water and CO_2 in Supercritical Geothermal System: A Case Study Larderello Geothermal Field in Italy. Master's Thesis, Jilin University, Changchun, China, 2023.
30. Sun, Z.X.; Xu, Y.; Lv, S.H.; Xu, Y.; Sun, Q.; Cai, M.Y.; Yao, J. A thermo-hydro-mechanical coupling model for numerical simulatio of enhanced geothermal systems. *J. China Univ. Pet. (Ed. Nat. Sci.)* **2016**, *40*, 109–117.
31. Xiao, Y. Study on THMC Coupling of Hydro shearing in Hot Dry Rock In enhanced Geothermal System. Ph.D. Thesis, Southwe Petroleum University, Chengdu, China, 2017.
32. Guo, Z.P.; Bu, X.B.; Li, H.S.; Gong, Y.L.; Wang, L.B. Numerical simulation of heat extraction in single-well enhanced geotherm system based on thermal-hydraulic-chemical coupling model. *Chem. Ind. Eng. Prog.* **2023**, *42*, 711–721.
33. Arash, K.A.; Ehsan, G.; Nicolas, P.; Nicolas, B. Experimental study of fracture response in granite specimens subjected t hydrothermal conditions relevant for enhanced geothermal systems. *Geothermics* **2018**, *72*, 205–224.
34. Zeng, Y.C.; Zhan, J.M.; Wu, N.Y.; Luo, Y.Y.; Cai, W.H. Numerical investigation of electricity generation potential from fracture granite reservoir through a single vertical well at Yangbajing geothermal field. *Energy* **2016**, *114*, 24–39. [CrossRef]
35. Pruess, K. Enhanced geothermal system (EGS) using CO_2 as working fluid: A novel approach for generating renewable energ with simultaneous sequestration of carbon. *Geothermics* **2006**, *35*, 351–367. [CrossRef]
36. Sun, B.J.; Wang, J.T.; Sun, W.C.; Wang, Z.Y.; Sun, J.S. Advances in fundamental research of supercritical CO_2 fracturing technolog forum conventional natural gas reservoirs. *J. China Univ. Pet. (Ed. Nat. Sci.)* **2019**, *43*, 82–91.
37. Gong, F.C. Multi-field Coupling Numerical Simulation of Enhanced Geothermal System and Development Model Optimizatio Master's Thesis, China University of Petroleum (East China), Dongying, China, 2020.
38. Li, P. Study on Effect of Thermo-Hydro-Mechanical-Chemical Coupling in CO_2-EGS of Hot Dry Rock. Ph.D. Thesis, Chin University of Mining and Technology, Xuzhou, China, 2020.

energies

MDPI

Article

Optimization of Nanofluid Flow and Temperature Uniformity in the Spectral Beam Splitting Module of PV/T System

Liwei Lu [1], Rui Tian [1,2,*] and Xiaofei Han [1]

[1] College of Energy and Power Engineering, Inner Mongolia University of Technology, Hohhot 010051, China; 20171000025@imut.edu.cn (L.L.); shawnhan@nuaa.edu.cn (X.H.)
[2] Inner Mongolia Key Laboratory of Renewable Energy, Hohhot 010051, China
* Correspondence: tianr@imut.edu.cn

Abstract: The mass fraction of 0.01 wt% ZnO nanofluid was prepared via the two-step method. The measurement verifies that ZnO nanofluids have better transmission characteristics in the frequency division window range of 400–1200 nm. At the same time, it has good absorption characteristics in ultraviolet and near-infrared bands, which meets the application conditions of the spectral beam-splitting module of the PV/T system. A spectral beam-splitting module of the PV/T system was designed. The simplified physical model was established in ANSYS 14.0. The flow field and convective heat transfer were simulated for different arrangements of the interlayer inlet to obtain a more ideal and uniform temperature distribution to improve the system's comprehensive efficiency. The results show that the fluid flow in the interlayer under case II is more uniform, and the temperature field distribution is better than other arrangements. Hence, this work could provide a reference for optimising nanofluid flow within a spectral beam-splitting module.

Keywords: spectral beam splitting; nanofluid; PV/T; velocity distribution; temperature field

1. Introduction

The rapid development of human society, the sharp decline of traditional fossil energy reserves and the adverse impact on the environment in the application process have posed new challenges to the sustainable development of human beings. The development of environmentally friendly, clean, and renewable energy sources and their further increase in utilization have become a consensus and development strategy in many countries [1]. Solar energy is an ideal alternative to traditional energy sources because of its cleanliness, accessibility, and abundance [2,3]. How to obtain and utilize solar energy more efficiently has been a hot research topic for scholars in recent years [4,5].

At present, solar energy utilization methods mainly include photovoltaic conversion and solar thermal conversion. Solar photovoltaic/thermal (PV/T) utilization systems that combine the two conversions can increase the efficiency of integrated solar energy utilization [6,7]. This system usually uses a cooling medium (e.g., water, air, etc.) to flow over the surface or backside of the cell to lower the cell temperature and remove and utilize the excess heat that affects power generation efficiency [8]. Due to the mutual thermal coupling between PV modules and the cooling medium in the system, the power generation efficiency of PV cells is directly affected by the temperature of the cooling medium. To ensure the conversion efficiency of photovoltaic devices, the cooling medium temperature is not high, and the effective utilization rate of heat is low. The proposed spectral beam splitting technology can avoid the problems of thermal coupling of PV and solar thermal utilization in PV/T systems [9,10]. Spectral beam splitting technology aims to use different wavelengths of light from the sun separately: Allocate the energy in solar radiation corresponding to the response band of photovoltaic cells to solar cells for photoelectric conversion; allocate solar radiation with higher thermal effect to solar thermal receivers for photothermal conversion [9,11]. Solar energy spectral beam splitting

Citation: Lu, L.; Tian, R.; Han, X. Optimization of Nanofluid Flow and Temperature Uniformity in the Spectral Beam Splitting Module of PV/T System. *Energies* **2023**, *16*, 4666. https://doi.org/10.3390/en16124666

Academic Editor: Artur Bartosik

Received: 18 April 2023
Revised: 10 June 2023
Accepted: 11 June 2023
Published: 12 June 2023

technology, while avoiding overheating caused by solar radiation energy projected onto PV cells in the PV response band, can independently convert solar radiation energy in these bands for photothermal conversion, thus obtaining a higher overall efficiency [12,13].

Nanofluid-based spectral beam-splitting techniques have become popular for research in recent years. When the particle size of the material reaches the nanometer level, it has a small size effect, quantum size effect, surface effect and macroscopic quantum tunneling effect, and its behavior is closer to that of liquid molecules. Meanwhile, certain metal oxide nanomaterials strongly absorb solar radiation in the corresponding wavelength range due to their specific photonic forbidden band values, allowing enhanced heat and mass transfer and spectral beam splitting utilization.

The effect of nanofluids (such as CuO [14], Al_2O_3 [15,16], TiO_2 [17], and ZnO [18,19]) on PV/T systems has been investigated in numerical and experimental studies.

Wu [20] investigated the impact of air-cooling channel location on heat transfer characteristics and PV panel efficiency. Abdollahi [21] experimentally investigated the effect of natural nanofluid circulation and PCM on PV panels. In this work, the heat will be transferred from the nanofluid used to cool the PV panel to the phase-change material tank. They found that the power generation using natural nanofluid circulation and PCM at solar intensities (410, 530 and 690 W/m^2) was increased by 44.74%, 46.63% and 48.23%, respectively, compared to the conventional case.

Tarun Mitta [22] presented a nanofluid-based dual-channel PV/T system and simulated it numerically. The results show a slight decrease in the electrical efficiency part, but the thermal efficiency is higher than that of the single-channel split-frequency PV/T system, making the dual-channel system a better choice.

Samir [23] proposed a new cascading nanofluid-based PV/T configuration with separate channels. The performance of the PV/T with separate channels (D-1) was compared with the double-pass channel (D-2). The simulation results show that the separate channel system (D-1) outperformed the double-pass design (D-2) by ~8.6%.

After reviewing references [20–23], it is not difficult to find that most studies focused on optical filtration effects and investigated the overall efficiency, including electrical and thermal efficiencies. There are few studies on fluid flow and temperature distribution in channels in PV/T systems. The uniform temperature distribution in the channel affects the overall efficiency of the PV/T system to some extent. Optimizing the arrangement of the interlayer inlet of frequency division components and uniform temperature distribution in the interlayer is beneficial to improve the photothermal conversion efficiency of the PV/T system.

In this paper, a PV/T integrated utilization module is designed and processed based on the already produced ZnO nanofluid. A simulation study of the flow characteristics of the ZnO nanofluid in the spectral beam splitting module was done, and an ideal flow method was identified to ensure a uniform temperature distribution on the surface of the PV cell while obtaining considerable heat and improving the integrated utilization efficiency.

2. Preparation and Optical Properties Characterization of ZnO Nanofluids

2.1. Preparation of Nanofluids

There are two main methods for preparing nanofluids: one-step and two-step methods. The one-step method disperses the particles into the base fluid while the particles are being prepared. The two-step method includes the selection of nanoparticles, the selection of dispersant and the ultrasonic dispersion of the suspension. The advantages of preparing nanofluids by the two-step method are that it is relatively simple and easy to control the concentration of the suspension, which helps further research work.

The ZnO nanofluid with a mass fraction of 0.01% was prepared by the two-step method. The material reagent parameters used for the preparation are shown in Table 1. Figure 1 shows the preparation process of ZnO nanofluids.

Table 1. Materials and reagents.

Name	Parameter
Nano ZnO powder	Purity: 99.9% Average particle size: 30 nm
SDBS (Sodium dodecylbenzene sulfonate)	Purity: >90%

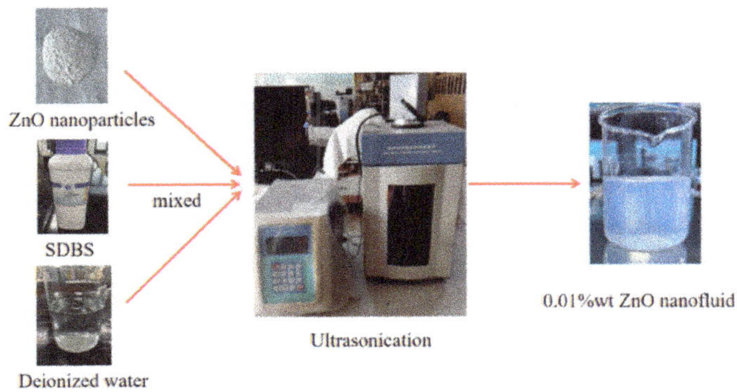

Figure 1. Process of ZnO nanofluids preparation.

2.2. *Optical Properties of Nanofluids*

The band gap of ZnO is 3.2 eV, corresponding to an absorption wavelength of 388 nm. Due to the quantum size effect, the band gap increases to 4.5 eV for a ZnO nanoparticle size of 10 nm. By scattering and absorption of UV light, ZnO nanomaterials have excellent UV shielding effects.

To meet the characteristics of ZnO nanofluid as a spectral beam-splitting heat collector with a certain absorbance while ensuring a high transmittance, the final ZnO nanofluid of 0.01 wt% was made through experiments. The fabricated nanofluid was tested for optical properties by UV-3600 UV-Vis NIR spectrophotometer, mainly for transmittance and absorbance measurements. The test results are shown in Figures 2 and 3.

As shown in Figure 2, compared with DI (Deionized water), the transmittance of the resulting ZnO nanofluid varies consistently with a wavelength in the visible light range of 400–800 nm, and the degree of attenuation is within 10%. This indicates that the ZnO nanoparticles do not have strong absorption of solar radiation in the visible wavelength range, which can ensure that the irradiation in the effective wavelength range that is beneficial to the power generation of PV panels is irradiated to the surface of PV panels through the fluid. The transmission of the visible ZnO nanofluid decreases steeply around 380 nm, indicating that it has very strong absorption of UV light near this wavelength.

Compared with DI (deionized water), the absorbance of the resulting ZnO nanofluid in the visible light range of 400–800 nm showed the same trend with wavelength, and the degree of attenuation was within 0.1, as illustrated in Figure 3. This means that the absorption of solar radiation in the visible wavelength range of ZnO nanoparticles is not much different from that of DI, which can ensure that the irradiation in the effective wavelength range beneficial to the power generation of PV panels can be irradiated to the surface of PV panels through the fluid. The visible ZnO nanofluid has an obvious absorption peak of around 380 nm. Still, it maintains an absorption value of about 1.25 between 230 and 380 nm, indicating a strong absorption effect on UV light in this wavelength band.

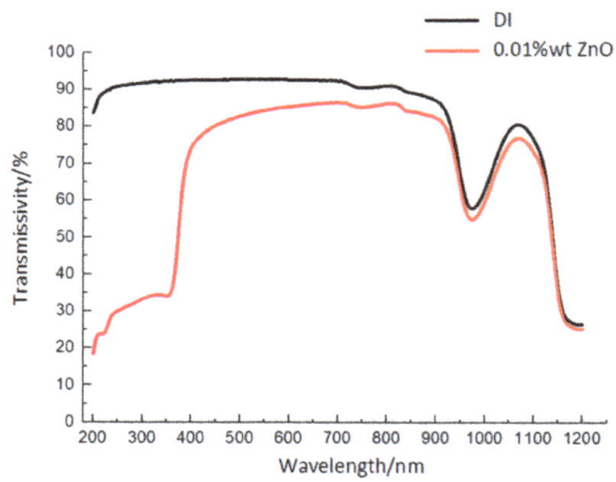

Figure 2. Curves of 0.01 wt% ZnO nanofluid transmittance as a function of wavelength.

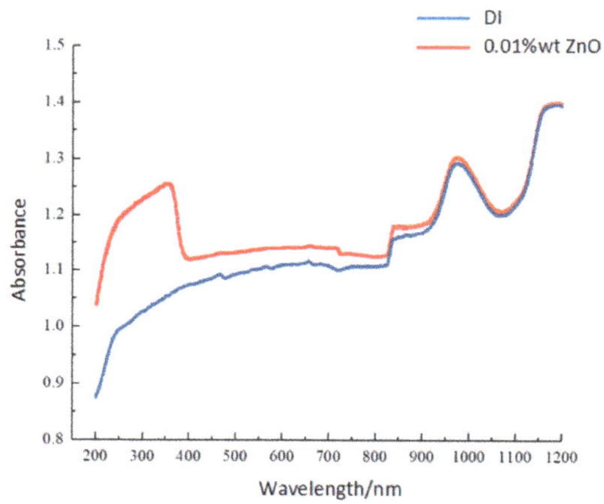

Figure 3. The curve of absorbance of 0.01 wt% ZnO nanofluid as a function of wavelength.

3. Spectral Beam Splitting Module and Physical Model Settings

3.1. Spectral Beam Splitting Module and Parameters

The physical PV/T spectral beam splitting module is shown in Figure 4, and th dimensions are shown in Table 2. To ensure the nanofluid's effective absorption of sola radiation and to consider the physical strength and economy of the module, the surface i encapsulated with an ultra-white glass of 5 mm thickness.

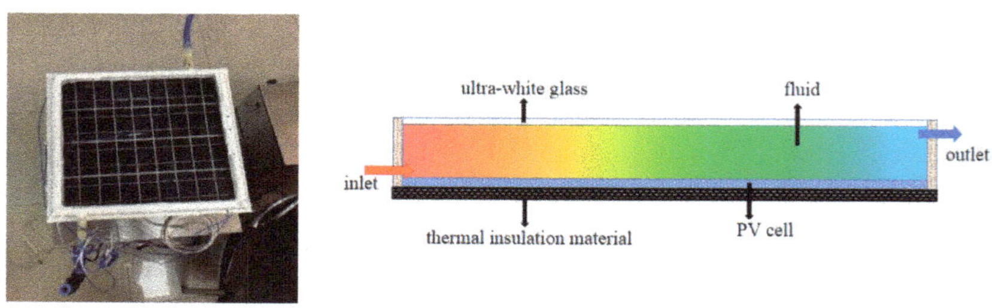

Figure 4. Schematic diagram and picture of PV/T spectral beam splitting module.

Table 2. Geometric Size of spectral beam splitting module.

Geometric Parameters/mm	Values
Length/L	350
Width/W	330
Height/H	25
Interlayer thickness/M	20
Inlet and outlet pipe diameter/D	10
Pipe thickness/d	1

3.2. *Physical Model and Boundary Condition Settings*

To simulate the flow state and heat transfer of a nanofluid in the interlayer of the splitting module more accurately and conveniently, this experiment simplifies the actual module into a three-dimensional rectangular interlayer with an inlet and outlet. The geometric dimensions are kept the same as the actual ones. The computational domai diagram is shown in Figure 5a. The six boundary surfaces are set adiabatically, excluding other variables, and the influence of the inlet end on the flow state and heat transfer uniformity of the fluid inside the interlayer is studied purely. Assume that the angle between the light-receiving surface and gravity direction is 130°, and the nanofluid inside the interlayer is simplified to a liquid with the same thermal conductivity and viscosity as the actual fluid. The inlet flow rate is set to 0.533 m/s, the inlet temperature is 313 K, and the initial temperature is 298 K, which is imported into ANSYS for transient simulation. The walls are set to be insulated and have no-slip boundary conditions. The simulation variables are the length of the inlet pipe inserted into the cavity, the position, and the distribution of the water divider.

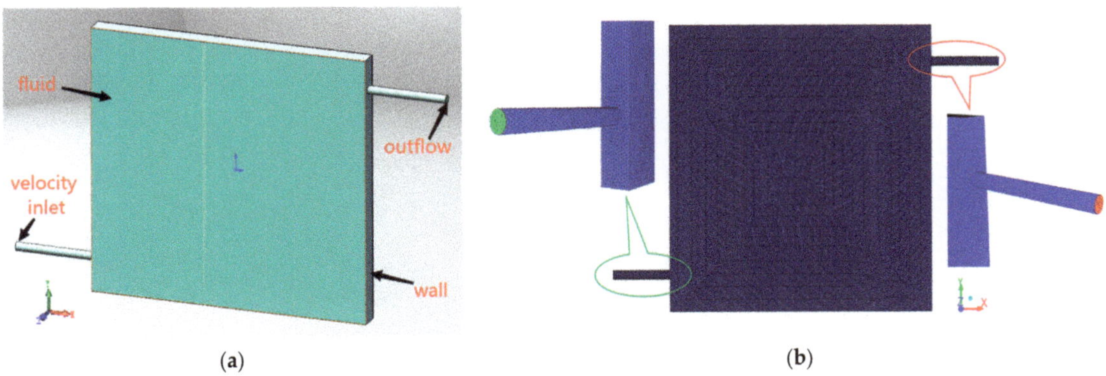

Figure 5. Computational domain and mesh for the PVT system. (**a**) Computational domain; (**b**) mesh.

The nanofluid is considered to be incompressible Newtonian fluid. The 3-D model of the system was numerically simulated by ANSYS Fluent. The grid is shown in Figure 5b. The control conservation equation is discretized by the finite volume method. The pressure basis scheme is used to solve the conservation equations. The SIMPLE method considers the coupling of pressure and velocity, and the convection terms are discretized by second-order upwind [24,25]. The inlet mass flow rate is 0.042 kg/s, and the Reynolds number is 5273. Turbulence model using k-ε standard model.

3.3. *Mathematical Model*

Mass equation

$$\frac{\partial \rho}{\partial t} + \frac{\partial(\rho u)}{\partial x} + \frac{\partial(\rho v)}{\partial y} + \frac{\partial(\rho w)}{\partial z} = 0 \tag{1}$$

Momentum equation

$$\frac{\partial(\rho u)}{\partial t} + \frac{\partial(\rho uu)}{\partial x} = \frac{\partial}{\partial x}\left(\mu\frac{\partial u}{\partial x}\right) - \frac{\partial P}{\partial x} \quad (2)$$

$$\frac{\partial(\rho v)}{\partial t} + \frac{\partial(\rho vv)}{\partial y} = \frac{\partial}{\partial y}\left(\mu\frac{\partial v}{\partial y}\right) - \frac{\partial P}{\partial y} \quad (3)$$

$$\frac{\partial(\rho w)}{\partial t} + \frac{\partial(\rho ww)}{\partial z} = \frac{\partial}{\partial z}\left(\mu\frac{\partial w}{\partial z}\right) - \frac{\partial P}{\partial z} \quad (4)$$

Energy equation

$$\rho C_p\left(u\frac{\partial T}{\partial x} + v\frac{\partial T}{\partial y} + w\frac{\partial T}{\partial z}\right) = k\left(\frac{\partial^2 T}{\partial x^2} + \frac{\partial^2 T}{\partial y^2} + \frac{\partial^2 T}{\partial z^2}\right) \quad (5)$$

where ρ represents fluid density, u, v, and w are velocity components in x, y and z, respectively, P is pressure, μ is dynamic viscosity, Cp is specific heat capacity, and k is thermal conductivity.

3.4. Grid and Time Independence Study and Model Validation

The grid independence study of the module is conducted by generating different grid numbers of 500, 1000, 2000, 2600, 3700, 4300, and 5000 thousand. Grids are three-dimensional and unstructured. A detailed grid diagram is shown in Figure 5b. The outlet temperature for different mesh numbers is presented in Figure 6a. The reason for choosing 2.6 million is that the PV/T outflow temperature is almost the same when the number of grids is greater than 2.6 million. The minimum grid size of 0.5 mm. The suitable time step is also determined according to the outlet temperature difference of the system for seven-time steps of 0.1 s, 0.2 s, 0.5 s, 1 s, 1.5 s, 2 s and 2.5 s. Figure 6b illustrates the outlet temperature difference in the simulations with various time step sizes. Therefore, the step size of 0.5 s is chosen.

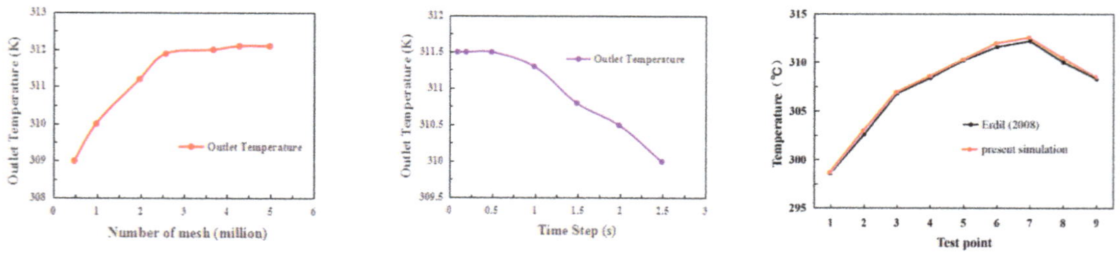

Figure 6. Grid and time independence study and mathematical model validation. (**a**) Grid independence study; (**b**) Time independence study; (**c**) mathematical model validation [26].

To prove the accuracy of this mathematical model, the related parameters are set equally with that in the reference [26]. For example, the mass flow rate is 18 L/h, the inlet water temperature is 297 K, the collector slope is 45, and the interlayer thickness is 5 mm. The calculated outlet temperature matched the tested results in Ref. [26] (Figure 6c).

4. Analysis of Simulation Results

For the convenience of description, the following simulation results are labeled as case I, case II, case III, case IV, case V and case VI, according to the different arrangements at the inlet end. The details of the arrangement are as follows.

Case I: The inlet end is set in the lower left of the interlayer, and the outlet end is upper right, and the schematic diagram of the length of the inlet end is shown in Figure 5a.

Case II: The inlet end is set at the lower left of the interlayer, the outlet end is upper right, and the inlet pipe is inserted into the inner part of the interlayer, with nine water-dividing holes of 4 mm in diameter. The hole spacing is 30 mm, and the end of the pipe is sealed. The schematic diagram is shown in Figure 7a.

Case III: The inlet end is set at the lower left of the interlayer, the outlet end is at the upper right, the inlet pipe is inserted into the inner part of the interlayer, and the lower part is opened with nine 4 mm diameter water distribution holes with a hole spacing of 30 mm, and the end of the pipe is sealed. The schematic diagram is shown in Figure 7b.

Case IV: The inlet end is set at the bottom of the left side of the interlayer, with the outlet pipe at an angle of 90°, the outlet end is on the right, and the inlet pipe is inserted into the interior of the interlayer, and nine water distribution holes with a diameter of 4 mm are opened to the side of the outlet end, with a hole spacing of 30 mm, and the end of the pipe is sealed. The schematic diagram is shown in Figure 7c.

Case V: The inlet end is set at the lower left of the interlayer, the outlet end is at the upper right, and the inlet pipe is inserted into the inner part of the interlayer, with four water distribution holes of 4 mm diameter at the lower part and five water distribution holes of 4 mm diameter at the upper part, with a hole spacing of 30 mm, and the end of the pipe is not sealed. The schematic diagram is shown in Figure 7d.

Case VI: The inlet end is set at the lower left side of the interlayer, the outlet end is at the upper right side, and the inlet pipe is inserted into the inner part of the interlayer, with four water distribution holes of 4 mm diameter at the lower part and five water distribution holes of 4 mm diameter at the upper part, with a hole spacing of 30 mm, and the end of the pipe is sealed. The schematic diagram is shown in Figure 7e.

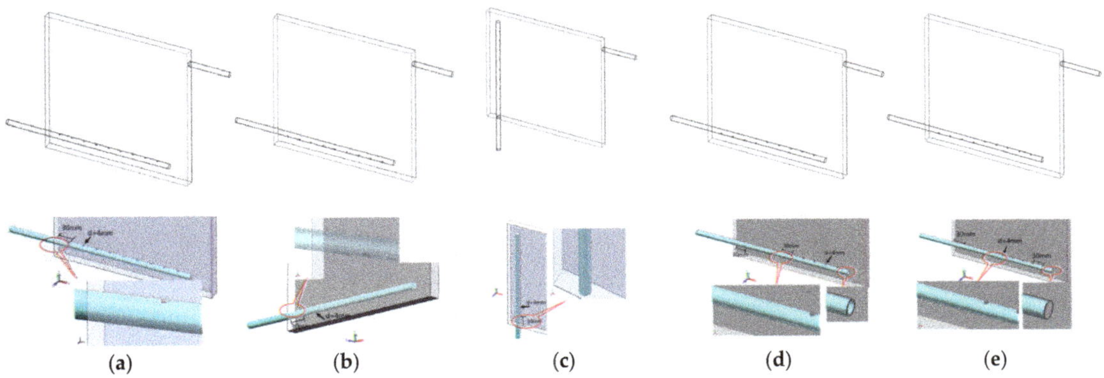

Figure 7. Diagram of the different arrangements at the inlet end of the interlayer (**a**) case II; (**b**) case III; (**c**) case IV; (**d**) case V; (**e**) case VI.

4.1. Analysis of Flow Characteristics under Different Arrangements at the Inlet End

For the sake of description, the central cross-section parallel to the supposed light-receiving surface is called the transverse cross-section, while the central cross-section perpendicular to the supposed light-receiving surface is referred to as the longitudinal cross-section.

Figure 8a shows the velocity distribution of the cross-section under case II. As shown in the figure, the velocity distribution in this plane does not change from 50 s to 250 s under the two-dimensional plane of flow, and it can be seen that the flow velocity is greater near the diversion orifice near the outlet because the local pressure is smaller there compared to that near the inlet section. The outlet end of the orifice suddenly contracted, so the fluid flow velocity increased. The velocity distribution under case III is similar to that of case II (Figure 8b), but the flow velocity near the diversion holes is more uniform than that of case II. As for the case IV in Figure 8c, even though there is gravitational influence, the direction of flow out of the diversion holes is still biased toward the outlet end, and th flow is stronger on the upper side of the diversion holes than on the lower side.

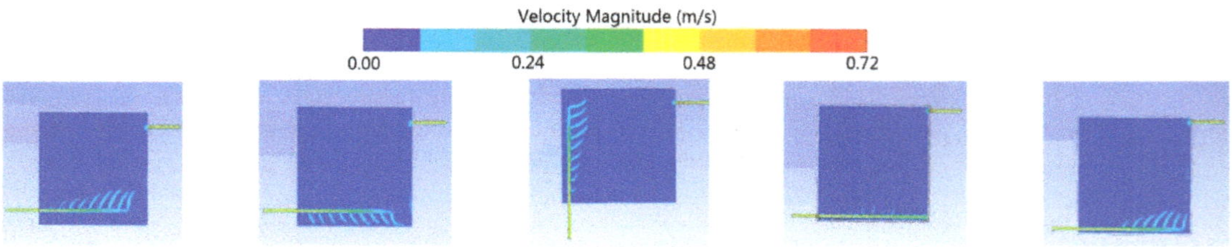

(a) (b) (c) (d) (e)

Figure 8. Velocity distribution of the cross-section (**a**) case II (**b**) case III (**c**) case IV (**d**) case (**e**) case VI.

As seen in Figure 8d, unlike case III, the splitting pressure of the splitting holes i smaller because the end of the inlet pipe is not sealed. The velocity distribution is no obvious, and the first and last few diversion holes are smaller than the middle wate flow. Due to the sealing of the end of the inlet pipe under the case VI (Figure 8e), th diversion hole partial pressure increases, and the end diversion hole water flow rat increases compared to case V.

A more apparent counterclockwise backflow is formed in the interlayer space at th upper end of the inlet pipe under case II (Figure 9a). Analysis of the reason, the right outle suddenly contracted, part of the fluid flow velocity increased to discharge the interlaye and the other part went back along the inner wall of the interlayer, forming a gyratory. Th gyratory center fluid flow rate is slower. Inlet pipe with the pipe length flow rate slowe down, but the back end of the water splitting hole flow rate is greater than the front, mainl because of the difference in pressure near the water splitting hole.

Figure 9b shows the cross-sectional velocity vector diagram in case III. The circulatio in the central space of the interlayer is not apparent, and the fluid flows out more alon the lower wall through the right wall. The velocity distribution is denser along the righ wall. The lower opening of the diversion hole makes the fluid flow out and back in othe directions after flowing against the lower side wall.

As shown in Figure 9c, the fluid forms a clockwise return flow in the center space c the mezzanine, which is more similar to the case II. A small vortex with counterclockwis flow is formed in the upper left corner at the end of the inlet pipe due to the blockage c the wall.

As shown in Figure 9d, the vortex in the upper left corner gradually decreases wit time. The lower right corner forms a delta of velocity vector distribution with more obviou demarcation because the pressure on both sides is different due to the unsealed enc resulting in different flow field distributions, and a region of slower flow velocity is forme between the diagonals.

Compared with case V, the velocity vector in case VI (Figure 9e) almost does no change with time. The same gyration is formed in the center space of the interlayer, and th velocity distribution is denser along the right side of the wall. The direction of the wate outlet from the diversion hole is biased towards the direction of the outlet end. The reaso is more similar to case II and will not be repeated.

The above comparison shows that the fluid velocity distribution in the interlayer tend to stabilize after a short time of formation. The inlet end arrangement has a more obviou effect on the flow field distribution, and various arrangements have produced differen sizes of gyration and slower flow areas.

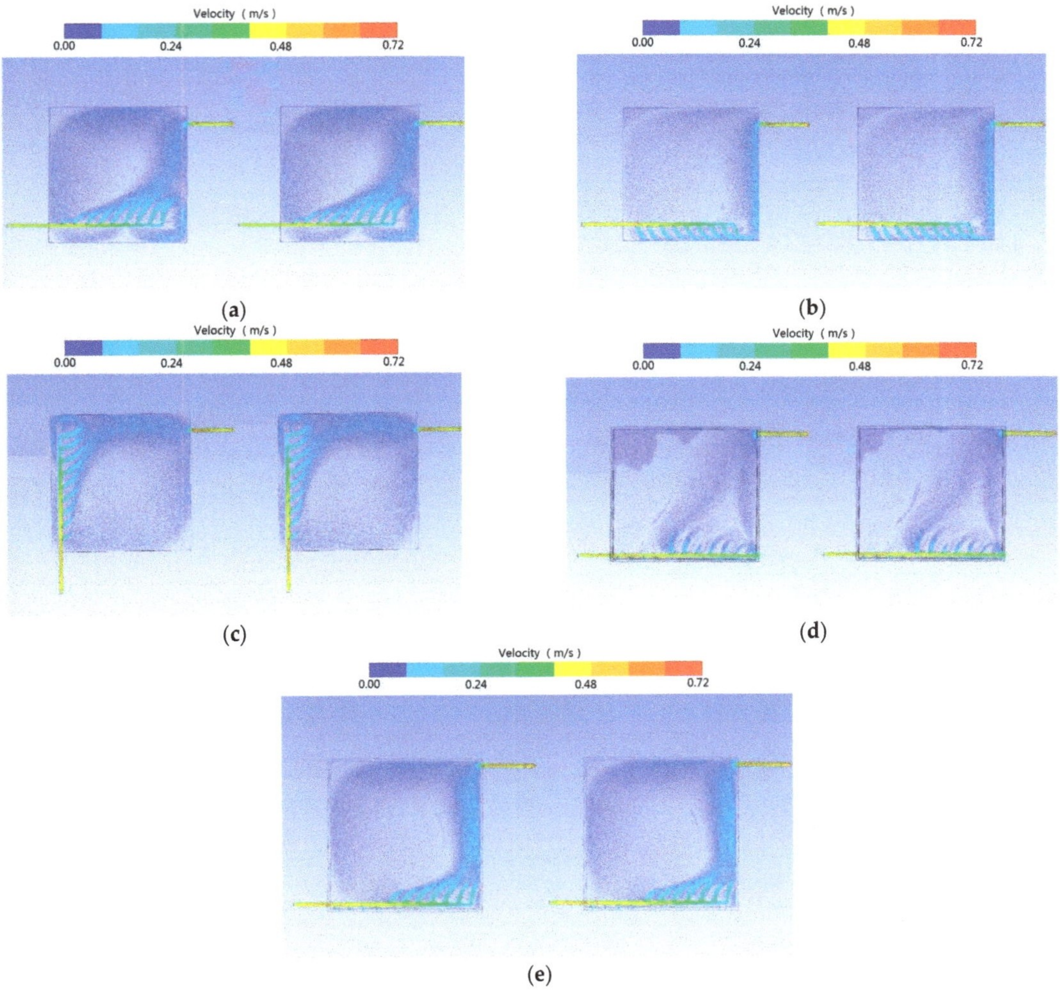

Figure 9. Velocity vector diagram of the cross-section at the same time node (**a**) case II (**b**) case III (**c**) case IV (**d**) case V (**e**) case VI.

4.2. Convective Heat Transfer Characteristics Analysis

For solar thermal utilization, the uniformity of the temperature field is crucial. On the one hand, a better temperature distribution can improve the thermal efficiency of the collector system. On the other hand, it can also avoid damage to the modules due to thermal stresses caused by temperature differences. For fractional frequency modules, the uniformity of the temperature distribution also affects the circulation performance of the fractional nanofluid and the photovoltaic conversion efficiency of the PV panel.

The temperature distribution of the cross-section at different time nodes under case I is shown in Figure 10. The lower side and right-side walls reach the highest temperature in the interlayer first. Because of the rotation in the center of the interlayer, the fluid flow in the center is slow, compared to the surrounding convective heat transfer is very weak, so there is a "leafy" low-temperature zone, and the temperature distribution is not uniform. The upper left corner is also a small area of low-temperature zone formed due to the slow flow and weak heat transfer. After 400 s, the maximum temperature difference reaches about 9 °C.

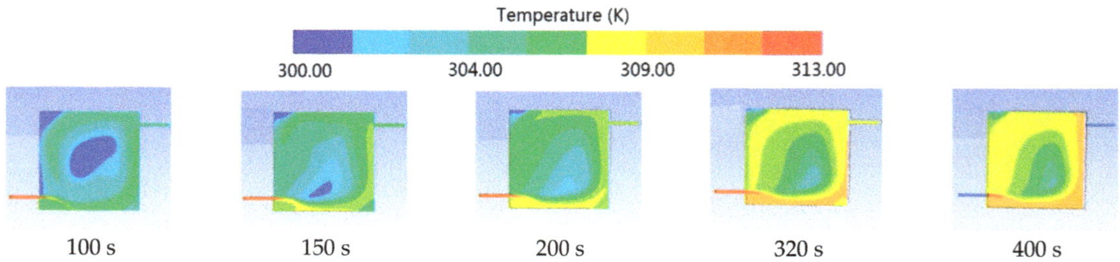

Figure 10. Temperature distribution of cross-section at different time nodes under Case I.

As illustrated in Figure 11, the velocity field is more uniformly distributed than that of case I. Hence, the "leafy" area formed in the center of the interlayer is smaller than that of case I. The maximum internal temperature difference is only 2 °C after 450 s of flow. However, the temperature is more likely to converge over time because of the stronger convective heat transfer around it. As shown in Figure 12, the distribution is more uniform in the longitudinal depth and close to the symmetric distribution.

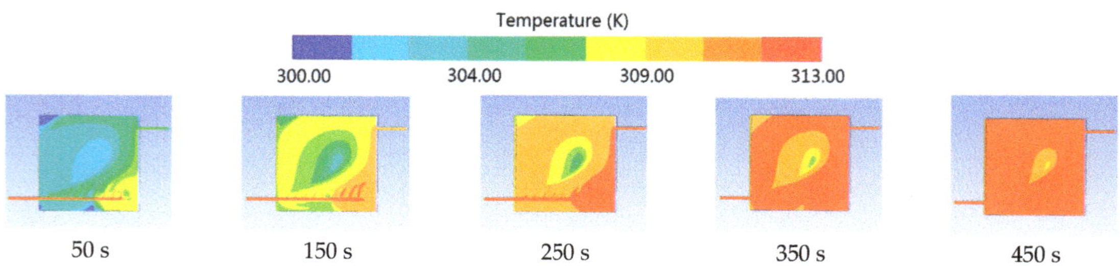

Figure 11. Temperature distribution of cross-section at different time nodes under case II.

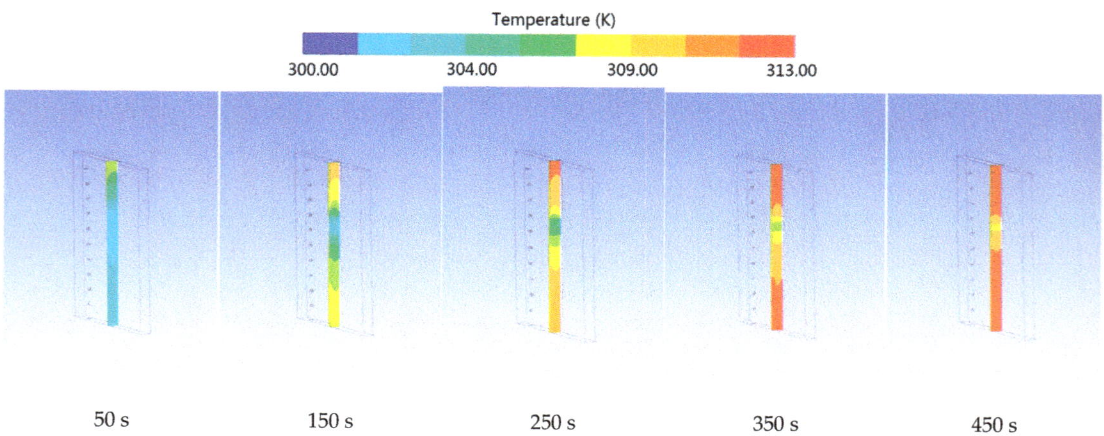

Figure 12. Temperature distribution of longitudinal section at different time nodes under case II.

Compared with the arrangement of case II, the formation of gyration in case III at the same time node is relatively late, and the area of the "leafy" low-temperature zone formed in the center is smaller with time, but there is an obvious heat transfer hysteresis zone in the upper left corner (as exhibited in Figure 13). Referring to Figure 14, the temperature distribution in the longitudinal section is more uniform than that in case II. However, the temperature in the lower part of the longitudinal section is lower in the early stage of convection heat transfer.

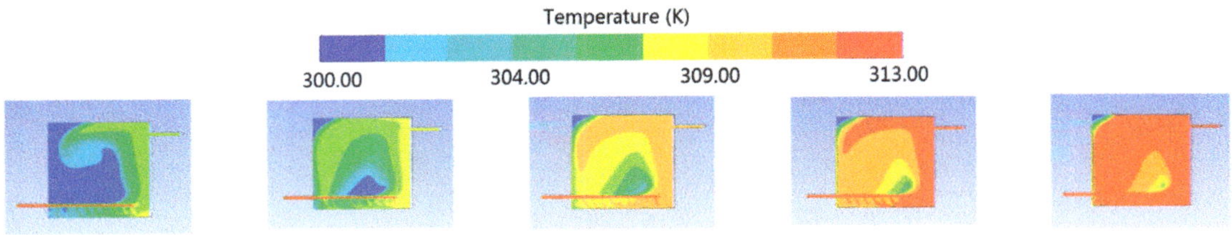

50 s 100 s 250 s 350 s 400 s

Figure 13. Temperature distribution of cross-section at different time nodes under case III.

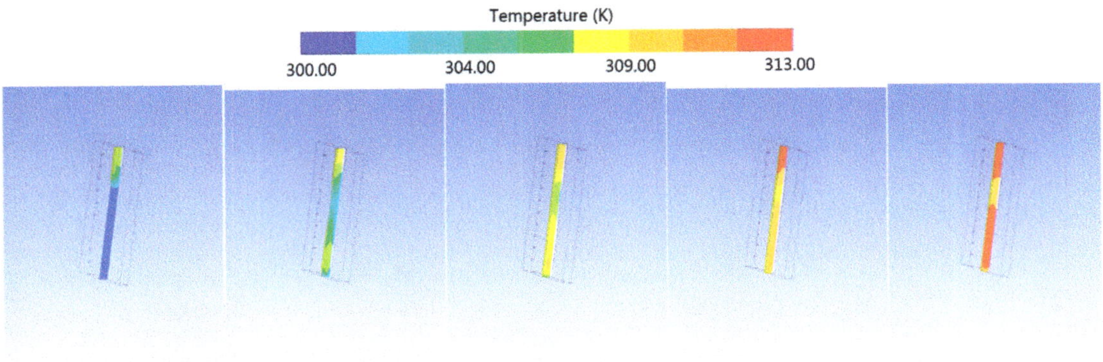

50 s 100 s 250 s 350 s 400 s

Figure 14. Temperature distribution of longitudinal section at different time nodes under case III.

The temperature distribution at 50 s in Figure 15 is similar to case III, but the temperature gradient is obviously smaller. The area of the lowest temperature point in the center of the "leafy" low-temperature zone formed in the center is smaller. In contrast, the four corners do not form a heat transfer hysteresis zone, and the temperature distribution is more uniform. Combined with Figure 16, it is similar to case III, and the distribution is more ideal.

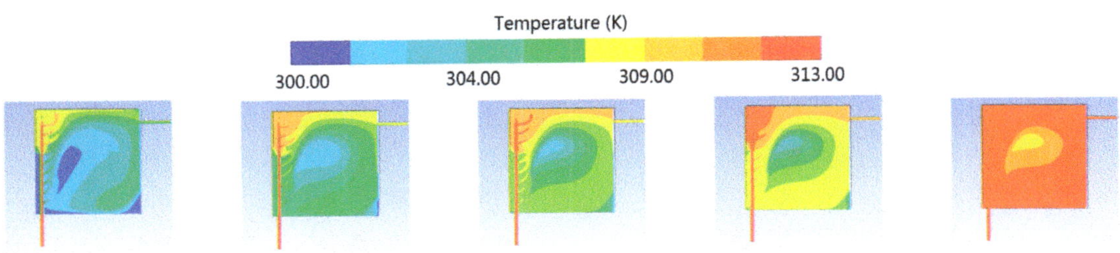

50 s 100 s 150 s 300 s 450 s

Figure 15. Temperature distribution of cross-section at different time nodes under case IV.

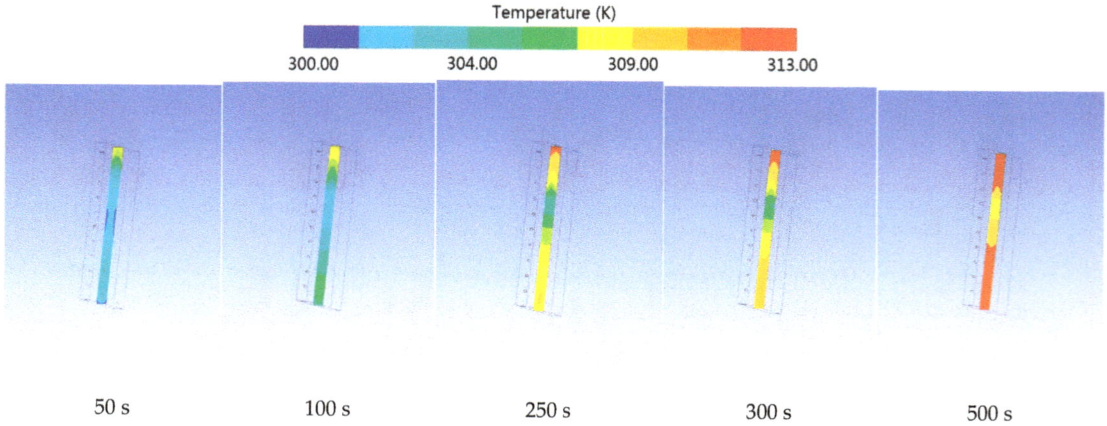

Figure 16. Temperature distribution of longitudinal section at different time nodes under case IV.

Since the end of case V inlet pipe is not sealed, convective heat exchange from the beginning of the flow is mainly concentrated on the side of the end opening. In contrast, the relative side maintained the initial temperature after the flow proceeded to 400 s, indicating that convective heat exchange near this place is extremely weak (Figure 17). The temperature distribution in Figure 18 is stepped, with poor uniformity.

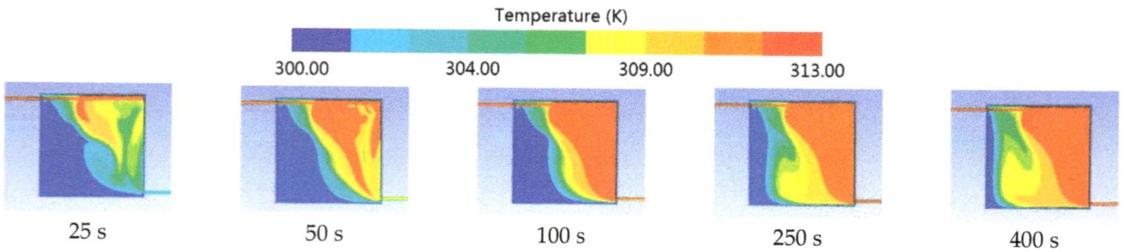

Figure 17. Temperature distribution of cross-section at different time nodes under case V.

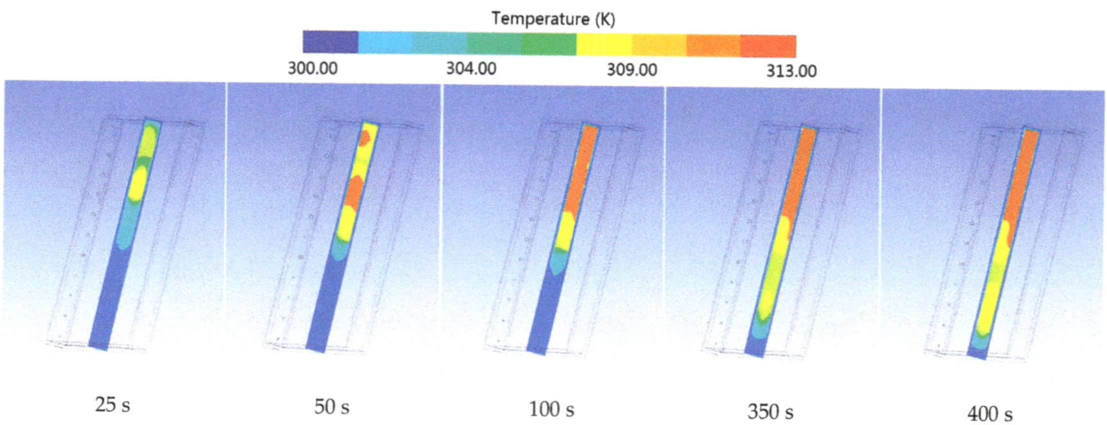

Figure 18. Temperature distribution of longitudinal section at different time nodes under case V.

As shown in Figure 19, compared with case V, an obvious " leafy" low-temperature zone is formed. Compared with the previous arrangements, case VI has a wider range of lower temperatures in the center, and the adjacent temperature difference is not large, but the heat transfer efficiency is lower, and the overall temperature at the same time node is lower than the other arrangements. Similar conclusions can be drawn by referring to Figure 20.

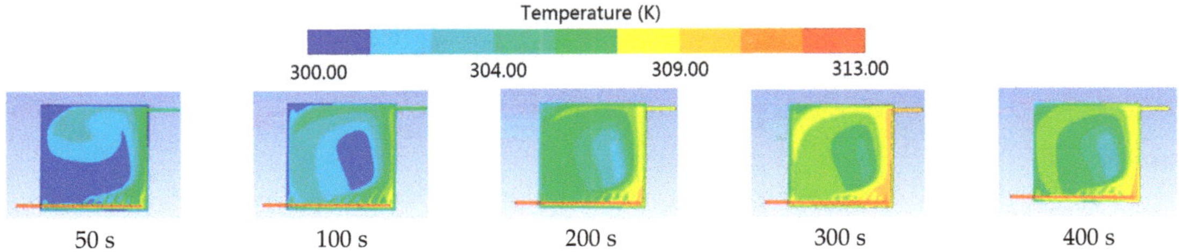

Figure 19. Temperature distribution of cross-section at different time nodes under case VI.

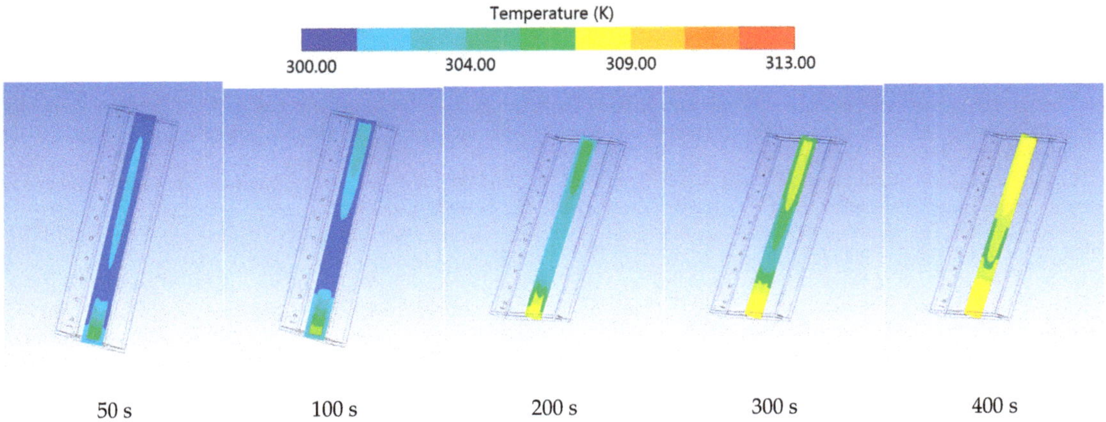

Figure 20. Temperature distribution of longitudinal section at different time nodes under case VI.

5. Conclusions

Increasing energy utilization and comprehensive efficiency of conventional PV/T integrated utilization systems using nanofluid have been hot research topics in recent years. The use of frequency-divided nanofluid to improve the efficiency of PV/T integrated utilization system has high requirements on the flow characteristics of nanofluid and the arrangement of frequency-divided components. In this paper, we simulated the arrangement of the nanofluid inlet of the split-frequency module based on the existing experimental components to obtain a more optimal arrangement. The following conclusions are obtained from the simulations.

1. The shorter the inlet end is, the more uneven the velocity distribution of the interlayer fluid is, which leads to the smaller intensity of convective heat transfer within the interlayer and the uneven distribution of the temperature field.
2. Case II and case III are more ideal, of which case II is the most ideal. Analysis of the reasons, case II velocity field distribution, is more uniform. It indicates that the fluid flow state and the flow velocity distribution play a decisive role in the uniformity of temperature distribution.
3. The cross-sectional area of the central "leafy" low-temperature region resulting from different arrangements is not the same, indicating that the central cross-section's temperature field distribution is not symmetric. The temperature field distribution at the cross-section is similar, but the temperature field is different in the direction of interlayer thickness.

Author Contributions: Conceptualization, R.T.; Software, X.H.; Data curation, L.L.; Writing—original draft, X.H.; Writing—review & editing, L.L.; Supervision, R.T.; Funding acquisition, R.T. All authors have read and agreed to the published version of the manuscript.

Funding: This work was supported by the Inner Mongolia Science and Technology Major Project (NO. 2019ZD014).

Data Availability Statement: The data presented in this study are available on request from the corresponding author.

Conflicts of Interest: The authors declare no conflict of interest.

References

1. Kuşkaya, S.; Bilgili, F.; Muğaloğlu, E.; Khan, K.; Hoque, M.E.; Toguç, N. The role of solar energy usage in environmental sustainability: Fresh evidence through time-frequency analyses. *Renew. Energy* **2023**, *206*, 858–871. [CrossRef]
2. He, Q.; Wang, S.; Tong, M.; Liu, Y. Experimental study on thermophysical properties of nanofluids as phase-change material (PCM) in low temperature cool storage. *Energy Convers. Manag.* **2012**, *64*, 199–205. [CrossRef]
3. Chen, N.; Ma, H.; Li, Y.; Cheng, J.; Zhang, C.; Wu, D.; Zhu, H. Complementary optical absorption and enhanced solar thermal conversion of CuO-ATO nanofluids. *Sol. Energy Mater. Sol. Cells* **2017**, *162*, 83–92. [CrossRef]
4. Sajid, M.U.; Bicer, Y. Nanofluids as solar spectrum splitters: A critical review. *Sol. Energy* **2020**, *207*, 974–1001. [CrossRef]
5. Otanicar, T.P.; Phelan, P.E.; Prasher, R.S.; Rosengarten, G.; Taylor, R.A. Nanofluid-based direct absorption solar collector. *J. Renew. Sustain. Energy* **2010**, *2*, 33102. [CrossRef]
6. Zondag, H. Flat-plate PV-Thermal collectors and systems: A review. *Renew. Sustain. Energy Rev.* **2008**, *12*, 891–959. [CrossRef]
7. Kumar, R.; Rosen, M.A. A critical review of photovoltaic–thermal solar collectors for air heating. *Appl. Energy* **2011**, *88*, 3603–3614. [CrossRef]
8. Huaxu, L.; Fuqiang, W.; Dong, Z.; Ziming, C.; Chuanxin, Z.; Bo, L.; Huijin, X. Experimental investigation of cost-effective ZnO nanofluid based spectral splitting CPV/T system. *Energy* **2020**, *194*, 116913. [CrossRef]
9. Blocker, W. High-efficiency solar energy conversion through flux concentration and spectrum splitting. *Proc. IEEE* **1978**, *66*, 104–105. [CrossRef]
10. Jiang, S.; Hu, P.; Mo, S.; Chen, Z. Optical modeling for a two-stage parabolic trough concentrating photovoltaic/thermal system using spectral beam splitting technology. *Sol. Energy Mater. Sol. Cells* **2010**, *94*, 1686–1696. [CrossRef]
11. Imenes, A.; Mills, D. Spectral beam splitting technology for increased conversion efficiency in solar concentrating systems: A review. *Sol. Energy Mater. Sol. Cells* **2004**, *84*, 19–69. [CrossRef]
12. Hu, P.; Zhang, Q.; Liu, Y.; Sheng, C.; Cheng, X.; Chen, Z. Optical analysis of a hybrid solar concentrating Photovoltaic/Thermal (CPV/T) system with beam splitting technique. *Sci. China Technol. Sci.* **2013**, *56*, 1387–1394. [CrossRef]
13. Liu, M.; Du, M.; Long, G.; Wang, H.; Qin, W.; Zhang, D.; Ye, S.; Liu, S.; Shi, J.; Liang, Z.; et al. Iron/Quinone-based all-in-one solar rechargeable flow cell for highly efficient solar energy conversion and storage. *Nano Energy* **2020**, *76*, 104907. [CrossRef]
14. Hajabdollahi, Z.O.; Mirzaei, M.; Kim, K.C. Effects of a mixture of cuo and AL_2O_3 nanoparticles on the thermal efficiency of a flat plate solar collector at different mass flow rates. *Heat Transf. Res.* **2019**, *50*, 945–965. [CrossRef]
15. Sardarabadi, M.; Passandideh-Fard, M. Experimental and numerical study of metal-oxides/water nanofluids as coolant in photovoltaic thermal systems (PVT). *Sol. Energy Mater. Sol. Cells* **2016**, *157*, 533–542. [CrossRef]
16. Hissouf, M.; Feddaoui, M.; Najim, M.; Charef, A. Numerical study of a covered Photovoltaic-Thermal Collector (PVT) enhancement using nanofluids. *Sol. Energy* **2020**, *199*, 115–127. [CrossRef]
17. Farajzadeh, E.; Movahed, S.; Hosseini, R. Experimental and numerical investigations on the effect of Al2O3/TiO2- H2O nanofluids on thermal efficiency of the flat plate solar collector. *Renew. Energy* **2018**, *118*, 122–130. [CrossRef]
18. Wole-Osho, I.; Adun, H.; Adedeji, M.; Okonkwo, E.C.; Kavaz, D.; Dagbasi, M. Effect of hybrid nanofluids mixture ratio on the performance of a photovoltaic thermal collector. *Int. J. Energy Res.* **2020**, *44*, 9064–9081. [CrossRef]
19. Salari, A.; Kazemian, A.; Ma, T.; Hakkaki-Fard, A.; Peng, J. Nanofluid based photovoltaic thermal systems integrated with phase change materials: Numerical simulation and thermodynamic analysis. *Energy Convers. Manag.* **2020**, *205*, 112384. [CrossRef]
20. Wu, S.-Y.; Wang, T.; Xiao, L.; Shen, Z.-G. Effect of cooling channel position on heat transfer characteristics and thermoelectric performance of air-cooled PV/T system. *Sol. Energy* **2019**, *180*, 489–500. [CrossRef]
21. Abdollahi, N.; Rahimi, M. Potential of water natural circulation coupled with nano-enhanced PCM for PV module cooling. *Renew. Energy* **2019**, *147*, 302–309. [CrossRef]
22. Mittal, T.; Saroha, S.; Bhalla, V.; Khullar, V.; Tyagi, H.; Taylor, R.A.; Otanicar, T.P. Numerical Study of Solar Photovoltaic/Thermal (PV/T) Hybrid Collector Using Nanofluids. In Proceedings of the ASME 2013 4th International Conference on Micro/Nanoscale Heat and Mass Transfer, Hong Kong, China, 11–14 December 2013.
23. Hassani, S.; Taylor, R.A.; Mekhilef, S.; Saidur, R. A cascade nanofluid-based PV/T system with optimized optical and thermal properties. *Energy* **2016**, *12*, 963–975. [CrossRef]

24. Khatibi, M.; Nemati-Farouji, R.; Taheri, A.; Kazemian, A.; Ma, T.; Niazmand, H. Optimization and performance investigation of the solidification behavior of nano-enhanced phase change materials in triplex-tube and shell-and-tube energy storage units. *J. Energy Storage* **2021**, *33*, 102055. [CrossRef]
25. Kazemian, A.; Taheri, A.; Sardarabadi, A.; Ma, T.; Passandideh-Fard, M.; Peng, J. Energy, exergy and environmental analysis of glazed and unglazed PVT system integrated with phase change material: An experimental approach. *Sol. Energy* **2020**, *201*, 178–189. [CrossRef]
26. Erdil, E.; Ilkan, M.; Egelioglu, F. An experimental study on energy generation with a photovoltaic (PV)–solar thermal hybrid system. *Energy* **2008**, *33*, 1241–1245. [CrossRef]

MDPI AG
Grosspeteranlage 5
4052 Basel
Switzerland
Tel.: +41 61 683 77 34

Energies Editorial Office
E-mail: energies@mdpi.com
www.mdpi.com/journal/energies